An Introduction to Underwater Acoustics

Principles and Applications

Springer
Berlin
Heidelberg
New York
Hong Kong
London
Milan
Paris
Tokyo

Xavier Lurton

An Introduction to Underwater Acoustics

Principles and Applications

Springer

Published in association with

Praxis Publishing

Chichester, UK

Dr Xavier Lurton
Institut Français de Recherche pour l'Exploitation de la Mer (IFREMER)
Plouzané
France

SPRINGER–PRAXIS BOOKS IN GEOPHYSICAL SCIENCES
SUBJECT *ADVISORY EDITOR*: Dr Philippe Blondel, C.Geol., F.G.S., Ph.D., M.Sc., Senior Scientist, Department of Physics, University of Bath, Bath, UK

ISBN 3-540-42967-0 Springer-Verlag Berlin Heidelberg New York

Die Deutsche Bibliothek – CIP-Einheitsaufnahme
  Lurton, Xavier:
  An introduction to underwater acoustics: principles and applications/
  Xavier Lurton. – London; Berlin; Heidelberg; New York; Barcelona;
  Hong Kong; Milan; Paris; Santa Clara; Singapore; Tokyo: Springer;
  Chichester, UK: Praxis Publ., 2002
    ISBN 3-540-42967-0
  0101 deutsche buecherei

Library of Congress Cataloging-in-Publication Data
  Lurton Xavier
    An introduction to underwater acoustics : principles and applications /
    Xavier Lurton.
      p. cm.
    Includes bibliographical references and index.
    ISBN 3-540-42967-0 (alk. paper)
    1. Underwater acoustics.   I. Title.
    QC242.2 .L88 2002
    620.2′5—dc21   2002030560

Reprinted 2004
Printed in Germany

Translated by Philippe Blondel, University of Bath, UK
Cover design: Jim Wilkie
Project management: Originator, Gt Yarmouth, Norfolk, UK

Printed on acid-free paper

Contents

Foreword

This book is based on my experience teaching in engineering schools (mainly at ENSIETA[1]) for the last decade, together with my activity as a research engineer in acoustical oceanography at IFREMER.[2] As indicated by its title, the first objective of this book is to present the reader keen to learn about underwater acoustics with an overall, up-to-date panorama of the field, as easily accessible as possible, with a clear emphasis on applications. The main subjects treated here are: physical processes (sound wave propagation in the ocean, influence of interfaces and targets, noise and fluctuations) and their basic modelling; specific technologies (transducers, array processing, signal processing) and assessment of their performance; and finally the different underwater acoustic systems and their practical applications (with a particular emphasis on mapping systems).

Going further than a descriptive introduction to these various topics, this book also aims at providing the interested student, engineer or scientist with the fundamental basics necessary for academic or practical solutions to current problems. To this effect, I tried as much as possible to present the different themes at a homogeneous level (Chapter 8, presenting seafloor mapping sonars in detail, is the exception to the rule), in order to present the reader with a more balanced and coherent approach. I deliberately focused the content on the underwater domain of acoustics, without repeating the fundamental principles (which can already be found in general textbooks and are reduced here to the minimum necessary), and without going either into too much detail on particular subjects, for which there are already monographs targeted toward the specialist.

Finally, I hope this book will arouse vocations for acoustics, both in students and in scientists working or studying geology, oceanography, biology … May it help to facilitate comprehension between engineering and the various fields of ocean sciences.

---

[1] Ecole Nationale Supérieure des Ingénieurs des Etudes et Techniques d'Armement, Brest, France.
[2] Institut Français de Recherche pour l'Exploitation de la Mer, Plouzané, France.

This Foreword is a nice opportunity for me to thank all the colleagues, teachers, scientists and engineers who have largely influenced me for years by their professional and personal qualities. The list is (fortunately for me) very long, but (unfortunately for them) far too long for this Foreword. So I shall only name, on several accounts: Michel Bruneau, Jean Kergomard, Claude Leroy, Jean-Yves Bervas, Bernard Leduc, Eric Pouliquen, Jean-Pierre Allenou, Michel Voisset, and Jean-Marie Augustin; I think they are worthy representatives of all the others.

As far as the writing of this book is concerned, I want to thank more particularly Yves Le Gall and Valérie Mazauric (IFREMER), together with Gérard Lapierre (GESMA), for their helpful reading and suggestions – and also for drawing some inspiration from their ENSIETA lecture notes.

My heartfelt thanks go to Philippe Blondel, the dynamic Editor of the Praxis Series in Geophysical Sciences, who was blinded by his passion for sonars and seafloor mapping to the point of proposing me to write this book. All my friendly thanks for his patience and understanding as editor, and for the quality of his translation.

Finally, and although they are only marginally concerned with sonars, I would like to dedicate this book to my wife Françoise, and our children Thibaut, Jean-Baptiste, Bérénice and Gauthier.

Xavier Lurton
Brest, September 2002

Abbreviations

ADC, A/D	Analogue to Digital Conversion
AGC	Automatic Gain Control
ADCP	Acoustic Doppler Current Profiler
APL	Applied Physics Laboratory (University of Washington, Seattle), USA
ASDIC	Anti-Submarine Detection Investigation Committee, UK
ASW	Anti-Submarine Warfare
ATOC	Acoustical Thermometry of Ocean Climate
AUV	Autonomous Underwater Vehicle
BCS	Backscattering Cross-Section
BPSK	Binary Phase Shift Keying
BFSK	Binary Frequency Shift Keying
BS	Backscattering Strength
BT	Bandwidth–Time product
BTR	Bearing Time Record
CRB	Cramer–Rao Bound
CTD	Conductivity–Temperature–Depth probe
CW	Continuous Wave
DI	Directivity Index
DSC	Deep Sound Channel
DEMON	Demodulated Noise
DICASS	Directional Command Activated Sonobuoy System
DIFAR	Direction Finding Acoustic Receiver
DSL	Deep Scattering Layer
EL	Echo Level
FFP	Fast Field Program
FFT	Fast Fourier Transform
FM	Frequency Modulation
FOM	Figure Of Merit
FSK	Frequency Shift Keying

GESMA	Groupe d'Etudes Sous-Marines de l'Atlantique, France
GLORIA	Geological Long-Range Inclined Asdic
GPS	Global Positioning System
HF	High Frequency
IHM	Instituto Hidrográfico de la Marina, Spain
IHO	International Hydrography Organisation
IHPT	Instituto Hidrográfico, Portugal
IOSDL	Institute of Oceanographic Sciences, Deacon Laboratory, UK
IFREMER	Institut Français de Recherche pour l'Exploitation de la Mer, France
JASA	*Journal of the Acoustical Society of America*
LOFAR	Low Frequency Analysis and Recording
MBES	Multibeam Echosounder
MCM	Mine Counter Measures
MLBS	Maximum Length Binary Sequence
NAVOCEANO	Naval Oceanographic Office, USA
NDRC	National Defence Research Committee, USA
NIMA	National Imagery and Mapping Agency, USA
NL	Noise Level
NOAA	National Oceanographic and Atmospheric Administration, USA
NUSC	Naval Underwater Systems Center, USA
OAT	Ocean Acoustic Tomography
ONR	Office of Naval Research, USA
PDF	Probability Density Function
PG	Processing Gain
PSK	Phase Shift Keying
PVDF	Polyvinylidene Bifluoride
PVDS	Propelled Variable Depth Sonar
PZT	Lead and Titanium Zirconate ceramic
QPSK	Quadrature Phase Shift Keying
RANHS	Royal Australian Navy Hydrographic Service
R&D	Research and Development
ROC	Receiver Operational Characteristics
ROV	Remote Operated Vehicle
R/V	Research Vessel
Rx	Receiver
SACLANTCEN	Supreme Allied Commander, Atlantic NATO Undersea Research Center, La Spezia, Italy
SAR	Synthetic Aperture Radar
SAS	Synthetic Aperture Sonar
SBL	Short Baseline
SBP	Sub-Bottom Profiler
SHC, CHS	Service Hydrographique Canadien, Canadian Hydrographic Service

SHOM	Service Hydrographique et Océanographique de la Marine, France
SI	Système International
SIO	Scripps Institution of Oceanography, USA
SL	Source Level
SNR	Signal to Noise Ratio
SOC	Southampton Oceanography Centre, UK
SOFAR	SOund Fixing And Ranging
SONAR	SOund Navigation And Ranging
SOSUS	SOund SUrveillance System
SPB	Signal Processing Board
SPL	Sound Pressure Level
SSN	Nuclear-Powered Attack Submarine
SSBN	Submarine Submersible Ballistic Nuclear
SSP/SVP	Sound Speed Profile/Sound Velocity Profile
TASS	Towed Array Sonar System
TL	Transmission Loss
TOBI	Towed Ocean-Bottom Instrument
TS	Target Strength
TVG	Time Varying Gain
Tx	Transmitter
UKHO	United Kingdom Hydrographic Office
USBL	Ultra Short Baseline
USGS	US Geological Survey
UWA	Underwater Acoustics
VDS	Variable Depth Sonar
VLF	Very Low Frequency
WHOI	Woods Hole Oceanographic Institution, USA
XCTD	Expendable Conductivity–Temperature–Depth probe
XBT	Expendable Bathy-Thermograph

Figures

Tables

Colour plates (between pages 166 and 167)

1

The development of underwater acoustics

1.1 RATIONALE

1.1.1 Exploring the underwater environment

One of the major technological achievements of modern history has been the design and implementation of means of communication and transmission of information over large distances (telephone, radio, television). The social and cultural implications of this development are huge and far-reaching. Most of these systems use electromagnetic waves, similar in nature to light. The properties of these waves were discovered in the 19th century, and they were first used from the beginning of the 20th century. Ever since, they revealed themselves as a formidable vector for the transmission of information and, used in radar systems, as a powerful tool for the exploration and monitoring of our environment as well. Since electromagnetic waves can propagate both in a vacuum and in the atmosphere, their domain of application has spread following man's entry into space, in particular with the generalised use of telecommunication and remote sensing satellites.

But a large portion of the Earth remains inaccessible to electromagnetic waves. The underwater realm, which accounts for more than 70% of the Earth's surface, does not allow for propagation of radio and radar waves. Indeed, water, and especially salt water, exhibits a strong conductivity, and is therefore highly dissipative. This means that electromagnetic waves are attenuated extremely rapidly, which limits their range and usefulness. Acoustic waves are today the only practical way to carry information underwater. They consist in mechanical vibrations of their propagation medium. They can propagate very easily in sea water, where their use might compensate to some degree for the strongly limited capabilities of electromagnetic and light waves in the ocean.

Sound waves are an integral part of our natural or artificial aerial environment. Through our physiological senses, we intuitively master their physical characteristics and use them constantly, mostly for communication, although they also participate to some degree in our faculties of orientation. In water, things are somewhat different. Acoustic waves have "better" transmission characteristics than in air. Their propagation speed is four to five times higher in water, they can reach much higher levels and most of all they undergo less attenuation and thus can propagate over large distances.[1] These advantages are mitigated by other constraints: useful signals are in particular perturbed by a higher ambient noise level and unwanted echoes taking advantage of the same conditions for easy propagation. The acoustic environment potentialities are therefore much different underwater than in the air. For example, cetaceans – the most evolved beings in the ocean – make abundant use of sound waves to communicate between each other, to explore their environment and to locate their prey. Such faculties are without parallel in terrestrial animals (except in bats).

The ease with which acoustic waves propagate in water was noticed long ago. However, the actual use of underwater sound is very recent. The first practical realisations came to light at the beginning of the 20th century, as permitted by technological progress. The number and type of applications has grown since, and it can be said, with some degree of simplification, that underwater acoustics today plays in the ocean the same essential role played by radar and radio waves in the atmosphere and in space (although with lower performances). Acoustic waves are used underwater to:

- *detect* and *locate* obstacles and targets; this is the primary function of *sonar* systems, mostly for military applications such as anti-submarine warfare and minehunting, but also used in fisheries;
- *measure* either the characteristics of the marine environment (seafloor topography, living organisms, currents and hydrological structures, …) or the location and velocity of an object moving underwater;
- *transmit* signals, which may be data acquired by underwater scientific instrumentation, messages between submarines and surface vessels, or commands to remotely operated systems.

These systems are for the most part *active systems*; that is, they transmit a characteristic signal, and this signal will be reflected on a target or transmitted directly to a receiver. But there are also *passive systems*, designed to intercept and exploit underwater sounds coming from the target itself.

1.1.2   Influence of the propagation medium

One of the main differences between using electromagnetic waves in the air and using acoustic waves in water resides in the constraints brought about by the propagation

[1] While propagation ranges under current aerial conditions hardly exceed a few kilometres, sound propagation in the ocean can currently be observed at ranges of up to thousands of kilometres.

medium. Sea water is relatively favourable to the propagation of acoustic waves, but still brings many limitations:

- *Attenuation* of the signals transmitted, due in particular to the *absorption* of sound waves in water, limiting the ranges reachable with a particular system.
- The relatively small propagation speed (1,500 m/s, compared with 300,000 km/s for radar waves in space).
- *Perturbations* of propagation by the variations in sound speed and reflection on seafloor and sea surface interfaces, resulting in (a) *inhomogeneous insonification* of the propagation medium (acoustic waves do not propagate in straight lines); (b) *multiple paths*, generating delayed parasite echoes and interference.
- Deformation of the signals transmitted – the fluctuations are related to hetero-geneities of the medium, interferences between multiple paths, reflections on the seafloor and at the sea surface and frequency changes (*Doppler effect*) due to relative movements of sonars and targets.
- *Ambient noise* in the ocean, whose level tends to mask the useful part of the signal – this noise comes from movements of the sea surface, volcanic and seismic activity, shipping, living organisms, rain, etc., to which one must add the *self-noise* characteristic of the acoustic system and its platform (surface vessel or submarine vehicle).

The characteristics of the propagation medium are also extremely varied, and vary both in space and in time. Environmental fluctuations include, but are not limited to, geographical and seasonal variations in temperature and salinity, seabed relief, swell, currents, tides, internal waves, movements of the acoustics systems and their targets ... All this concurs to give underwater acoustic signals a mostly *random fluctuating* character.

In all applications of current systems, used for detection and location, the transmitted signal is sent back by sonar to a target. This *backscattering* process is always very complicated; it depends on the physical structure of the target and its dimensions, as well as the angle of arrival and frequency of the signal. The echo from the target is usually associated with a continuum of signals scattered back by objects present in the propagation medium and at the interfaces, masking the actual signal; this is the *reverberation* phenomenon.

The marine environment therefore plays a very complex role in the transmission of signals. This may be the main problem encountered in the use of underwater acoustics. This justifies the important efforts undertaken over the last half-century in studying and modelling *underwater acoustic propagation*, currently one of the most active branches of underwater acoustics. Comparing again with electromagnetic waves, one notes that electromagnetic wave propagation (far less critical in radio and radar performances) has been the subject of comparatively fewer publications, accounting for the huge importance of their use compared with underwater acoustics and the relative ease of experimentation that they afford.

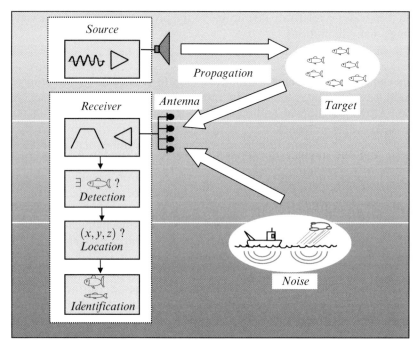

Figure 1.1. Generic organisation of a sonar system.

1.1.3   Structure of sonar systems

Target and obstacle detection/location systems using acoustic signals are commonly called *sonars* (acronym of *SOund Navigation And Ranging*). This acronym makes a clear acoustic parallel with the electromagnetic *radar* (*RAdio Detection And Ranging*). To detect and locate a particular target, a sonar system relies on the reception:

● either of the echo of a signal transmitted by the system and sent back by the target (*active sonar*);
● or the acoustic noise directly radiated by the target (*passive sonar*).

In the first case (Figure 1.1), the sonar must be able to transmit high-power acoustic signals. The signals transmitted, and finally reflected by the target, will propagate in the ocean, both on the way in and on the way out; they will be attenuated, deformed and overlaid with noise. They will be received by an antenna, possibly an *array* made of several transducers (*hydrophones*), whose outputs are combined in a suitable fashion. The signals are then fed into a dedicated *processing chain*, to increase the signal-to-noise ratio by amplifying and filtering, prior to deciding whether the object has been detected or not. They are then used for the measurement (e.g. range and angle estimations) and identification or characterisation of the target.

1.1.4   Signal processing

Together with the constraints of the physical environment, the design of signals and their processing directly affects the performance of underwater acoustics systems. The main sonar signal-processing functions are (Figure 1.1):

- *detection* (identification, with enough confidence, of the presence of a signal buried in noise);
- *estimation* of specific parameters (usually the propagation time of the acoustic signal and the angular direction of arrival), mostly for tracking the target relative to sonar; and
- *identification* (target recognition) or *characterisation* (estimation of some of its intrinsic parameters).

Detection and spatial location are made possible using *antennas*, whose purpose is to increase the signal-to-noise ratio and to measure angles. To improve the performances of these different functions, a very large number of theoretical and applied studies have been pursued over the last few decades, in the various domains of "sonar signal processing". Many of the studies concerning active sonars have been conducted along parallel lines to earlier or present work on active radars. Conversely, passive sonar processing techniques are very specific and have required the development of original theories and concepts. In both cases, progress has been clearly technology-related, in particular by generalisation of digital signal processing.

1.2   HISTORICAL HIGHLIGHTS

1.2.1   The pioneers

The possibility of using sound to detect distant ships, by simply listening to the noise they radiate into water, has apparently been known for a very long time. Leonardo da Vinci is often mentioned as the first to propose this. But practical applications have been more recent, and the first underwater acoustic devices used efficiently were the passive detection systems developed by the Allies during World War I, to counter the then new threat of German submarines. Steerable earphones were submerged, enabling an operator to detect noise sources and locate their direction through binaural receiving.[2]

However, the idea that obstacles to navigation and targets could be detected by *active acoustic systems* had been studied from the beginning of the 20th century. This research effort had been spurred, in particular, by the loss of the *Titanic* in 1912. The same year, Fessenden built a first prototype of underwater electro-acoustic sources, based on an electrodynamic transducer. A major breakthrough came from Paul Langevin, a French physicist mainly known for his work in relativity: during

---

[2] In parallel, aerial acoustic systems were developed to detect aircraft, using the same angular location techniques.

history-making experiments on the River Seine and at sea, between 1915 and 1918, he demonstrated that it was possible to transmit signals, and to actively detect submarines, giving both their angles and distances from the receiver. His decisive innovation consisted in using a *piezoelectric transducer*, working at 38 kHz. This concept would be adopted by the majority of subsequent applications. But these discoveries would occur too late to be of use in World War I.

1.2.2   World War II

Sonar technology improved considerably between the two world wars. It benefited from the emergence of first-generation electronics and from progress in the newborn radio industry. At the beginning of World War II, the technology of active sonar was advanced enough to be used on a large scale by the Allied navies (these were the famed ASDIC systems of the Royal Navy). Sonar played a major role in the Battle of the Atlantic, pitching Allied convoys against German submarines. As soon as the USA entered the war in 1941, they made a huge effort in sonar research and development (as they did in radar and nuclear R&D). This greatly improved the performance of active sonar systems, as well as the understanding of underwater acoustic propagation, or the theories associated with detection and measurement of signals buried in noise. A large portion of the basic sonar knowledge still used today dates back to this time.[3]

1.2.3   After the War

Although the War ended in 1945, the "Cold War" between the Western and Eastern blocks ensured the efforts in scientific and technological research would continue. The strategic nuclear weapon race only slowed down with the breakdown of the USSR at the beginning of the 1990s. In the intervening decades, this situation justified huge efforts in developing underwater acoustic systems. In the West and in the Soviet Union, large programmes of research and experimentation were started. At the end of the 1950s, a new impulse was given by the appearance of nuclear submarines capable of launching strategic missiles (SSBNs), followed swiftly by the appearance of attack nuclear submarines (SSNs). This led to a complete review of underwater warfare strategies. Until then, sonar was used locally to monitor ship convoys or shipping corridors. Now, it had to be able to monitor vast areas of the ocean. In the 1960s, priority was thus given to the study of passive detection techniques, capable of achieving much larger ranges than active acoustics. A technological revolution occurred at the end of the 1960s, with the introduction of digital signal processing. This led to the dramatic increase in capabilities and versatility of sonar systems, particularly as computer performance was quickly evolving. The extreme degree of sophistication reached at this time by passive sonar was, however, countered by progress in lessening acoustic noise radiated by submarines. So the trend was again reversed in the 1990s, with a

---

[3] These studies were synthesised in *Physics of Sound in the Sea* (NRDC, 1945).

return of active sonar techniques, extended to lower frequencies (to reach larger ranges). Those conflicts at the end of the 20th century with a maritime component confirmed the importance of mastering sonar techniques to counter the threats of attack submarines (Falklands War) or of mines (Gulf War).

1.2.4   Civilian developments

In parallel with military developments, oceanography and industry were able to profit from the development of underwater acoustics. *Acoustic sounders* quickly replaced the traditional lead line to measure the water depth below a ship or to detect obstacles. It became widely used between the two world wars; these systems are today both scientific instruments and indispensable navigation tools. These same systems began to be used to detect fish shoals early in the 1920s; underwater acoustics has since progressively become one of the main elements of sea fishing and scientific monitoring of the biomass. The use of *sidescan sonars* to obtain "acoustic images" from the seabed became one of the major tools of marine geology after their invention in the early 1960s. The efficiency of seabed acoustic mapping increased dramatically with the emergence in the 1970s of *multibeam echosounders*, allowing multiplication of the number of simultaneous soundings. Merged with sidescan imagery at the end of the 1980s, this concept today allows the collection of maps of remarkable quality, which measure both the topography of the seabed and its *acoustic reflectivity* (i.e., yielding insights into its nature). The offshore oil and underwater industries were also concerned with acoustic mapping developments. They initiated the development of specific acoustic techniques for the *positioning* of ships or underwater vehicles, or for *data transmission.* In the domain of physical oceanography, the propagation of acoustic waves has been used since the 1970s to measure hydrological perturbations locally (*acoustic Doppler current profiling*) or on a medium scale (*ocean acoustic tomography*). Techniques of acoustic monitoring have even been suggested to monitor the evolution of the average temperature of large ocean basins on a permanent basis, as part of global climate studies.

1.3   OUTLINE OF UNDERWATER ACOUSTICS APPLICATIONS

Underwater acoustics is now part and parcel of most human activities at sea. Its technology is essential in mastering the oceans, scientifically, militarily or industrially. Underwater acoustics allows several kinds of application (detection, tracking, imaging, transmission and measurement), at a large range of scales. Despite the high constraints of both the natural and artificial environment, it answers most of today's needs. We will briefly show here the main modern applications of underwater acoustics. These will be detailed further in Chapters 7 and 8.

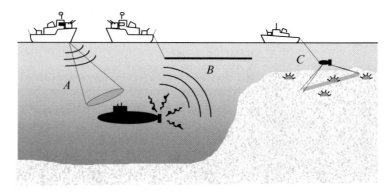

Figure 1.2. Examples of military applications: active sonar (*A*); passive sonar (*B*); minehunting active sonar (*C*).

1.3.1   Military applications

We have seen that most of the research and industrialisation effort in underwater acoustics was long linked to military applications. These systems are therefore mostly aimed at detecting, locating and identifying two types of target: submarines and mines.

Military sonars are classified into two main categories, depending on their mode of functioning (Figure 1.2):

- *Active sonars* – transmitting a signal and receiving echoes from a target (usually a submerged submarine). The measured time delay is used to estimate the distance between the sonar and its target, and receiving the signal on a suitable antenna completes the measurement with a determination of the angle of arrival of the signal. Further analysis of the echo allows identification of more characteristics of the target (e.g., its speed using the *Doppler effect*). *Minehunting* sonars are a particular type of active sonar with very high resolution – they are designed to detect and identify mines laid (or buried) on the seabed in coastal areas.
- *Passive sonars* – with no civilian equivalent. They are designed to intercept noises (and possible active sonar signals) radiated by a target vessel. Their main interest lies in their total stealth; they can be used on submarines as well as on the ships hunting them. Acquiring the target-radiated noise allows not only its detection, but also its location (from analysis of the spatial structure of the acoustic field received on a long enough antenna), and its identification (from analysis of its *acoustic signature*).

1.3.2   Civilian applications

Civilian underwater acoustics is a more modest sector of industrial and scientific activity, but it is highly varied and growing. It has been invigorated by the needs for

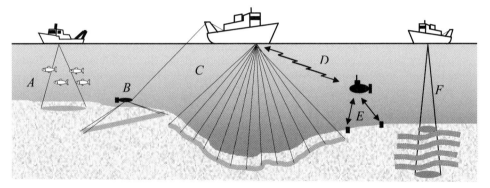

Figure 1.3. Examples of civilian applications: bathymetry or fishery sounder (*A*); sidescan sonar (*B*); multibeam sounder (*C*); data transmission system (*D*); acoustic positioning system (*E*); sediment profiler (*F*).

scientific instrumentation raised by large scientific programmes of environment study and monitoring, as well as developments in offshore engineering and industrial fishing. The main categories of systems are as follows (Figure 1.3):

- *Bathymetric sounders* – sonars specialised in the measurement of water depth. They transmit a signal downward, vertically, inside a narrow beam. They measure the time delay of the seabed echo. These sounders are now very well distributed and are universally used, from professional navigation to leisure yachting.
- *Fishery sounders* – designed for the detection and localisation of fish shoals. They are similar to bathymetric sounders, but they support additional tools to detect and process the echoes coming from the entire water column.
- *Sidescan sonars* – used for the acoustic imaging of the seabed. They allow high-accuracy observations. Placed on a platform towed close to the bottom, the sonar transmits, in a direction very close to the horizontal, a short pulse that sweeps the bottom. The signal, reverberated as a function of time, yields an image of the irregularities, obstacles and changes in structures. These systems are used in marine geology, or for the detection of mines and shipwrecks.
- *Multibeam sounders* – used for seafloor mapping. They are installed aboard oceanographic or industrial survey vessels to map the topography of the seabed accurately. A fan of elementary beams, transmitting athwartship, rapidly sweeps a large swathe of the seabed and measures its relief. If the angular aperture is large enough, the sounder can also provide acoustic images, like sidescan sonars.
- *Sediment profilers* – used for the study of the stratified internal structure of the seabed. These are single-beam sounders, similar to those used in bathymetry. But their frequency is much lower, enabling penetration of tens and even hundreds of metres, depending on the type of seafloor. In the same domain, *seismic* systems use explosive or percussive sources and long receiving antennas.

They can explore the seabed down to several kilometres deep, and are widely used in oil and gas exploration, and in geophysics.

- *Acoustic communication systems* (e.g., underwater telephone). Beyond their primary use as a phone link, they are also used for the transmission of digital data (e.g., remote control commands, images, results from measurements). Their performance is limited by the small bandwidths available, and by the difficulties inherent in underwater propagation. Rates of several kilobits per second are, however, achievable at distances of several kilometres.
- *Positioning systems* – used, for example, for the dynamic anchoring of oil drilling vessels or the tracking of submersibles or towed platforms. The mobile target is often located by measuring the time delays for signals coming from several fixed transmitters placed on the bottom. Different geometries are in fact achievable.
- *Acoustic Doppler systems* – they use the frequency shift of echoes to measure the speed of the sonar (and its supporting platform) relative to a fixed medium (*Doppler log*, used for navigation), or the speed of water relative to a fixed instrument (*Acoustic Doppler Current Profilers or ADCP*, used for scientific measurements and monitoring).
- *Acoustic tomography networks* – they use fixed transmitters and receivers. They measure propagation times (over large distances), or amplitude fluctuations (over small distances), to assess the structure of hydrological perturbations, using speed variations estimates.

2

Underwater acoustic wave propagation

The first aspect of underwater acoustics that we shall investigate is related to the propagation of acoustic waves in water: how can a signal go from one point to another? And what are the constraints and transformations it undergoes by so doing? The chapter will start with reminders about the fundamental notions associated with the physical nature of acoustic waves, and the orders of magnitude of their properties; in particular, the logarithmic notation (in decibels) used throughout this book is explained. The first effect of propagation is to decrease the amplitude of the signal, by geometrical effects on one part and by absorption on the other; the latter is linked to the chemical properties of sea water, and is a definite factor in the propagation of underwater acoustic waves, limiting their range at high frequencies. The estimation of propagation losses is an important factor in the evaluation of sonar system performance. As the underwater propagation medium is limited by two well-marked interfaces (the bottom and the sea surface), the propagation of a signal is often accompanied by a series of multiple paths generated by unwanted reflections at these two interfaces. Practically, these multiple echoes show up as bursts, or series, of replicas of the signal transmitted (at high frequencies), or as the setting of a spatial field of stable interferences (at low frequencies). Both are usually sources of trouble for the reception and exploitation of useful signals. Furthermore, the velocity of acoustic waves varies spatially in the ocean, mostly with depth, because of temperature and pressure constraints. The paths of sound waves are thus deviated, depending on the velocity variations encountered, and of course complicate the modelling and interpretation of the sound field spatial structure. The easiest and most efficient modelling technique uses geometrical acoustics, relating the local values of wave propagation direction and velocity (the Snell–Descartes law is the best known). The fundamental principles and formulas of this technique will be provided here, with the main formulas yielding the geometrical descriptions and durations of these paths. The ease of use of this technique and its very intuitive results make it the most popular approach in underwater acoustic

propagation modelling. Using this model, we shall introduce a few archetypal propagation configurations, corresponding to simple but typical variations in sound
velocity. The wave methods, such as normal modes or parabolic equation, are
more rigorous than the geometric approach. They aim at solving directly the
equation of wave propagation in a medium with non-constant velocity. At the end
of the chapter we shall introduce the fundamental principles of these approaches,
whose application is mostly useful for very low-frequency signals.

2.1   ACOUSTIC WAVES

2.1.1   Acoustic pressure

Acoustic waves originate from the propagation of a mechanical perturbation. Local
compressions and dilations are passed on from one point to the surrounding points,
because of the medium's elastic properties. From one point to the next (Figure 2.1),
this perturbation will propagate away from its source. The propagation rate of this
perturbation of the medium will be called *velocity*.

The same event $s(x, t)$ will be observed at different points in the medium at
different times. If the propagation only occurs along one spatial dimension (as in
Figure 2.1), one can write the equation:

$$s(x_1, t_1) = s(x_2, t_2) = s(x_3, t_3) \tag{2.1}$$

The observation times t_1, t_2, ... are linked to the locations x_1, x_2, ... and to the
propagation speed c:

$$\begin{cases} x_2 - x_1 = c(t_2 - t_1) \\ x_3 - x_2 = c(t_3 - t_2) \end{cases} \tag{2.2}$$

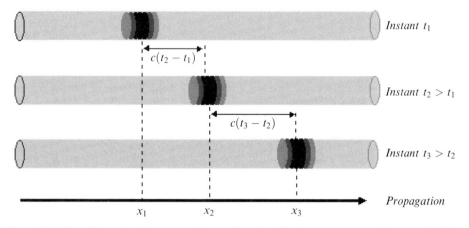

Figure 2.1. One-dimensional propagation of a localised overpressure in a cylindrical
waveguide, as a function of time.

An acoustic wave requires an *elastic material* support to propagate (i.e., gas, liquid or solid). The mechanical properties of this supporting medium dictate the propagation speed, or *velocity*.

The acoustic wave is characterised: by the amplitude of the "local movement" of each particle in the propagation medium around its position of equilibrium; by the corresponding particle velocity (not to be confused with the wave propagation velocity); and by the resulting acoustic pressure (variation around the average hydrostatic pressure).

Practically, *acoustic pressure* is the most often used quantity in underwater acoustics. Marine equivalents of aerial microphones, *hydrophones* used as receivers are in fact pressure transducers. The difference between the maximal and minimal values, or dynamics, of acoustic pressures is extremely high. Background noise measured in a narrow frequency band under quiet conditions may be only a few tens of microPascals (μPa), whereas the sound level close to a high-power source may reach $10^{12}\,\mu$Pa (i.e., more than 10 orders of magnitude higher!).

2.1.2 Velocity and density

The propagation velocity of an acoustic wave is imposed by the characteristics of the propagation medium: it depends on the density ρ and the elasticity modulus E:[1]

$$c = \sqrt{\frac{E}{\rho}} \tag{2.3}$$

In sea water, the velocity of the acoustic wave is close to $c = 1{,}500\,$m/s (usually between $1{,}450\,$m/s and $1{,}550\,$m/s, depending on pressure, salinity and temperature). The density of sea water is approximately $\rho = 1{,}030\,\mathrm{kg\,m^{-3}}$.

In a marine sediment (considered as a fluid medium, in the first approximation), the density ranges between $1{,}200\,\mathrm{kg\,m^{-3}}$ and $2{,}000\,\mathrm{kg\,m^{-3}}$. In a saturated sediment (where velocity is proportional to the velocity in water), the sound velocity ranges between $1{,}500\,$m/s and $2{,}000\,$m/s.

And, of course, in air the respective values of sound velocity and density are approximately $340\,$m/s and $1.3\,\mathrm{kg\,m^{-3}}$.

2.1.3 Frequency and wavelength

Acoustic signals are generally not instantaneous perturbations, but maintained vibrations. They are characterised by their frequency f (the number of vibrations per second, expressed in Hertz) or by their period T (period of an elementary vibration, linked to frequency by the relation $T = 1/f$). The frequencies used in

---

[1] The elasticity modulus quantifies the relative variation of volume or density due to pressure variations: $E = \left[-\dfrac{1}{V}\dfrac{\partial V}{\partial P}\right]^{-1} = \left[\dfrac{1}{\rho}\dfrac{\partial \rho}{\partial P}\right]^{-1}$. Water is much less compressible than air, hence its larger acoustic velocity.

underwater acoustics range roughly from 10 Hz to 1 MHz, depending on the application (i.e., periods of 0.1 s to 1 µs).

The wavelength is the spatial interval between two points in the medium, undergoing the same vibration with a phase shift of 2π. Stated differently, this is the
distance travelled by the wave during one period of the signal with velocity c. It
verifies therefore:

$$\lambda = cT = \frac{c}{f} \tag{2.4}$$

For a sound velocity of 1,500 m/s, the underwater acoustic wavelengths will be 150 m
at 10 Hz, 1.5 m at 1 kHz and 0.0015 m at 1 MHz.

These very diverse values of frequencies and wavelengths will of course correspond to different physical processes, both for the propagation of waves in water and
for the characteristics of the underwater acoustic system itself. The main constraints
on the frequencies usable for a particular application are:

- the dampening of sound waves in water, limiting the maximum usable range,
  whose effect increases very rapidly with frequency;
- the size of the sound sources, which increases with lower frequencies, for a given
  transmission power;
- the spatial selectivity related to the directivity of the acoustic sources and
  receivers, improving (for a given transducer size) as frequency increases;
- the acoustic response of the target, depending on frequency – a target will reflect
  less energy as its dimensions are small relative to the acoustic wavelength.

All of these points must be taken into consideration when choosing the frequency for
a particular application; the choice will always result in a compromise. Table 2.1
shows the frequency ranges commonly used for the different types of underwater
acoustic systems.

2.1.4   The wave equation and its elementary solutions

Acoustic waves propagating in gases and liquids follow the laws of fluid mechanics
(e.g., Medwin and Clay, 1998). It can be shown that their propagation is described
by the Helmholtz equation:

$$\Delta p = \frac{\partial^2 p}{\partial x^2} + \frac{\partial^2 p}{\partial y^2} + \frac{\partial^2 p}{\partial z^2} = \frac{1}{c^2(x, y, z)} \frac{\partial^2 p}{\partial t^2} \tag{2.5}$$

where p is the pressure of a wave propagating in the space (x, y, z) as a function of
time t, and $c(x, y, z)$ is the local propagation velocity of the wave. Δ is the Laplacian
operator.

In a medium where the velocity is constant, $c(x, y, z) = c$, and where the propagation is restricted to the single direction $\mathbf{x}$, Equation (2.5) becomes:

$$\frac{\partial^2 p}{\partial x^2} = \frac{1}{c^2} \frac{\partial^2 p}{\partial t^2} \tag{2.6}$$

Table 2.1. Frequency ranges of the main underwater acoustic systems and orders of magnitude of the maximum usable ranges (the latter are not valid for passive sonars, seismic sounders and sediment profilers).

Frequency (kHz)	0.1	1	10	100	1,000
Maximum ranges (km)	1,000	100	10	1	0.1
Multibeam sounders					
Sidescan sonars					
Transmission and positioning					
Active military sonars					
Passive military sonars					
Fishery echo sounders and sonars					
Acoustic Doppler current profilers					
Sediment profilers					
Seismics					

For a sinusoidal wave of frequency f_0, it can be shown that the solution of Equation (2.6) will be a pressure wave of the type:

$$p(t) = p_0 \exp\left(2j\pi f_0 \left(t - \frac{x}{c} \right) \right) \tag{2.7}$$

With a constant amplitude, and a phase depending on a single Cartesian space coordinate, the surfaces of constant phases associated with this wave will be planes. This type of wave is therefore called a *plane wave*.

The particle velocity **v** is related to the acoustic pressure by the equation:

$$\nabla p = -\rho \frac{\partial \mathbf{v}}{\partial t} \tag{2.8}$$

where ρ is the density and ∇ is the gradient operator.

For a plane wave propagating in the x direction, this relation simplifies into:

$$\frac{\partial p}{\partial x} = -\rho \frac{\partial v}{\partial t} \quad \Rightarrow \quad p_0 = \rho c v_0 = \rho c \omega a_0 \tag{2.9}$$

where $\omega = 2\pi f_0$ is the angular frequency.

The product ρc is the *characteristic impedance* of the propagation medium. It relates the average acoustic pressure to the corresponding movement of particles. The SI-related unit for acoustic impedance is the Rayleigh.[2] In a high-impedance medium like water ($\rho c \approx 1.5 \times 10^6$ rayl), a particle movement with a given amplitude

---

[2] In honour of the English physicist John William Strutt, Lord Rayleigh (1842–1919), one of the most famous names of acoustics, and author of *The Theory of Sound*.

will yield an acoustic pressure level much higher than in a low-impedance medium like air ($\rho c \approx 0.5 \times 10^3$ rayl).

When considering the propagation in the three directions of a space assumed to be isotropic, one can show that the solution to the wave equation is a *spherical wave*:

$$p(t) = \frac{p_0}{R} \exp\left(2j\pi f_0\left(t - \frac{R}{c}\right)\right) \tag{2.10}$$

The space variable considered here is the distance R to the source. The wave fronts are now spheres centred on the source ($R = 0$), and the wave's amplitude decreases in $1/R$, from its value p_0 considered 1 metre away from the source.

Plane and spherical waves are the two basic tools for modelling the propagation of acoustic waves. The former are the easiest to use; they are used when modelling can approximate the amplitude to a constant and the wave fronts show a negligible curvature. These conditions are satisfied far enough away from the sound source, when looking at the modelling of local processes. Spherical waves are descriptive of a field transmitted by a well-localised source and when one must account for the amplitude decreasing with propagation.

When the particles move in the direction of propagation, the waves are called *longitudinal waves* (or *compressional waves*). They are encountered in fluids (liquids and gases). Solid materials are also subject to *transverse waves* (or *shear waves*), in which displacement and propagation direction are perpendicular. They are characterised by two velocity values (longitudinal and transverse), which of course makes the models more complicated. In underwater acoustics, most propagation processes can be described by compressional waves. Shear waves are taken into account when describing propagation in consolidated sediments and scattering by solid targets.

2.1.5   Intensity and power

The propagation of a sound wave is associated with an *acoustic energy*. This energy can be decomposed into a kinetic part (corresponding to particle movements) and a potential part (corresponding to the work done by elastic pressure forces).

The acoustic intensity I is the mean value of the energy flux by unit of surface and time. One can show it equals the average of the product of acoustic pressure by particle velocity. For a plane wave of amplitude p_0, this yields:

$$I = \frac{p_0^2}{2\rho c} \quad \text{(in Watts/m}^2\text{)} \tag{2.11}$$

The acoustic power P received by a surface S is the intensity, corrected for the surface considered. For the plane wave, this would be:

$$P = I \times S = \frac{p_0^2 S}{2\rho c} \quad \text{(in Watts)} \tag{2.12}$$

Like acoustic pressure, intensity and power can vary enormously. A high-power sonar transmitter may deliver acoustic power of several tens of kilowatts, whereas a nuclear submarine in silent mode might radiate only a few milliwatts!

2.2 LOGARITHMIC NOTATION: DECIBELS AND REFERENCES

2.2.1 The decibel

Because of their huge dynamics, acoustic values like pressure or energy are usually quantified on a logarithmic scale, and noted in *decibels* (dB). By definition, the decibel corresponds to ten times the base-10 logarithm of the ratio of two powers. For example, $10 \log_{10}(P_1/P_2)$ quantifies the difference between two powers P_1 and P_2, and a 10-dB difference means that P_1 is in fact 10 times higher than P_2.

The decibel can be adapted to other physical values; for example, the acoustic pressure. As the power P_i is proportional to the square of pressure p_i^2, the same ratio can be expressed as:

$$10 \log\left(\frac{P_1}{P_2}\right) = 10 \log\left(\frac{p_1^2}{p_2^2}\right) = 20 \log\left(\frac{p_1}{p_2}\right) \tag{2.13}$$

We shall therefore remember to use $10 \log_{10}(X_1/X_2)$ for quantities akin to energies (e.g., power, intensity) and $20 \log_{10}(x_1/x_2)$ for quantities akin to acoustic pressure.

A difference of 3 dB between two signals will therefore correspond to an energy ratio of 2, and an amplitude ratio of $\sqrt{2}$, of their acoustic pressures. A 10-dB difference will correspond to an energy ratio of 10 and an amplitude ratio of $\sqrt{10} \approx 3.1$. A 1-dB difference (which one can consider as a practically achievable limit of the accuracy of current underwater acoustic measurements) corresponds to around 10% variation in acoustic pressure.

One should also remember that the product of several variables translates into the sum of their values in dB ($[AB]_{dB} = A_{dB} + B_{dB}$), because of the properties of logarithmic operations. But the sum in dB of physical values does not equal the sum of the individual values in dB. Table 2.2 shows the increase of the level of a signal A when adding (in energy terms, i.e. $10 \log$) a second signal B, depending on their level differences. For example, if the level of B is larger than the level of A by 5 dB, the level of the sum $A + B$ will be 6.2 dB larger than the level of A alone.

Table 2.2. Correspondence between the level difference (in dB) between two quantities (akin to energies), and the level increase obtained by their summations.

$B_{dB} - A_{dB}$	0	1	2	3	4	5	7	10	12	15	20
$[A + B]_{dB} - A_{dB}$	3.0	3.5	4.1	4.8	5.5	6.2	7.8	10.4	12.3	15.1	20.0
$[A + B]_{dB} - B_{dB}$	3.0	2.5	2.1	1.8	1.5	1.2	0.8	0.4	0.3	0.1	0.0

2.2.2   Absolute references and levels

A reference level is necessary if one is to give absolute pressure or intensity levels in dB. In underwater acoustics, the pressure reference is usually the microPascal ($p_{ref} = 1\,\mu Pa$):[3]

$$p_{dB} = 20\log\left(\frac{p}{p_{ref}}\right) \tag{2.14}$$

The absolute acoustic pressure is therefore expressed in dB relative to $1\,\mu Pa$, noted dB/1 μPa or dB re $1\,\mu$Pa. The reference must be explicitly given, as the definition of the reference level for underwater acoustic pressure can be a source of confusion:

- On one part, this absolute pressure reference of $1\,\mu$Pa is different from the reference in air, $20\,\mu$Pa (average human hearing threshold at 1 kHz). A similar absolute pressure level will therefore be expressed as "absolute decibels", with a value higher by $20\log(20) = 26\,dB$, if considered in water rather than in air.
- On the other part, one must know that the underwater acoustic reference for pressure has long been the microbar, worth $10^5\,\mu$Pa. Absolute levels in dB/1 μPa are higher by $20\log(10^5) = 100\,dB$ than those expressed in dB/1 μbar.

The relation between the expressions of a signal pressure in dB (re $1\,\mu$Pa) and of its intensity (re $1\,W/m^2$) are easy to deduce for a plane wave. As $I = p^2/2\rho c$, with I in Watts/m^2 and p in Pascals, one gets:

$$10\log\left(\frac{I}{1\,W/m^2}\right) = 20\log\left(\frac{p}{1\,Pa}\right) + 10\log\left(\frac{1}{2\rho c}\right)$$

$$\approx 20\log\left(\frac{p}{1\,\mu Pa}\right) + 20\log\left(\frac{1\,\mu Pa}{1\,Pa}\right) - 64.8$$

$$\approx 20\log\left(\frac{p}{1\,\mu Pa}\right) - 184.8 \tag{2.15}$$

2.3   BASICS OF PROPAGATION LOSSES

When acoustic waves propagate, the most visible process is their loss of intensity, because of geometric spreading (*divergence* effect) and absorption of acoustic energy by the propagation medium itself. This *propagation loss* (or *transmission loss*) is a key parameter for acoustic systems, as it will constrain the amplitude of the signal received, hence the receiver's performance, directly dependent on the signal-to-noise ratio.

---

[3] The Pascal (Pa) is the SI-derived unit for pressure; it is defined as $1\,N/m^2$. Correspondences with other pressure units are given in A.1.1.

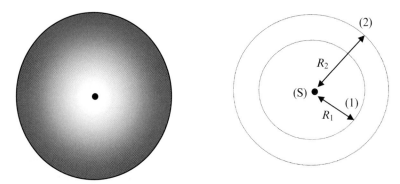

Figure 2.2. Spherical spreading: the acoustic intensity decreases with distance from the source, in inverse proportion to the sphere's surface.

2.3.1 Geometrical spreading losses

An acoustic wave propagating from a sound source will spread the transmitted acoustic energy on a larger and larger surface. As energy is conserved, the intensity will decrease proportionally to the inverse of the surface. This process corresponds to the *geometric spreading loss*.

The simplest (and most useful) case is the homogeneous, infinite medium, with a small-dimension source radiating in all directions (point source). The energy transmitted is conserved, but will be spread over spheres of larger and larger radii (Figure 2.2).

The decrease in local acoustic intensity between points (1) and (2) (Figure 2.2) is inversely proportional to the ratio of the surfaces S_1 and S_2 of the spheres:

$$\frac{I_2}{I_1} = \frac{S_1}{S_2} = \left(\frac{4\pi R_1}{4\pi R_2}\right)^2 = \left(\frac{R_1}{R_2}\right)^2 \tag{2.16}$$

where R_i is the radial distance from the source. Intensity decreases in $1/R^2$, and pressure decreases in $1/R$. This is similar to the amplitude variation law for spherical waves. The spreading transmission loss considered from the reference unit distance $R_{1m} = 1m$ can be expressed in dB:

$$TL = 20 \log\left(\frac{R}{R_{1m}}\right) \tag{2.17}$$

Common usage is to write $TL = 20 \log R$, with no reference to the unit distance. Although incorrect, this usage is convenient and widespread.

2.3.2 Attenuation losses

Sea water is a *dissipative* propagation medium; it absorbs part of the energy of the transmitted wave, which is dissipated through viscosity or chemical reactions. The local amplitude decrease is proportional to the amplitude itself; the acoustic pressure then decreases exponentially with distance. This will add to spreading losses. For example, for a spherical wave, the pressure becomes:

$$p(R,t) = p_0 \exp(-\gamma R) \frac{\exp\left(2j\pi f_0\left(t - \frac{R}{c}\right)\right)}{R} \tag{2.18}$$

Attenuation is quantified by the parameter γ (expressed in Neper/m). This additional exponential decrease of pressure corresponds to a decrease proportional to the distance, expressed in dB. This results in an *attenuation coefficient* α expressed in decibels per metre (dB/m), related to γ by $\alpha = 20\gamma \log e \approx 8.686\gamma$. Note that, in fact, attenuation of sound in sea water is most often expressed in dB/km.

Attenuation is often the most limiting factor in acoustic propagation. Its amount depends strongly on the propagation medium and the frequency. In sea water, attenuation comes from:

- the viscosity of pure water;
- the relaxation of magnesium sulphate ($MgSO_4$) molecules above 100 kHz;
- the relaxation of boric acid ($B(OH)_3$) molecules above 1 kHz.

Molecular relaxation (see, e.g., Medwin and Clay, 1998) consists in the dissociation of some ionic compounds in solution (e.g., $MgSO_4$ and $B(OH)_3$), due to local pressure variations created by the acoustic wave. This process is dominant for sound absorption in sea water. If the period of the local pressure variation is longer than the time necessary for the molecule to recompose itself (*relaxation time*), the process is reproduced at every change and permanently dissipates energy. The attenuation due to this process appears at frequencies lower than the characteristic relaxation frequency of the relevant compound. Ever since the beginnings of underwater acoustics, a high amount of attention has been brought to the modelling of absorption coefficients, and many models have been proposed. The most recent ones are under the form:

$$\alpha = C_1 \frac{f_1 f^2}{f_1^2 + f^2} + C_2 \frac{f_2 f^2}{f_2^2 + f^2} + C_3 f^2 \tag{2.19}$$

The first two terms of this equation show the contributions from the two relaxation processes; the third term corresponds to the viscosity of pure water. The relaxation frequencies f_i and coefficients C_i depend on temperature, pressure and salinity; they are determined by experiments in the laboratory or at sea.

The model most used today was developed by Francois and Garrison (1982a, b). This model is shown in Figure 2.3; it explicitly uses temperature, salinity and hydrostatic pressure, as well as frequency. Based upon a large number of previous experimental results and theoretical studies, it is very complete and precise; its use is highly recommended.

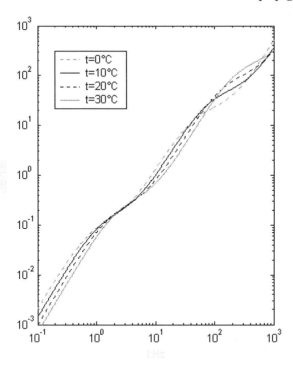

Figure 2.3. Sound absorption coefficient in sea water, as a function of frequency, for a salinity of 35 p.s.u., a depth of zero and different values of temperature.

Model of Francois–Garrison

The absorption coefficient is decomposed into three terms, corresponding to the contributions of boric acid, magnesium sulphate and pure water:

$$\alpha = A_1 P_1 \frac{f_1 f^2}{f_1^2 + f^2} + A_2 P_2 \frac{f_2 f^2}{f_2^2 + f^2} + A_3 P_3 f^2 \tag{2.20}$$

where α is the attenuation, in dB/km; z is the depth, in m; S is the salinity, in p.s.u. (practical salinity units); T is the temperature, in °C; and f is the frequency, in kHz.

The contribution of boric acid $B(OH)_3$ reads:

$$\begin{cases} A_1 = \dfrac{8.86}{c} 10^{(0.78pH-5)} \\[2mm] P_1 = 1 \\[2mm] f_1 = 2.8 \sqrt{\dfrac{S}{35}} 10^{\left(4 - \frac{1245}{T+273}\right)} \\[2mm] c = 1412 + 3.21T + 1.19S + 0.0167z \end{cases} \tag{2.21}$$

The contribution of magnesium sulphate $Mg(SO)_4$ reads:

$$\begin{cases} A_2 = 21.44 \dfrac{S}{c}(1 + 0.025T) \\[2mm] P_2 = 1 - 1.37 \times 10^{-4}z + 6.2 \times 10^{-9}z^2 \\[2mm] f_2 = \dfrac{8.17 \times 10^{\left(8 - \frac{1990}{T+273}\right)}}{1 + 0.0018(S - 35)} \end{cases} \qquad (2.22)$$

The contribution of pure water viscosity reads:

$$\begin{cases} \qquad\qquad P_3 = 1 - 3.83 \times 10^{-5}z + 4.9 \times 10^{-10}z^2 \\[2mm] T < 20°C \quad \Rightarrow \quad A_3 = 4.937 \times 10^{-4} - 2.59 \times 10^{-5}T \\[2mm] \qquad\qquad\qquad\quad + 9.11 \times 10^{-7}T^2 - 1.5 \times 10^{-8}T^3 \\[2mm] T > 20°C \quad \Rightarrow \quad A_3 = 3.964 \times 10^{-4} - 1.146 \times 10^{-5}T \\[2mm] \qquad\qquad\qquad\quad + 1.45 \times 10^{-7}T^2 - 6.5 \times 10^{-10}T^3 \end{cases} \qquad (2.23)$$

Figure 2.3 shows that attenuation increases very rapidly with frequency, and that the orders of magnitude are highly variable. For frequencies of 1 kHz and less, sound attenuation is of a few hundredths of dB per km, and is therefore not a limiting factor. At 10 kHz, a coefficient of around 1 dB/km forbids ranges of more than tens of kilometres. At 100 kHz, the attenuation coefficient reaches several tens of dB/km and the range cannot extend further than 1 km. Underwater systems using frequencies in the MHz range will be limited to ranges of less than 100 m, with an attenuation of several hundreds of dB/km.

Depth dependence

Depth is another very important factor for the attenuation coefficient. It can strongly influence local system performance (e.g., sidescan sonar or data transmission). At a lesser degree, it will influence the signals transmitting throughout the entire water column (e.g., bathymetric systems).

 If the frequency is high enough that the relaxation effect of $MgSO_4$ is predominant, it is often accurate enough to multiply the attenuation coefficient by the coefficient P_2 of Francois and Garrison's model (Equation 2.22). Table 2.3 gives the evolution of P_2 as a function of depth.

 For example, at 100 kHz, if the attenuation is 40 dB/km at the surface of, say,

Table 2.3. Evolution of the depth-correction coefficient P_2 with depth (Francois and Garrison's model).

Z(m)	0	500	1,000	1,500	2,000	2,500	3,000	3,500	4,000	4,500	5,000	5,500	6,000
P_2	1.00	0.93	0.87	0.81	0.75	0.70	0.64	0.60	0.55	0.51	0.47	0.43	0.40

Table 2.4. Evolution of the absorption correction term $A(H)$ as a function of the water depth.

H(m) 0	500	1,000	1,500	2,000	2,500	3,000	3,500	4,000	4,500	5,000	5,500	6,000
$A(H)$ 1.00	0.97	0.93	0.90	0.86	0.83	0.79	0.76	0.73	0.69	0.66	0.62	0.59

Mediterranean-type water ($S = 38.5$ p.s.u., $T = 14°C$), it decreases to 30 dB/km at 2,000 m depth, and 22 dB/km at 4,000 m. These variations, of course, create very important differences in the maximum ranges attainable.

For particular applications concerned with total attenuation over the entire water column (e.g., for the signal of a depth sounder, propagating from the ship to the bottom and back to the ship again), one will look at the integrated attenuation:

$$\hat{\alpha}(H) = \frac{1}{H} \int_0^H \alpha(z)\, dz \qquad (2.24)$$

Using the same assumptions, and assuming as well that the variations of salinity and temperature with depth z are negligible, the mean coefficient now reads:

$$\hat{\alpha}(H) = \alpha(0)A(H) = \alpha(0)\left[1 - 1.37\,10^{-4}\frac{H}{2} + 6.2 \times 10^{-9}\frac{H^2}{3}\right] \qquad (2.25)$$

where $\alpha(0)$ is the value of attenuation at the surface. The values of $A(H)$ are given in Table 2.4.

Once again, these values only give orders of magnitude showing the effect of depth on attenuation, and they cannot describe every specific case. For accuracy in practical applications, it is recommended to use complete equations, using the temperature and salinity profiles neglected here.

2.3.3 Conventional propagation loss

Spherical spreading, corrected from attenuation, is systematically used as a first approximation when evaluating the propagation loss and the performance of underwater acoustic systems. In dB, the transmission loss reads:

$$TL = 20 \log R + \alpha R \qquad (2.26)$$

Variations of TL with range and frequency are presented in Figure 2.4. All systems using the echo from a target are undergoing propagation losses on the outgoing *and* returning paths; the total loss is therefore:

$$2TL = 40 \log R + 2\alpha R \qquad (2.27)$$

One must pay attention to the units used in Equations (2.26) and (2.27). R is expressed in metres, and the exact expression should in fact be $20 \log(R/R_{1m})$. But α is most often expressed in dB/km, and the units should be converted appropriately.

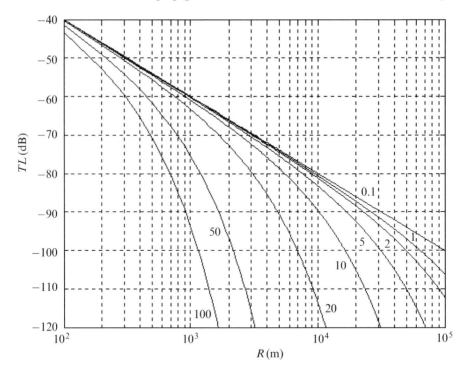

Figure 2.4. Conventional propagation loss $-TL = -20 \log R - \alpha R$, as a function of range, for frequencies of 0.1, 1, 2, 5, 10, 20, 50 and 100 kHz, respectively. The absorption coefficient α is calculated for ambient conditions: $T = 10°C$; $S = 35$ p.s.u.; $z = 10$ m.

   This simple formula is often sufficient to evaluate the performance of underwater acoustic systems. However, some applications require specifically adapted propagation models (geometric rays, normal modes, parabolic approximation). This is in particular necessary when spatial velocity variations induce refraction of wave paths, or when some interfaces induce multiple concurrent paths, as geometric spreading cannot be approximated by spherical spreading alone.

2.3.4   Effect from air bubbles

Air bubbles are created by sea surface movements, and in some cases by the displacement of a ship. They form an inhomogeneous layer, close to the surface, whose dual-phase mixing (water and gas) modifies locally and strongly the acoustic characteristics of the propagation medium (velocity and attenuation). This process decreases in importance as depth increases, because the hydrostatic pressure increases. Below 10 to 20 m, the effect of bubbles can be neglected.

   This local perturbation of the propagation medium has several effects on the functioning of hull-mounted sonar:

- additional attenuation to that of sea water, and resulting in the absorption of the signals transmitted, or the masking of receiving transducers;
- local modification of sound velocity, inducing refraction;
- parasite backscattering, which can be observed, for example, in the first moments of echoes recorded by a sounder.

These processes can degrade the performance of an acoustic instrument until it fails completely, as commonly observed in practice. According to their amplitude, they can simply decrease slightly the levels of echoes, but they can also degrade the quality of the measurements, and finally induce detection loss or parasite echoes.

It is of course helpful to control as much as possible the many parameters acting on these performance-limiting processes. The presence of bubble clouds depends for one part on factors linked to the surrounding environment, and is therefore uncontrollable (as are weather-induced surface agitation, biological activity, ship wakes, etc.). But the shape of the hull and the speed of the instrument-bearing platform, as well as the position and geometry of the transducers, can be controlled. One must therefore be very careful, when studying the design and installation of sonar antennas on the hull, that antennas are placed away from shallow water-streaming areas. This is done by studying the hydrodynamic behaviour of the hull, using numerical simulations or in a test tank, and usually leads to the installation of the transducers in the forward part of the hull.

Surface bubbles also modify the reflection characteristics of waves at the sea surface, which is masked by absorbing layers. This effect needs to be accounted for when modelling sonar signals propagating along multiple paths and reflecting at the surface.

Many theoretical studies and experimental work have been conducted to look at bubbles and their acoustic behaviour. Each bubble acts as a spherical obstacle with a very high acoustic impedance contrast, and therefore as a scatterer of the incident sound wave. The scattering effect will be maximum around the intrinsic resonance frequency of the bubble (Medwin and Clay, 1998):

$$f_R = \frac{3.25}{a} \sqrt{1 + 0.1z} \qquad (2.28)$$

where f_R is the resonance frequency (in Hertz), z is the depth (in m) and a is the radius of the gas sphere (in m).

Visco-thermal absorption by the walls of the air bubble adds to the scattering effects. When the bubble population increases, the intensity of the transmitted sound wave is decreased by the sum of the scattering processes (dominated by the contribution from resonating bubbles) and the absorption processes. The loss in transmitted intensity can be modelled with an equivalent absorption coefficient, depending on the individual characteristics of bubbles and their statistical size distribution.

The individual behaviour of the bubbles, or the behaviour of populations with simple statistical distributions, is therefore well understood. For a really useful practical application, the problem lies in the prediction of the perturbations provoked by evolving, inhomogeneous bubble populations. The models available from the literature are generally quite complex, and difficult to summarise here. For a first approach, it is possible to use the following simple formula (APL, 1994), quantifying the additional loss for a path reflecting on the surface and crossing the entire surface bubble layer:

$$
\begin{cases}
TL_b = \dfrac{1.26 \times 10^{-3}}{\sin \beta} \nu_w^{1.57} f^{0.85} & \text{for } \nu_w \geq 6\,\text{m/s} \\[2mm]
TL_b = TL_b(\nu_w = 6)\, e^{1.2(\nu_w - 6)} & \text{for } \nu_w < 6\,\text{m/s}
\end{cases}
\tag{2.29}
$$

where TL_b (in dB) is the two-way transmission loss coming from the bubbles, ν_w (in m/s) is the wind speed 10 m above the surface, f (in kHz) is the signal frequency and β is the grazing angle. The value of $\nu_w = 6$ m/s corresponds to the threshold for the existence of crashing waves that generate bubbles.

Developments in the theory of bubble acoustics can be found in Medwin and Clay (1998) or Leighton (1994). For more practical models taking into account the depth of the sonar, velocity variations and the depth-velocity profile, see Hall (1989), or APL (1994).

2.4   MULTIPLE PATHS

2.4.1   Notion of multiple paths

Because the propagation medium is limited by the sea surface and the seabed, the signals transmitted undergo successive reflections at the interfaces. Variations in sound velocity within the medium may also deform the paths of sound waves. Due to these processes, a given signal can therefore propagate from a source to a receiver along several distinct paths, corresponding to distinct directions and durations. The main, "direct" signal arrives along with a series of echoes, of amplitudes decreasing with the number of reflections undergone. The time structure of the signal to be processed is of course somewhat affected, and the performance of a system can be highly degraded by these parasite signals, in particular for data transmission applications. The number of noteworthy multiple paths is highly variable, depending on the configuration; none in the best case, several tens or even hundreds in long-distance propagation cases (these multiple paths are then impossible to distinguish individually, and show up as a continuous signal trail).

Multiple paths may be considered from two points of view, depending on the configuration. In high frequency, for short signals (shorter than the typical delay between path arrivals), their effect is observable in the time domain, with typical sequences of multiple echoes (Figure 2.5). Whereas, for low-frequency permanent signals at fixed frequencies, the contributions add together permanently; this creates a stable interference pattern, with strong variations in the field amplitude (a simple case is described in Section 2.4.2).

These multiple paths are typical of underwater acoustics, and can be very penalising. Their effects will be different from multiple paths encountered in electro-

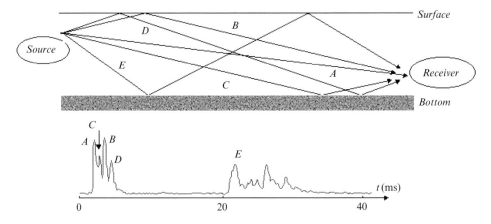

Figure 2.5. (*Top*) multiple paths: (*A*) direct path; (*B*) reflection on the surface; (*C*) reflection on the bottom; (*D*) reflection on the surface and bottom; (*E*) reflection on the bottom and surface. (*Bottom*) multiple paths as visible in a time domain signal.

magnetic propagation, even if they both correspond to similar physical processes. The multiple paths of radio or radar waves interfere and cause *fading*, but the delayed time echo is not always troublesome when decoding the received signal (because the speed of light induces very short delays). In underwater acoustics, on the contrary, the propagation velocity is low, and the time delays are much more important (Figure 2.5). They create distinguishable echo and reverberation effects. The physical anisotropy of the waveguide (velocity variations, influence of the interfaces) is also much more marked than in electromagnetic propagation.

2.4.2   Sea-surface interference

We shall see in Chapter 3 that, when a sound wave hits the sea surface from below, it can be reflected with a reflection coefficient $V \approx -1$ (no change in amplitude, and a phase shift of π). In low-frequency transmission configurations close to horizontal, this yields interesting effects since the surface-reflected signal interferes with the direct path signal. The coherent summation of the two signals produces interference fringes. Noting (Figure 2.6) that D represents the horizontal range, and z_s and z_r the respective depths of the source and the receiver, the direct spherical length reads:

$$R_d = \sqrt{D^2 + (z_s - z_r)^2} \qquad (2.30)$$

And the surface-reflected spherical length reads:

$$R_s = \sqrt{D^2 + (z_s + z_r)^2} \qquad (2.31)$$

Considering that z_s and z_r are small relative to D, these expressions can be approximated:

$$\begin{cases} R_d = D\sqrt{1 + \dfrac{(z_s - z_r)^2}{D^2}} \approx D + \dfrac{(z_s - z_r)^2}{2D} \\[4mm] R_s \approx D + \dfrac{(z_s + z_r)^2}{2D} \end{cases} \qquad (2.32)$$

This leads to:

$$R_s \approx R_d + \frac{2z_s z_r}{D} \qquad (2.33)$$

Disregarding the amplitude terms (since the spherical ranges are almost identical for the two paths), and accounting for the sign change of the reflected wave, the field at the receiver is proportional to the sum:

$$p \propto \exp(-jk R_d) - \exp(-jk R_s)$$

$$= \exp(-jk R_d)\left[1 - \exp\left(-2jk\frac{z_s z_r}{D}\right)\right] \qquad (2.34)$$

The resulting intensity is thus proportional to:

$$|p|^2 \propto 1 - \cos\left(2k\frac{z_s z_r}{D}\right) \qquad (2.35)$$

It presents a series of maxima and minima as a function of range. For instance, the minima are given by:

$$2k\frac{z_s z_r}{D} = 2n\pi, \text{ with } n = 1, 2, 3 \ldots \quad \Rightarrow \quad D_n = k\frac{z_s z_r}{n\pi} = 2\frac{z_s z_r}{n\lambda} \qquad (2.36)$$

The interference nulls get closer in range as n increases. The longest range with a pressure minimum is given by $n = 1$; that is:

$$D_1 = 2\frac{z_s z_r}{\lambda}$$

Figure 2.6 shows interference fringe patterns as a function of range alone, then range and frequency.

2.4.3   An ideal model of multipath propagation

If the propagation medium is stratified (i.e. the bottom is horizontal) and characterised by a constant velocity, the signal transmitted by the source arrives at the receiver by a series of rectilinear paths undergoing multiple reflections on the surface and on the bottom. These different paths correspond to the transmission from image sources obtained by successive symmetries of the source relative to the surface and bottom (Figure 2.7).

Image sources are defined by the difference z_{ij} between their ordinates and the ordinate z_r of the receiver. Four kinds of source exist, corresponding to different

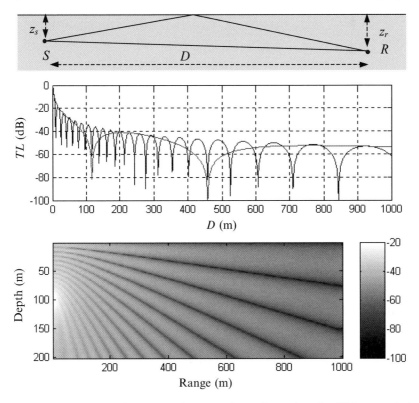

Figure 2.6. Sea surface interference: (*top*) geometric configuration; (*middle*) transmission loss as a function of horizontal range, for two frequencies ($f = 10$ and $100\,\mathrm{Hz}$, $z_s = z_r = 200\,\mathrm{m}$); (*bottom*) pressure field in dB, as a function of range and depth ($f = 100\,\mathrm{Hz}$, $z_s = 100\,\mathrm{m}$).

recurrence relations for the successive reflections, and using the water height H and the depths of the transmitter z_s and the receiver z_r:

$$
\begin{cases}
z_{1i} = 2(i-1)H + z_s - z_r \\
z_{2i} = 2(i-1)H + z_s + z_r \\
z_{3i} = 2iH - z_s - z_r \\
z_{4i} = 2iH - z_s + z_r
\end{cases}
\tag{2.37}
$$

with $i = 1, 2, 3, \ldots$

The oblique distance corresponding to a source ji equals $R_{ji} = \sqrt{z_{ji}^2 + D^2}$, D being the horizontal distance from source to receiver. The grazing angle of the corresponding path is $\beta_{ji} = \arctan\left(\dfrac{z_{ji}}{D}\right)$, and the travel time is $\tau_{ji} = \dfrac{R_{ji}}{c}$.

The contribution of each image source can be identified with the transmission of a spherical wave, thus with a propagation loss in $\exp(-\gamma R)/R$ (γ being the water absorption coefficient). Losses and phase shifts at the reflections on the surface and

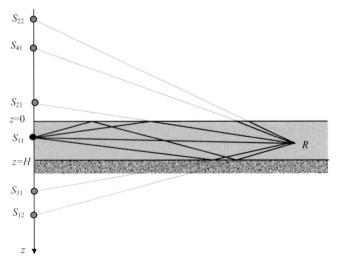

Figure 2.7. Image sources in a constant velocity water layer.

at the bottom (for an image source S_{ij}) are first calculated from the (complex) reflection coefficients of the plane waves $V_s(\beta)$ and $V_f(\beta)$ for $\beta = \beta_{ji}$. The characteristics of these reflection coefficients will be detailed further in Chapter 3. The number of reflections at the surface and at the bottom is a function of the orders i and j:

$$\begin{cases} N_{S_{1j}} = i - 1; N_{b_{1j}} = i - 1 \\ N_{S_{2i}} = i; N_{b_{2i}} = i - 1 \\ N_{S_{3i}} = i - 1; N_{b_{3i}} = i \\ N_{S_{4i}} = i; N_{b_{4i}} = i \end{cases} \tag{2.38}$$

If $S(t)$ is the transmitted signal, the contribution of each image source can be written as:

$$p_{ji}(t) = S(t - \tau_{ji}) \frac{\exp(-\gamma R_{ji})}{R_{ji}} V_s^{N_{s_{ji}}}(\beta_{ji}) V_s^{N_{b_{ji}}}(\beta_{ji})$$

$$= S(t - \tau_{ji}) A_{ji} \tag{2.39}$$

And the resulting signal is:

$$p(t) = \sum_{i=1,2,\dots} \sum_{j=1}^{4} p_{ji}(t) \tag{2.40}$$

The impulse response of the medium can be written with this model:

$$H(t) = \sum_{i=1,2,\dots} \sum_{j=1}^{4} A_{ji} \delta(t - \tau_{ji}) \tag{2.41}$$

where δ is the Dirac distribution.

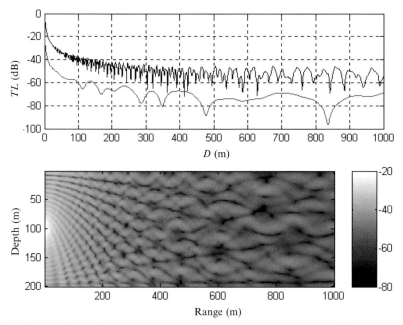

Figure 2.8. Transmission loss TL computed from image sources ($H = 200$ m; $z_s = 100$ m; $c_2 = 1,700$ m/s; $\rho_2 = 1.8$; $\alpha_2 = 0.5$ dB/λ_2): (*top*) as a function of the range ($z_r = 100$ m) for frequencies of 500 Hz and 50 Hz, respectively (the latter lowered by 20 dB for the drawing); (*bottom*) pressure field in dB as a function of range and depth at 100 Hz, to be compared with Figure 2.6.

This ideal model can be used as a first approximation for shallow water cases, or deep water/small ranges. It yields easy formulations of the energy, time and angular variations of the field; an example of energy transmission loss is given in Figure 2.8. This model can be used, for example, to simulate signals under multiple path conditions, in order to assess the performance of particular processing techniques.

2.4.4   Average energy flux in a waveguide

In specific configurations, for example at large ranges in shallow water, the transmitted field is composed of many paths propagating by successive reflections on the surface and on the bottom. The acoustic energy remains trapped between these two boundaries. If the signal frequency is high enough that the field oscillations can be considered as random, it can be interesting to model the average energy flux. The basic principle is to consider, as a function of the cylindrical distance r:

- The spherical propagation from the source, until the entire channel depth is insonified, defining a transition range r_0 (Figure 2.9). This implies the existence of a limit angle, caused by either reflection on the channel boundaries or refraction effects.
- Cylindrical propagation from r_0 onwards. Instead of spherically diverging, the

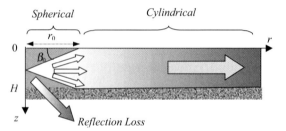

Figure 2.9. The energy flux model: transition at r_0 from spherical to cylindrical.

acoustic energy spreads on cylindrical surfaces limited by the channel bound-aries. Its intensity thus decreases in $10\log(r/r_0)$.

For instance, in a constant-velocity channel of height H, with a bottom imposing a maximum grazing angle β_0, the effective angular sector is defined by the interval $[-\beta_0; +\beta_0]$. Outside this interval, the sound waves quickly disappear because of the bottom reflection losses. Supposing the source is at mid-height, the transition range is obviously defined as (Figure 2.9):

$$r_0 = \frac{H}{2\tan\beta_0} \tag{2.42}$$

The average transmission loss (in dB) then reads:

$$\begin{cases} TL = 20\log r + \alpha r & r < r_0 \\ TL = 20\log r_0 + 10\log\dfrac{r}{r_0} + \alpha r = 10\log(rr_0) + \alpha r & r < r_0 \end{cases} \tag{2.43}$$

This approach is relevant for rapid calculations of the transmission losses under propagation conditions dominated by numerous boundary-reflected multipaths, and when only the coarse characteristics of the field are needed. Figure 2.10, for example, presents the average intensity of the acoustic field, corrected for the standard deviation of a Rayleigh distribution[4] of pressure, and compared with a full coherent calculation obtained by image sources. As a first approximation, it gives very simply a fair estimate of the average field and its fluctuation rate (except, for the latter, at short ranges).

This concept may be extended to propagation media where the sound velocity is not constant, or where there are slight losses at the boundaries. The transmitted field cannot then be taken as whole: it has to be decomposed into its angular components, and the cyclic characteristics of the various beams must be detailed (see Appendix A.2.3 for details).

---

[4] See Section 4.6.3.2. The standard deviation of a Rayleigh distribution is 0.522 times its average. Therefore, these values in dB are $20\log(1.522) = 3.6\,\mathrm{dB}$ and $20\log(0.478) = -6.4\,\mathrm{dB}$.

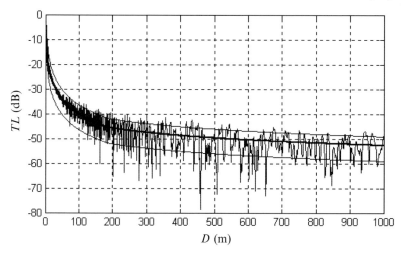

Figure 2.10. Comparison of the average intensity flux (monotonous curves) with a complete computation (image sources in an isovelocity medium, fluctuating curve). The thick line is the average intensity level, the thin ones account for $+/-$ the standard deviation of a Rayleigh amplitude distribution (see Section 4.6.3.2) ($f = 1,000\,\text{Hz}$; $H = 200\,\text{m}$; $z_s = 100\,\text{m}$; $z_r = 100\,\text{m}$; $c_2 = 1,700\,\text{m/s}$; $r_0 = 187\,\text{m}$).

2.4.5   General sound field prediction

The elementary solutions presented earlier (image sources, losses in $20\log R + \alpha R$, average energy flux) are quite simple, but they should not mask the fact that solving the generalised Helmholtz equation is one of the main theoretical problems in acoustic propagation. It has motivated researchers for many decades. The complexity of the solutions, and the tools necessary to obtain them, depends on the velocity field $c(x, y, z)$. The techniques used change with the frequency range considered:

- At high frequencies, the Helmholtz equation can be approximated (*eikonal equation*) to a formula describing trajectories in the sound ray space (similar to optics); its behaviour follows local velocity variations. With this approach, one can follow energy travelling in the waveguide, and reconstruct the field at each point as a sum of multiple paths (similarly to the source–image method, but accounting for path refractions). The time and angle domain characteristics of the signals thus predicted are excellent, and the predictions of energy levels are generally acceptable.
- At low frequencies, one would use modal field analysis when the velocity only depends on depth, or the parabolic equation approximation when the velocity also depends on distance. In both cases, the monochromatic coherent field is calculated very accurately in each point of the waveguide.

These approaches will be briefly discussed in the rest of this chapter, along with some details about the ray-tracing equations, since they are the most used in practice.

Much more detailed descriptions can be found in specialist books (e.g., Brekhov-skikh and Lysanov, 1992; Frisk, 1994; Jensen *et al.*, 1994).

2.5   OTHER DEFORMATIONS OF UNDERWATER ACOUSTIC SIGNALS

2.5.1   Doppler effect

The Doppler effect corresponds to a shift of the apparent signal frequency after propagation, due to a change in the duration of the source–receiver paths during transmission time, caused by the relative displacement of the source and the receiver, or the source and the target.

Let us consider a source transmitting Dirac-like pulses with a regular period T, towards a receiver at distance D. If D does not vary with time, the receiver gets each pulse after a time $t = D/c$. The period of the signal is therefore not modified, and the apparent frequency remains $f_0 = 1/T$.

Conversely, if D decreases (for example) with time as $D(t) = D - v_r t$, because of the relative speed v_r between the source and the receiver, the length of time between the pulses received will change too.

If the first pulse (transmitted at $t = 0$) arrives at time $t_1 = D(t_1)/c$, the second pulse (transmitted at $t = T$) arrives at time:

$$t_2 = T + \frac{D(t_2)}{c} = T + \frac{D(t_1) - v_r T}{c} \tag{2.44}$$

The time lag between the two successive receptions is then:

$$t_2 - t_1 = T - \frac{v_r T}{c} = T\left(1 - \frac{v_r}{c}\right) \tag{2.45}$$

instead of T. The apparent frequency of the signal is then modified:

$$f = \frac{1}{T\left(1 - \dfrac{v_r}{c}\right)} = \frac{f_0}{\left(1 - \dfrac{v_r}{c}\right)} \approx f_0\left(1 + \frac{v_r}{c}\right) \tag{2.46}$$

The frequency variation δf created by the Doppler effect is thus:

$$\delta f = f_0 \frac{v_r}{c} \tag{2.47}$$

where v_r is positive when moving closer, negative when moving away. For an echo on the target, as the sound travels both ways, we get:

$$\delta f = 2 f_0 \frac{v_r}{c} \tag{2.48}$$

The Doppler effect complicates the processing of signals, particularly in communication and data transmission applications. But it can be put to good use for some applications. For example, its measurement can be used to determine the speed of a ship relative to the bottom or the water column (Doppler log, see Section 7.1.3 for

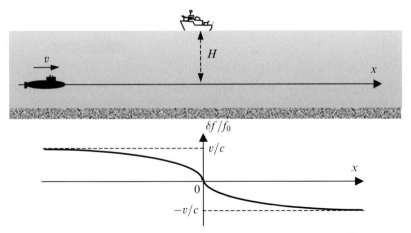

Figure 2.11. Effect of a Doppler shift between a mobile source and a fixed receiver.

details), or to measure the characteristics of a marine current (see Section 7.5.3 for details).

In Anti-Submarine Warfare (ASW), the Doppler shift is useful to track a target, and even identify it by measuring its speed. Let us consider an acoustic source passing by a fixed receiver (Figure 2.11). The source has a linear speed v, and the minimum range between the source and the receiver is represented by H. It can be shown that the radial speed component equals:

$$v_r = \frac{xv}{\sqrt{x^2 + H^2}} \tag{2.49}$$

Or, in terms of frequencies:

$$\frac{\delta f}{f_0} = \frac{v}{c} \frac{x}{\sqrt{x^2 + H^2}} \tag{2.50}$$

This last expression shows that the Doppler shift goes to zero and changes sign at the closest point of approach $x = 0$.

The relative frequency variations observed using sonar can be large: $\delta f/f_0 \approx 0.7\%$ for a relative speed of 10 knots (18.5 km/h), much larger than the same variations encountered with radar (0.0002% for a plane travelling at 1,000 km/h, the velocity of radar waves being 3×10^8 m/s).

2.5.2   Time characteristics of the echoes

The duration of the signal transmitted strongly influences the shape and the level of the signals received. The concept of "point" target (see Section 3.2.1) assumes the signal is long enough to insonify the target totally at a given time. Long signals allow the maximum instantaneous energy of the response. Conversely, short signals will

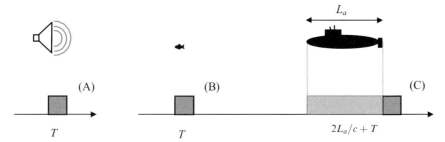

Figure 2.12. Lengthening of a time signal: signal transmitted (A), backscattered by a point target (B) and by a long target (C).

allow separation of the different components of the global echo (e.g., to distinguish close targets).

The signal scattered back is lengthened by δt. This lengthening is a function of the apparent length L_a of the target insonified along the propagation direction (Figure 2.12). It reads:

$$\delta t = \frac{2L_a}{c} \qquad (2.51)$$

A similar effect of signal lengthening occurs during reflections on the interfaces; the signal transmitted intercepts part of the surface or the seabed, and the signal reflected reproduces the extension of this reflector. Finally, as we saw earlier, the global time structure of the signal received depends on the multiple paths.

In all cases, the modelling of underwater acoustic signals must be complemented with the fluctuating and random characteristics of the propagation medium. This leads one to consider statistically the physical processes, and construct models accordingly. These processes are discussed in the second part of Chapter 4.

2.6   SOUND VELOCITY IN THE OCEAN

2.6.1   Velocity parameters

Sound velocity depends on temperature, salinity and depth, all at the same time, and varies with them as well.

- *Temperature.* Globally, seawater temperature decreases from the surface to the seabed. But there are many local variations. The time and space variations are maximal in the shallower layers (surface mixing, solar heating, currents, external inputs), but decrease with depth. Beyond a limit depth (usually around 1,000 m in open oceans, but shallower in closed seas: e.g., 100–200 m in the Mediterranean), the average temperature remains stable, decreasing very slowly with depth and not varying much from one place to another. A fine temperature microstructure gets added to the average temperature profile, and creates local random fluctuations.

- *Depth*. Hydrostatic pressure[5] makes the sound velocity increase with depth, because of variations in the elasticity modulus. This increase is linear as a first approximation, of around 0.017 m/s per metre down.
- *Salinity*. Sea water is made of a mix of pure water and dissolved salts (NaCl, $MgSO_4$, ...). The latter's percentage mass defines salinity, expressed in practical salinity units[6] or p.s.u.

Salinity in the large ocean basins (Atlantic, Pacific and Indian Oceans) is on average about 35 p.s.u., but locally it can vary as a function of the hydrological conditions. In closed seas, the average value of salinity can be quite different from 35 p.s.u., depending on whether evaporation is more important (e.g., 38.5 p.s.u. in the Mediterranean) or freshwater input is more important (e.g., 14 p.s.u. in the Baltic). At a given point, salinity will usually only slightly vary with depth (1 to 2 p.s.u.). But there might be stronger variations with depth in local freshwater input configurations: river estuaries, ice melt, etc.

2.6.2 Velocity models

The impact of sound velocity variations on acoustic propagation was first noticed and has been systematically studied from the 1940s (NDRC, 1945). However, local velocity measurements are quite difficult to perform accurately, whereas its constitutive parameters (temperature, salinity, depth) are much more easily quantified. Parametric models have been extensively studied, and several sound velocity models are now available in the literature (e.g., Wilson, 1960; Leroy, 1969; Del Grosso, 1974; Chen and Millero, 1977).

As a first approximation, one can use the formula proposed by Medwin (1975), which is quite simple but limited to a 1,000-m depth:

$$c = 1449.2 + 4.6t - 0.055t^2 + 0.00029t^3 + (1.34 - 0.01t)(S - 35) + 0.016z \quad (2.52)$$

where c is the sound velocity, in m/s; t is the temperature, in °C; z is the depth, in m; and S is the salinity, in p.s.u.

More recent and accurate models are, however, recommended. One such model, proposed by Chen and Millero (1977), is now widely used. It is endorsed by UNESCO, and used as a standardized reference model:

$$c = c_0 + c_1 P + c_2 P^2 + c_3 P^3 + AS + BS^{3/2} + CS^2 \quad (2.53)$$

---

[5] Hydrostatic pressure in sea water is given precisely by the Leroy formula (Leroy, 1968):

$$P = [1.005\,240\,5(1 + 5.28 \times 10^{-3} \sin^2 \phi)z + 2.36 \times 10^{-6}z^2 + 10.196] \times 10^4$$

in which the pressure P is in Pa, ϕ is the latitude (in °) and z is the depth (in metres).

[6] Salinity was traditionally expressed as per thousandths of the seawater mass, noted as ‰ or p.p.t. (part per thousand). It is now expressed in practical salinity units. The numerical values do not change from one system to the other.

The first four terms correspond to the contribution of pure water, and the next three terms to salinity, all being influenced by pressure (e.g., depth). c is the sound velocity, in m/s; P is the hydrostatic pressure, in bars ($P = 0$ at the surface, although the actual pressure there equals 1 bar); t is the temperature in °C; and S is the salinity in p.s.u.

The parameters of the model are:

$$
\begin{cases}
c_0 = 1,402.388 + 5.037\,11t - 5.808\,52 \times 10^{-2}t^2 + 3.3420 \times 10^{-4}t^3 \\
\qquad - 1.478\,00 \times 10^{-6}t^4 + 3.1464 \times 10^{-9}t^5 \\
c_1 = 0.153\,563 + 6.8982 \times 10^{-4}t - 8.1788 \times 10^{-6}t^2 + 1.3621 \times 10^{-7}t^3 \\
\qquad - 6.1185 \times 10^{-10}t^4 \\
c_2 = 3.1260 \times 10^{-5} - 1.7107 \times 10^{-6}t + 2.5974 \times 10^{-8}t^2 - 2.5335 \times 10^{-10}t^3 \\
\qquad + 1.0405 \times 10^{-12}t^4 \\
c_3 = -9.7729 \times 10^{-9} + 3.8504 \times 10^{-10}t - 2.3643 \times 10^{-12}t^2 \\
A = A_0 + A_1 P + A_2 P^2 + A_3 P^3 \\
A_0 = 1.389 - 1.262 \times 10^{-2}t + 7.164 \times 10^{-5}t^2 + 2.006 \times 10^{-6}t^3 - 3.21 \times 10^{-8}t^4 \\
A_1 = 9.4742 \times 10^{-5} - 1.2580 \times 10^{-5}t - 6.4885 \times 10^{-8}t^2 + 1.0507 \times 10^{-8}t^3 \\
\qquad - 2.0122 \times 10^{-10}t^4 \\
A_2 = -3.9064 \times 10^{-7} + 9.1041 \times 10^{-9}t - 1.6002 \times 10^{-10}t^2 + 7.988 \times 10^{-12}t^3 \\
A_3 = 1.100 \times 10^{-10} + 6.649 \times 10^{-12}t - 3.389 \times 10^{-13}t^2 \\
B = -1.922 \times 10^{-2} - 4.42 \times 10^{-5}t + (7.3637 \times 10^{-5} + 1.7945 \times 10^{-7}t)P \\
C = -7.9836 \times 10^{-6}P + 1.727 \times 10^{-3}
\end{cases}
$$

$$(2.54)$$

Sea water contains many inhomogeneities: bubble layers close to the surface, living organisms (e.g., fish, plankton), mineral particles in suspension, etc. These are as many potential scatterers of acoustic waves, especially at high frequencies, and cause perturbations of the acoustic characteristics (sound velocity and absorption). Bubble clouds are a dual-phase medium, in which sound velocity can be significantly different from the sound velocity in sea water, particularly if the resonant frequency of a portion of the bubble population is close to the frequency of the acoustic wave. The velocity c_{bw} in water filled with gas bubbles of identical radius a is related to the sound velocity c_w in normal sea water:

$$
c_{bw} = c_w \left[1 - \frac{aNc_w^2}{2\pi f^2} \frac{f_0^2/f^2 - 1}{(f_0^2/f^2 - 1)^2 + \delta^2} \right]
$$

$$(2.55)$$

N is the number of bubbles per unit volume, f_0 is their resonant frequency and δ is the attenuation constant (see Section 3.3.3 for details).

2.6.3 Sound velocity measurements

Two types of system are commonly used for measuring sound velocity *in situ*. The depth-velocimeter directly measures the sound velocity (usually from phase) for a high-frequency wave transmitted over a perfectly calibrated distance. This type of instrument is very delicate, as it must be insensitive to constraints caused by important variations in temperature and hydrostatic pressure. A more common and simple device is the bathythermograph, or XBT (eXpendable BathyThermo-graph) probe. This device only measures water temperature as a function of depth. To deduce sound velocity, this means accessing independently the salinity profile. The latter can be measured simultaneously by a conductance meter, inte-grated into the same device (XCTD probe). Most often, however, one refers to databases, giving the average values of salinity to a sufficient accuracy. The bathy-thermograph is the most used instrument; the most widespread models can be used on board a ship in motion, with expendable probes.

The accurate knowledge of sound velocity profiles over a particular area is necessary for many sonar applications. In ASW, general knowledge of sound velo-cities is necessary to optimise detection capabilities. In bathymetric surveys, it is imperative so that sound ray trajectories can be accurately compensated. These applications require the joint use of *in situ* measurements and statistical databases (e.g., Levitus, 1982).

2.6.4 Depth–velocity profiles

The environment can often be approximated as horizontally stratified; sound velocity only depends on depth, which greatly simplifies the modelling of propaga-tion processes. The resulting sound velocity profiles (SVP) are made up of several characteristic parts (Figure 2.13):

- A homogeneous layer (*mixed layer*) of constant sound velocity, often present in the first few metres. It corresponds to the mixing of shallow water through surface agitation.
- A *surface channel*, corresponding to a sound velocity increasing from the surface down. This channel is often due to a shallow isothermal layer appearing in winter conditions, but can also be created by water that is very cold at its surface (e.g., ice melting), or by an input of fresh water near river estuaries.
- A *thermocline*, monotonous variation of temperature with depth. It is most often negative (temperature decreasing usually from the surface to the bottom), and then induces a velocity decrease with depth. It can be seasonal (close to the surface) or permanent.
- A *deep channel*, corresponding to a sound velocity minimum (e.g., between a negative thermocline and a deep isothermal layer). The average depth–velocity

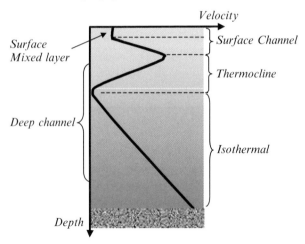

Figure 2.13. The different elements of a typical sound velocity profile. Sound velocity is represented on the *x*-axis and depth on the *y*-axis (going down).

profile of large ocean basins shows a deep channel between a few hundred metres and 2,000 metres.
- An *isothermal layer*, at constant temperature. Sound velocity increases linearly with depth, because of hydrostatic pressure. The deeper layers in the ocean are approximately isothermal. So are the sound velocity profiles in closed seas (e.g., the Mediterranean) or in shallow waters in winter conditions.

These different elements can combine to form very varied profiles, depending on local conditions. Local processes can also complicate matters:

- At intermediate spatial scales (tens of kilometres), the average velocity profiles are modified by the presence of currents, thermal fronts and gyres. These processes perturb propagation strongly; conversely, they can be studied by acoustic means (ocean acoustic tomography).
- At high latitudes, the presence at the surface of very cold water coming from the melting of the ice pack corresponds to a very noticeable minimum velocity.
- As an example of water mass exchange between ocean basins in the Gibraltar Strait, Mediterranean waters intrude into the North-East Atlantic, inducing there a warmer and more salty layer, with a marked maximum of velocity, between 1,000 and 2,000 m in depth.
- Near river estuaries, freshwater input creates a slower shallow layer, inducing important perturbations (the velocity difference between fresh water and sea water is ca. 40 m/s).

Other factors are the internal waves and tides, related to variations in water density with depth. When propagating, they create fluctuations of sound velocity profiles over large spatial scales. Their effect on acoustic propagation has been studied since the 1970s (see, e.g., Flatté, 1979, for a more detailed study).

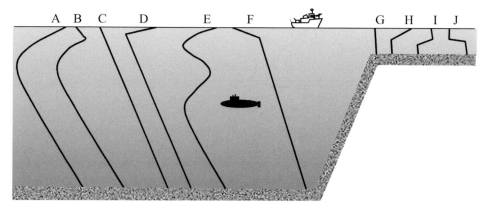

Figure 2.14. Generic sound velocity profiles. From *left* to *right*: (A) summer SOFAR channel[7]; (B) winter SOFAR; (C) winter Mediterranean (isothermal); (D) summer Mediterranean; (E) North-East Atlantic (with Mediterranean water layer); (F) polar; (G) winter shallow water; (H) summer shallow water; (I) autumn shallow water; (J) shallow water with fresh water at the surface.

Generic depth–velocity profiles are shown in Figure 2.14 for different types of water body.

2.7 GEOMETRICAL INVESTIGATION OF THE ACOUSTIC FIELD

In underwater acoustic modelling, it may be more convenient to deal with either grazing angles (referenced relative to the horizontal) or incidence angles (referenced relative to the vertical). Logically enough, the first convention is preferred in applications related to in-water "horizontal" propagation geometries involving multipaths (mainly military sonar applications, but also ocean acoustic tomography). And the second convention is chosen for vertical and oblique transmission applications (e.g., echo sounders and imaging sonars). In the following, we shall use both conventions, with this notation to avoid confusion:

- "horizontal" grazing angles will be represented by β; they will mainly be used in this chapter, devoted to propagation processes;
- "vertical" incident angles will be noted θ; they will also be used in Chapter 3 (refraction and backscattering) and in Chapter 8 (about seafloor mapping sonars).

2.7.1 Depth–velocity profile refraction

Let us now look at the modifications of the characteristics of a plane wave, at the interface between two homogeneous fluids with different sound velocities (c_1, c_2) (Figure 2.15). The change of sound velocity between the two media induces

---

[7] SOFAR: SOund Fixing And Ranging (see Section 2.8.3.1 for details).

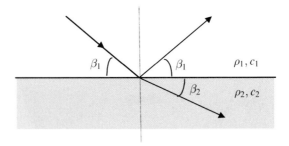

Figure 2.15. Reflection and refraction of a plane wave, due to the change in sound velocity at the interface.

specular reflection of the wave in the first medium (along a direction symmetrical to the normal at the incidence point), and refraction of the wave in the second medium, at an angle given by the famous law of Snell–Descartes:

$$\frac{\cos \beta_1}{c_1} = \frac{\cos \beta_2}{c_2} \tag{2.56}$$

Equation (2.56a) can only be written if $\cos \beta_2 \leq 1$ (i.e., for $\cos \beta_1 \leq c_1/c_2$). The limit angle given by the equation:

$$\beta_c = \arccos(c_1/c_2) \tag{2.57}$$

is the *critical angle* for the interface. For angles larger than β_c (high grazing incidence), there is *total reflection*; transmission into the second medium is then impossible.

   The Snell–Descartes law can be applied to a series of constant-velocity layers, located by the indices $i = 1, 2, \ldots$ The law then becomes:

$$\frac{\cos \beta_i}{c_i} = \frac{\cos \beta_{i+1}}{c_{i+1}} \tag{2.58}$$

It can be used to describe the behaviour of a wave propagating in a medium with non-constant velocity, along a coordinate z. The refraction relation is generalised into:

$$\frac{\cos \beta(z)}{c(z)} = \text{constant} \tag{2.59}$$

Continuous changes in the sound velocity of a propagation medium will progressively change the initial direction of the wave (Figure 2.16); the orientation of the wave vector depends on the local velocity. If the velocity gradient is vertical (which is most common), an increase in velocity will refract the wave towards the horizontal. If the velocity increases enough, an acoustic path inclined relative to the horizontal can undergo *total refraction*. Conversely, a decrease in velocity will increase the grazing angle of the wave. A wave path between two points will be called an *acoustic ray*.

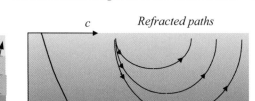

Figure 2.16. Refraction of a wave with a discontinuous (*left*) and continuous (*right*) change of the sound velocity with depth.

2.7.1.1 Case of a linear sound speed profile

If the depth–velocity law is linear ($c(z) = c_0 + g(z - z_0)$), the Snell–Descartes equation then becomes:

$$\cos \beta(z) = \frac{c(z)}{c_0} \cos \beta_0 = \left(1 + \frac{g}{c_0}(z - z_0) \right) \cos \beta_0 \qquad (2.60)$$

For any circle in the (x, z) plane, the general relations between the Cartesian coordinates of a point, the slope angle β and the circle radius R_c may be written as:

$$\begin{cases} x - x_0 = R_c(\sin \beta - \sin \beta_0) \\ z - z_0 = R_c(\cos \beta - \cos \beta_0) \end{cases} \qquad (2.61)$$

x_0, z_0 and β_0 define an arbitrary reference point of the circle. The cosine of the angle at a given point may then be linked with z:

$$\cos \beta(z) = \cos \beta_0 + \frac{z - z_0}{R_c} \qquad (2.62)$$

Formally, it is similar to equation (2.60). And therefore, in the case of a linear sound speed profile, the waves will follow circles of radii R_c depending on the input angle β_0:

$$R_c = \frac{c_0}{g \cos \beta_0} \qquad (2.63)$$

2.7.1.2 Generalisation

Geometric acoustics[8] aims at modelling the structure of the acoustic field as a set of paths (or rays) following the basic principles bulleted at the top of the next page:

---

[8] The theoretical bases of geometric acoustics are formally and rigorously derived from the wave equation. It can be shown that, as long as the frequency is high enough, the Helmholtz equation leads to an equation named "eikonal", linking the cosines of the plane wave vector to the local velocity conditions. This approach is detailed in many classical textbooks (e.g., Officer, 1958; Brekhovskikh and Lysanov, 1992; Tolstoy and Clay, 1987).

- refraction of the propagation direction by velocity changes, according to the law of Snell–Descartes;
- specular reflection at the interfaces;
- intensity losses along rays, through geometric divergence (spherical divergence being then modified by refraction), through absorption along the paths and through reflections on the interfaces.

It is therefore a generalisation of the basic model of image sources presented in Section 2.4.3, because it takes into account refraction by the sound velocity profile.

We saw that, for a linear profile, a very simple form of paths (arc circles) can be obtained. Other analytical solutions can be derived for velocity profiles different from the linear case. But they are of little practical interest for a general use.

In practice, the generalisation to more complicated velocity profiles (Figure 2.14) can be done by joining elementary layers with constant gradients. This allows tracing the paths for any velocity condition. One can remark that, beyond the simple tracing of acoustic paths, the *ray method* can provide the transmitted acoustic intensity (the spreading loss being estimated by calculating the distance between two neighbouring rays), and travel times (through integration along the paths). The main formulae of this approach are provided in the next section.

2.7.2   Sound ray calculations

Let $c(z)$ be the sound velocity profile. The profile is approximated by a finite number $N + 1$ (indices $n = 0, \ldots, N$) of points defining N layers (indices $n = 1, \ldots, N$), inside which the velocity is assumed to vary linearly (Figure 2.17).

In layer n, the sound velocity gradient reads:

$$g_n = \frac{c_n - c_{n-1}}{z_n - z_{n-1}} \tag{2.64}$$

The Snell–Descartes law then reads:

$$\frac{\cos \beta_n}{c_n} = \frac{\cos \beta_{n-1}}{c_{n-1}} = \text{constant} \tag{2.65}$$

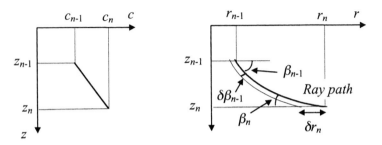

Figure 2.17. Geometry and notation of a ray in layer n.

Figure 2.18. Sign convention for the ray angles.

A ray with an entry angle β_{n-1} in the layer will undergo circular refraction with the curvature radius:

$$R_{cn} = \frac{c_{n-1}}{g_n \cos \beta_{n-1}} \tag{2.66}$$

Due to the cylindrical symmetry of the problem, in the following the horizontal coordinate is the cylindrical radius r. The location of the current point in the layer is given by:

$$\begin{cases} r - r_{n-1} = \dfrac{c_{n-1}}{g_n \cos \beta_{n-1}} (\sin \beta - \sin \beta_{n-1}) \\[2ex] z - z_{n-1} = \dfrac{c_{n-1}}{g_n \cos \beta_{n-1}} (\cos \beta - \cos \beta_{n-1}) \end{cases} \tag{2.67}$$

where β is the local angle, positive or negative depending on whether the ray is going up or down relative to the horizontal (Figure 2.18).

The travel time in layer n is given by:

$$t_n = \left| \frac{1}{g_n} \ln \left(\frac{\tan\left(\left| \dfrac{\beta}{2} \right| \right)}{\tan\left(\left| \dfrac{\beta_{n-1}}{2} \right| \right)} \right) \right| \tag{2.68}$$

One can remark that:

$$\tan \frac{\beta}{2} = \frac{\sin \beta}{1 + \cos \beta} \tag{2.69}$$

One can therefore deduce the equivalent expressions for the travel time:

$$t_n = \left| \frac{1}{g_n} \ln \left(\left| \frac{\sin \beta}{\sin \beta_{n-1}} \frac{1 + \cos \beta_{n-1}}{1 + \cos \beta} \right| \right) \right| = \left| \frac{1}{g_n} \ln \left(\frac{c_{n-1}}{c} \left| \frac{1 + \cos \beta_{n-1}}{1 + \cos \beta} \right| \right) \right| \tag{2.70}$$

The curvilinear path length is:

$$s_n = \frac{c_{n-1}}{g_n \cos \beta_{n-1}} (\beta - \beta_{n-1}) \tag{2.71}$$

This quantity is used to calculate the propagation loss by absorption.

At reflection on the interface, the ray is reflected specularly, with an angle $\beta_s' = -\beta_s$. It then follows a reflected path symmetrical to its incident path, relative to the impact point defined by $r = r_s$.

Across a gradient change, r, z and β must be continuous at the transition; the same follows for $\partial r / \partial \beta_0$ for geometric spreading (see below for more details). These conditions being fulfilled, all the above formulae are applicable layer by layer.

2.7.3   Losses from geometric spreading

Computing the losses from geometric spreading mainly consists in following the evolution of a section of geometric beam with an infinitesimal opening $\delta\beta_0$ at the source. The intensity loss at the point of abscissa r relative to the reference point r_0 reads:

$$\frac{I(r)}{I(r_0)} = \frac{dS(r_0)}{dS(r)} = \frac{\cos\beta_0}{r\left|\dfrac{\partial r}{\partial\beta_0}\right|\sin\beta} \tag{2.72}$$

Expressing this transmission loss in dB:

$$TL = 10\log\left[r\left|\frac{\partial r}{\partial\beta_0}\right|\frac{\sin\beta}{\cos\beta_0}\right] \tag{2.73}$$

The only difficulty of this expression lies in the term $|\partial r/\partial\beta_0|$, whose evolution must be followed layer after layer, with the formulae, respectively, giving the evolution of the beam aperture (Equation 2.74) and the evolution of the beam section projected along r (Equation 2.75):

$$\delta\beta_n = \frac{\tan\beta_{n-1}}{\tan\beta_n}\delta\beta_{n-1} \tag{2.74}$$

$$\delta r_n = \delta r_{n-1} + (r_n - r_{n-1})\frac{\delta\beta_{n-1}}{\cos\beta_{n-1}\sin\beta_n} \tag{2.75}$$

The above expression of the loss is not valid when $\sin\beta \to 0$ (ray observed horizontally). This is only because we chose to project the ray on the r axis. It would be enough to replace $|\partial r/\partial\beta_0|\sin\beta$ with $|\partial z/\partial\beta_0|\cos\beta$.

The geometric approach to transmission losses is no longer valid when $|\partial r/\partial\beta_0| \to 0$; the convergence of a geometric beam would create locally infinite intensities. In such areas of the sound field, called *caustics*, we are at the limits of the geometric method. Ideally, one should compute the exact solution of the propagation equation (e.g., Brekhovskikh and Lysanov, 1992), and apply locally the field thus computed.

2.7.4   Application of geometric acoustics

We have just seen that it is possible to build acoustic paths for rather complex velocity variations, just by using elementary geometric constructs, which can be easily implemented numerically on computers. We restricted ourselves to a variation of velocity with depth only, but it is perfectly possible to trace rays with velocities varying as a function of two or three spatial coordinates. The solution is then not to look for analytical forms (like the arc circles in Section 2.7.1), but solve directly and numerically the eikonal equation, following the ray path with a small numerical increment (the Runge–Kutta finite difference technique is often used).

Ordinarily, ray-tracing programmes calculate the paths of rays emitted in an angular sector (in the vertical plane) close to the source, with a relatively small increment (typically a tenth of a degree or less), up to the maximum range of interest. This first computation phase having been achieved, to ascertain the rays arriving at a particular reception point, one must determine the couples of rays emitted consecutively whose paths border the target point. Each couple thus determined corresponds to a different "history" of the ray (number of reflections and refractions). The exact characteristics of the rays arriving at the receiver (angle, time, phase, transmission loss) are finally computed by interpolating between the characteristics of the two rays of the selected couples.

The geometric description of the acoustic field is used in most applications aiming at the accurate determination of propagation times: bathymetry measurements by sounders, acoustic ocean tomography, etc. This technique has many advantages, which explains its predominance. First, its graphical representation of the acoustic field structure as paths is easily and intuitively understandable. Second, it describes accurately the structure of the field in a receiving point with propagation times and directions of arrival, which are key parameters for many applications, by numerical calculations that are relatively simple and rapid. Finally, although approximate, its prediction of intensity levels is good enough in most cases. It is only inadequate for very low frequencies (tens and hundreds of Hertz), at which wave processes dominate the purely geometric aspect of refractions.

2.8 UNDERWATER ACOUSTIC PROPAGATION – CASE STUDIES

We shall now look at a few typical propagation configurations imposed by the depth–velocity profile. These different configurations will be illustrated with acoustic ray tracing, and considerations about the time domain response.

2.8.1 Constant-velocity profile

In this configuration, the acoustic rays are rectilinear, and the tracing of the acoustic field is very easily obtained (Figure 2.19). It is not even necessary to trace rays *sensu stricto*, and the field can be practically described as a sum of contributions from image sources (cf. Section 2.4.3), corresponding to reflections at the interfaces. These contributions are obtained by successive symmetries of the actual source relative to the surface and the bottom.

Each image source corresponds to a particular ray of easily computable characteristics. The source and receiver being a distance D away, for a path defined by its source angle β determined in Section 2.4.3, the propagation time reads:

$$t(\beta) = \frac{D}{c \cos \beta}$$

The equivalent horizontal velocity of the path is:

$$V(\beta) = c \cos \beta \tag{2.76}$$

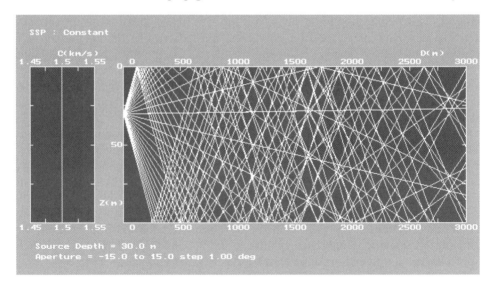

Figure 2.19. Ray tracing for a constant-velocity configuration.

The propagation loss is given by the oblique distance travelled $R(\beta) = D/\cos\beta$:

$$TL(\beta) = 20\log R(\beta) + \alpha R(\beta) \qquad (2.77)$$

It gets smaller as β decreases.

Reflection losses increase too as the grazing angle increases (see Chapter 3 for more details). The direct path, corresponding to the minimal distance of β, arrives first and has the maximum amplitude; the multiple reflected paths follow it with decreasing amplitudes (Figure 2.20).

Although clearly too idealised, the approximation of a constant-velocity profile is very useful in many configurations of shallow water propagation. It is used by default in the evaluation of sonar system performance. Finally, it is often used as a reference configuration, and is very useful for teaching purposes.

2.8.2   Isothermal profile

When the temperature does not vary with depth, the sound velocity increases with depth because of the hydrostatic pressure, with a gradient of around $0.017\,\text{m/s/m}$. The acoustic ray paths are therefore refracted upward, and propagate by successive bounces on the surface. This configuration is called a *surface channel*. This type of condition is rather common in shallow water; however, the acoustic field is then very close to the constant-velocity case, because the small velocity contrast between the surface and the bottom induces very few refractions. More significantly, one can also

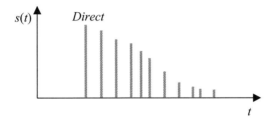

Figure 2.20. Impulse response for a constant-velocity channel: the direct path arrives first, the amplitude of multipaths decrease with their order (longer length, more reflections).

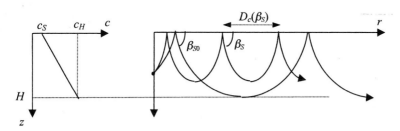

Figure 2.21. Ray geometry with isothermal sound velocity profiles.

find this condition of propagation in deep water: the isothermal profile can then correspond to a shallow layer or to the entire water column.[9]

The velocity variation with depth z is described by the linear law $c(z) = c_s + gz$, with $c(H) = c_H$. The rays are then arc circles, reflected by the surface (Figure 2.21). They show a periodic structure, with a horizontal cyclical distance $D_C(\beta_S)$ for a ray striking the surface with angle β_S:

$$D_C(\beta_S) = 2\frac{c_S}{g}\tan\beta_S \qquad (2.78)$$

The maximal angle possible corresponds to a ray whose turning point is just at the lower end $(z = H)$ of the channel; hence it is given by the ratio of velocities at each end of the profile:

$$\cos\beta_{S0} = \frac{c_S}{c_H} \quad \text{i.e., roughly} \quad \beta_{S0} = \sqrt{\frac{2gH}{c_S}} \qquad (2.79)$$

Such an insonification of the medium is globally homogeneous, but it favours upper layers if the acoustic source is shallow.

[9] This is, for example, the case in the Mediterranean, with a remarkably stable winter temperature around 13°C in the entire water column, or at high latitudes, where the velocity profiles, although not strictly isovelocity, increase monotonously from the surface down to the bottom, and create the same type of propagation.

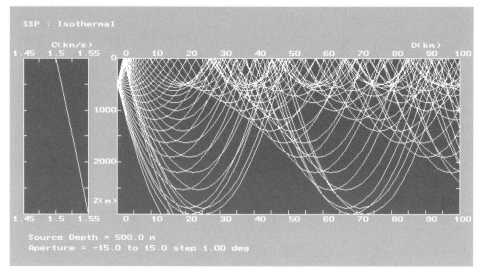

Figure 2.22. Ray tracing for an isothermal sound speed profile.

The propagation time over one cycle reads:

$$T_C(\beta_S) = \frac{2}{g \ln\left(\tan\left(\dfrac{\beta_S}{2} + \dfrac{\pi}{4}\right)\right)} \approx \frac{2}{g}\left(\beta_S + \frac{\beta_S^3}{6}\right) \tag{2.80}$$

And the corresponding mean horizontal velocity is:

$$V(\beta_S) = \frac{D_C}{T_C} \approx c_S\left(1 + \frac{\beta_S^2}{6}\right) \tag{2.81}$$

One can therefore see that the fastest paths correspond to the highest β_S (i.e., the largest arc circles, see Figure 2.22). This contradicts physical intuition, based on the constant-velocity case. It can be explained by considering that these rays have followed the longest curvilinear distance, but in regions with higher velocities: the increase in propagation velocity compensates the increase in distance.

Using the equations of Section 2.7.3, one can show that the intensity loss through geometric spreading between two points at a horizontal distance r from each other can be approximated as:

$$TL \approx 10\log\left[\frac{r^2 \sin\beta_0 \sin\beta}{\cos^2\beta_0 \sin^2\beta_S}\right] \tag{2.82}$$

The impulse response of the medium therefore has arrivals corresponding first to the highest angles (β_{S0}), and last to the most grazing angles (incidentally, with the highest amplitudes) (Figure 2.23). The time spreading of the signal is given by the velocity difference between the slowest (at $\beta_S = 0$) and the fastest (at β_{S0}) rays. Thus

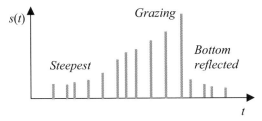

Figure 2.23. Impulse response for an isothermal channel: the steepest rays arrive first with the smallest amplitudes (longer paths, more surface reflections), and the grazing angles come last, with a maximum amplitude; they are followed by the bottom-reflected group.

it can be approximated as:

$$\Delta t = r\left(\frac{1}{V(0)} - \frac{1}{V(\beta_{S0})}\right) \approx r\frac{\beta_{S0}^2}{6c_S} \qquad (2.83)$$

Beyond β_{S0}, the rays can strike the bottom and be reflected. Their refracted character is less marked as the grazing angle increases, and their general behaviour resembles more closely the behaviour of rectilinear paths in isovelocity conditions.

2.8.3   Deep channel

The deep channel configuration corresponds to the case where the sound velocity decreases from the surface to a minimum value at depth, and then increases down to the bottom. This configuration exists in most deep oceanic basins; the temperature usually decreases from the surface to a deep value of 2°C to 4°C, and below this depth the velocity follows an almost isothermal profile. In the following, we will describe the sound velocity profile in the simplified form of two linear profiles, of gradients g_1 and g_2 with opposite signs. In the upper part ($z < z_A$), the velocity varies in $c(z) = c_S + g_1 z$, with $g_1 < 0$, decreasing from c_S to c_A. In the deep part ($z < z_A$), it varies in $c(z) = c_A + g_2(z - z_A)$, with $g_2 > 0$, increasing from c_A to c_F at the interface with the seafloor.

2.8.3.1   SOFAR propagation

If the source is close to the channel axis (minimum velocity in z_A), the rays are successively refracted by the two gradients, and can propagate without interacting with the interfaces (Figure 2.24). The condition of existence of this mode of propagation is that the angle β_A of the ray when crossing the axis verifies:

$$|\cos \beta_A| > \max\left(\frac{c_A}{c_S}, \frac{c_A}{c_F}\right) \qquad (2.84)$$

The cycle length of a ray then becomes:

$$D_C = 2c_A \tan \beta_A \left(\frac{1}{g_2} - \frac{1}{g_1}\right) \qquad (2.85)$$

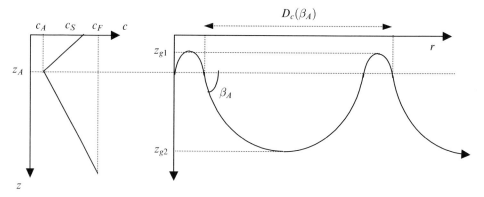

Figure 2.24. Deep channel ray geometry.

The depths at the turning points of the ray of angle β_A become:

$$\begin{cases} z_{g1} = z_A + \dfrac{c_A}{g_1}\left(\dfrac{1}{\cos\beta_A} - 1\right) \\[3mm] z_{g2} = z_A + \dfrac{c_A}{g_2}\left(\dfrac{1}{\cos\beta_A} - 1\right) \end{cases} \qquad (2.86)$$

The equations of the isothermal channel can be reused by writing $1/g_0 = 1/g_2 - 1/g_1$. For the ray of angle β_A in the axis, the cycle propagation time thus becomes:

$$T_C(\theta_A) = \frac{2}{g_0 \ln\left(\tan\left(\dfrac{\beta_A}{2} + \dfrac{\pi}{4}\right)\right)} \approx \frac{2}{g_0}\left(\beta_A + \frac{\beta_A^3}{6}\right) \qquad (2.87)$$

And the corresponding cycle horizontal distance is:

$$V(\theta_A) = \frac{D_C}{T_C} \approx c_A\left(1 + \frac{\beta_A^2}{6}\right) \qquad (2.88)$$

The propagation loss through spherical spreading is:

$$TL \approx 10\log\left[\frac{r^2 \sin\beta_0 \sin\beta}{\cos^2\beta_0 \sin^2\beta_A}\right] \qquad (2.89)$$

In this configuration, the insonification of the medium is quite homogeneous, especially close to the axis (Figure 2.25) because of the large number of rays interfering at any given point of the waveguide. The number of these paths is larger when closer to the axis of the channel, which corresponds to a configuration that is very favourable energetically. This type of propagation is called SOFAR (for SOund Fixing And Ranging). It allows very large transmission ranges. Because of the absence of energy loss by reflection at the interfaces on one hand, and because of the concentration of a large number of multiple paths on the other hand, the loss

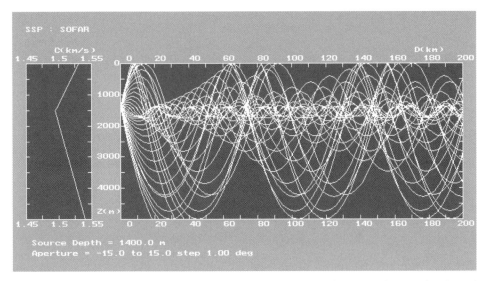

Figure 2.25. SOFAR propagation regime in a deep channel, with source close to the channel axis.

through geometric spreading is minimal. The absorption of sound in water is the only limitation. Ranges of several thousands of kilometres can be achieved by using frequencies that are suitably low. This process was discovered during Word War II and used for positioning at very large ranges. It is now used for acoustic oceanography experiments, transmitting signals to evaluate the variations of sound velocity at the scale of ocean basins (see Section 7.5 for more detail).

The impulse response of the SOFAR channel is very characteristic (Figure 2.26). It is an extension of the isothermal case: the most inclined rays arrive first, and are distinguishable if the signal resolution is fine enough. The arrivals are increasingly closer in time, with increasingly higher intensities. They end with a group arrival of indistinguishable paths corresponding to para-axial propagation. The time spread of the arrivals depends on the distance and the values of velocity at the interfaces and on the axis: from 2 s to 10 s at 1,000 km, depending on the depth–velocity profile. Evidently, this makes more complex the transmission of information other than simple detection or measurement of the arrival time of the end peak, unless one uses a large vertical antenna to filter the angles of the arrivals.

2.8.3.2 *Shadow and convergence zones*

Let us now look at a source close to the surface. The rays transmitted near the horizontal are rapidly refracted downward, because of the shallow gradient. The limit ray is tangential to the surface:

$$\cos \beta_0 = \frac{c_0}{c_S} \qquad (2.90)$$

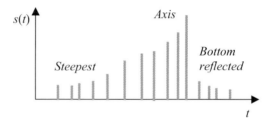

Figure 2.26. Impulse response for a SOFAR channel: the steepest rays arrive first with the smallest amplitudes (longer paths), and the subaxial rays are last, with a maximum amplitude due to the geometrical divergence; they may be followed by a bottom-reflected group.

This ray crosses again the depth of the source at the distance:

$$D_{S0} = \frac{2c_0}{g_1} \tan \beta_0 \tag{2.91}$$

The rays with grazing angles larger than the limit angle will strike the surface, and therefore cross the source depth at smaller distances.

The beam transmitted in the angular sector $[0, \beta_{S0}]$ crosses the depth z_A at angles, which one can show to be in the angular sector $[\arccos c_A/c_0, \arccos c_A/c_S]$, much smaller than the transmitted sector. These rays all follow more or less the same path, and therefore this ray bundle undergoes a very small geometric spread. They can be found again after refraction by the deep gradient, still grouped in a "convergence zone" (or "resurgence zone"), insonified very favourably (Figure 2.27). The empty space left by this beam group is called a "shadow zone"; it cannot be insonified, except by bottom reflections, and is therefore unfavourable energetically (at least in the strict frame of ray tracing, as diffraction outside the beam becomes important at very low frequencies).

The convergent beam continues to propagate with the same characteristics. The convergence zone is duplicated at a distance, somewhat widening with propagation, and becoming increasingly blurred (Figure 2.27). The size of shadow zones and the periodicity of convergence zones depend on the general bathymetric profile (typically 40 km in the Mediterranean, 60 km in the Atlantic).

This type of configuration is very important in ASW tactics, where submersibles and sonars move at relatively shallow depths. The shadow zones are used to hide, and the convergence zones are favoured places for detection (it is also possible to use several successive convergence zones).

2.9   WAVE CALCULATIONS OF THE ACOUSTIC FIELD

Wave techniques give the global solution of the propagation equation for the complex acoustic field. They are based on the direct solution, through different techniques, of the propagation equation.

Typically, these techniques require that the acoustic field be described with a

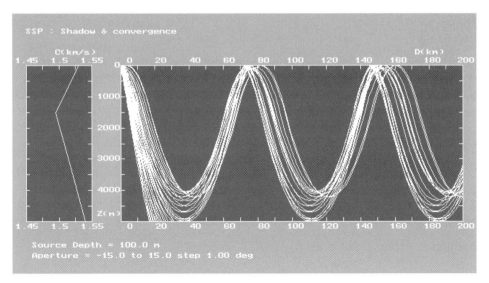

Figure 2.27. Propagation in a deep channel: shadow and convergence zones, with a shallow source. The narrowness of the ray bundle depends on the source depth.

resolution close to the wavelength, so that its wave behaviour can be analysed with high enough precision. These techniques are therefore mostly employed for very low frequencies,[10] where the oscillations/interferences of the field have a practical meaning. The geometric approach is preferred at high frequencies, where it is more accurate and faster.

Wave solutions are also a priori defined for monochromatic signals, and are badly suited to the description of transitory signals. They will therefore be kept for the study of permanent, narrowband signals (e.g., ship–noise propagation and passive sonars) or the very accurate description of active low-frequency signals when ray tracing becomes dubious. Geometric approaches using transitory processes (travel times) should therefore be reserved to active detection/measurement applications.

The following description of wave techniques is less detailed than the complexity of the subject ought to warrant. The interested reader is also referred to the more specialised books by Brekhovskikh and Lysanov (1992), Etter (1991) and Jensen *et al.* (1994).

2.9.1   Modal method

This method starts from the propagation equation for a stratified medium, with sound velocity $c(z)$:

$$\Delta p + k^2(z)p = 0 \qquad\qquad (2.92)$$

---

[10] Here, the notion of high or low frequencies is relative to the height of the waveguide.

By cylindrical symmetry, the solution $p(r, z)$ can be separated into a function depending on r only and a function depending on z only:

$$p(r, z) = \Gamma(r)\Phi(z) \tag{2.93}$$

We introduce a constant K^2 separating the variables in the propagation equation expressed in cylindrical coordinates:

$$\begin{cases} \dfrac{d^2\Gamma}{dr^2} + \dfrac{1}{r}\dfrac{d\Gamma}{dr} + K^2\Gamma = 0 \\[2mm] \dfrac{d^2\Phi}{dz^2} + (k^2(z) - K^2)\Phi = 0 \end{cases} \tag{2.94}$$

The first equation is a Bessel equation. For a wave spreading from the source, its solution is a Hankel function:

$$\Gamma(r) = H_0^{(1)}(Kr) \tag{2.95}$$

For "large" values of its argument (in fact, as soon as it is greater than 1), this function tends toward the asymptotic expression:

$$H_0^{(1)}(x) \rightarrow \sqrt{\frac{2}{\pi x}} \exp\left[j\left(x - \frac{\pi}{4}\right)\right] \tag{2.96}$$

The second equation must be verified simultaneously with the continuity conditions at the waveguide boundaries:

- the pressure at the air–water interface equals zero: $\Phi(0) = 0$;
- the condition at the water–bottom interface is of the general form:

$$F\left\{\Phi(H), \frac{d\Phi}{dz}(H)\right\} = 0$$

The differential equation in z associated with these conditions has a series of solutions expressed as $(K = k_n, \Phi(z) = \Phi_n(z))$, for $n = 1, 2, \ldots, N$. These are called *propagation eigenmodes*. Each mode n must verify the following system of equations:

$$\begin{cases} \dfrac{d^2\Phi_n}{dz^2} + (k^2(z) - k_n^2)\Phi_n = 0 \\[2mm] \Phi_n(0) = 0 \\[2mm] F\left\{\Phi_n(H), \dfrac{d\Phi_n}{dz}(H)\right\} = 0 \\[2mm] \displaystyle\int_0^{+\infty} \Phi_n(z)\Phi_m(z)\,dz = \delta_{mn} \end{cases} \tag{2.97}$$

The latter equation (where δ_{mn} is the Kronecker symbol) corresponds to the *eigenmode orthonormality*. The general solution can finally be expressed as:

$$p(r,z) = \sum_n \Phi_n(z)\Phi_m(z_0)H_0^{(1)}(k_n r) \qquad (2.98)$$

where z_0 is the depth of the source.

The main difficulty of this method therefore consists in determining the wave numbers of the eigenmodes (and thus the corresponding wave functions). In the simple case where the sound velocity is constant (see Appendix A.2.4), the wave functions are trigonometric functions that are easily calculable, of type $\Phi_n(z) \propto \sin(\sqrt{k^2 - k_n^2}z)$. It can be shown that the eigenmodes are separated into two categories:

- the *propagating modes*, whose wave numbers $\sqrt{k^2 - k_n^2}$ are real, and which carry most of the acoustic field's energy;
- the *attenuated modes*, with complex wave numbers, whose contribution is noticeable only at short ranges from the source.

An example of application is given in Figure 2.28, showing both the wave functions of the 24 modes computed, and the resulting field. The latter is to be compared with Figure 2.8, which was obtained geometrically for the same configuration.

The total number of modes (propagating and attenuated) N_m is approximately given by the ratio of the water depth to the average half-wavelength for the frequency considered:

$$N_m \approx \frac{2H}{\lambda} \qquad (2.99)$$

It is therefore proportional to the frequency. This number of modes is an interesting judge of the usefulness of the modal technique. A number of a few tens is optimal (Figure 2.28). Beyond a hundred, the use of geometric acoustics is undoubtedly justified. Conversely, if the number of modes is too small, a method of direct numerical integration (see Section 2.9.2) is preferable.

To understand this modal description physically (Brekhovskikh and Lysanov, 1992), one can decompose $\sin k_z z \propto \exp(jk_z z) + \exp(-jk_z z)$; that is, we can consider the wave function as the sum of an up-going wave and a down-going wave interfering in the waveguide. So:

- the existence of an eigenmode corresponds to constructive interference between the two plane waves;
- the propagating modes correspond to total reflection on the bottom, at over-critical incidence;
- the attenuated modes correspond to energy loss by reflection on the bottom, at subcritical incidence.

For real depth–velocity profiles $c(z)$, the wave functions have no more exact solution, as in the constant-velocity case, except for a few specific profiles. The problem can then be solved: either by cutting the sound speed profile into

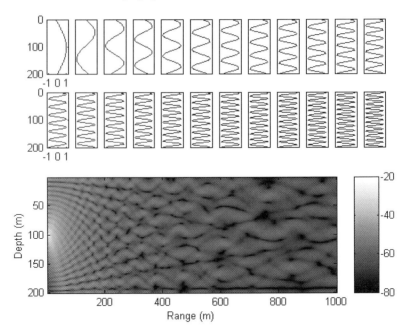

Figure 2.28. Normal mode depth functions $\Phi_n(z)$ in a constant-velocity channel at 100 Hz, with 24 modes computed, and resulting pressure field in dB. The configuration is the same as in Figure 2.8.

sections, which each admit an exact solution, and forcing continuity conditions between the layers; or by numerically solving the differential equation in z.

Eigenmodes in a non-stratified waveguide

The method of eigenmodes is by its very nature linked to the structure of a stratified waveguide. But it can be extended to media with a range dependence. The basic principle of these extensions is to consider the modal structure as a local notion, either discrete (the medium is divided into discontinuous sections in which one calculates the eigenmodes), or continuous. The problem then lies in joining the modal structures obtained. Two methods are often used:

- The method of *coupled modes* consists in imposing continuity conditions between the sections of the discretised medium, over the entire field structure. These conditions lead to coupling coefficients between the different sections. This method is theoretically rigorous, but quite heavy computationally.
- The method of *adiabatic modes* is lighter and more interesting physically. Assuming the medium varies slightly with range, this method supposes that each mode transforms with no energy loss into a corresponding mode (of the same order) in the neighbouring section. This implies that the transitions between sections are quite progressive, and this limits the calculation of the

number of modes to the smallest water depth. This method is frequently used, because of its relative simplicity. It is, however, increasingly regularly replaced by the parabolic equation technique (see Section 2.9.3).

2.9.2   Numerical solution of the wave equation

For a medium stratified with depth z, a very accurate solution of the acoustic field can be obtained numerically. It can be shown that the general solution to the Helmholtz equation can be written as the integral of the contributions from a spectrum of plane waves characterised by the horizontal component k_r of their wave number:

$$p(r, z, z_0) = \int_{-\infty}^{+\infty} G(z, z_0, k_r) H_0^1(k_r r) k_r \, dk_r \qquad (2.100)$$

H_0^1 is the Hankel function of first type and order 0. G is the Green function of the medium, obeying the following conditions:

- propagation equation, in z;
- limit conditions at the surface and the bottom;
- source conditions – continuity of G and discontinuity of dG/dz.

The system to be solved is similar to the system with eigenmode wave functions, with the addition of source conditions. The integral expression may indeed be interpreted with eigenmodes, when considering that the modes correspond to the poles of the integrand.

Numerical computation of the integral of the spectrum in k_r is made easier with the approximation of the Hankel function by its asymptotic form. This allows writing $p(r, z)$ as:

$$p(r, z, z_0) \approx \frac{1}{\sqrt{r}} \int_{-\infty}^{+\infty} G(z, z_0, k_r) \sqrt{k_r} \exp(j k_r r) \, dk_r \qquad (2.101)$$

This is in fact a Fourier transform, which can be computed with FFT algorithms. The main part of the calculation thus consists in the evaluation of $G(z, z_0, k)$ for a series of values sampling the domain $\{k_r\}$ of interest. The FFT results in a series of values of p as a function of the range, for a given couple (z, z_s).

The bases of this method (Jensen et al., 1994) date back to the 1970s. Its main interest is to provide a very accurate solution in the case of a stratified medium. Its main fault is that it does not give direct access to the physical interpretation of the problem. It is mostly used as a benchmark for calculations of the propagation of very low-frequency monochromatic signals, or their reflection, in stratified media.

2.9.3   Parabolic approximation

This method is used for environments slowly varying with range and azimuth. Let us consider the case of a depth–velocity profile $c(r, z)$. We have:

$$k(r, z) = k_0 n(r, z) = k_0 \frac{c_0}{c(r, z)} \qquad (2.102)$$

where c_0 is a reference average velocity. The Helmholtz equation writes:

$$\Delta p + k_0^2 n^2(r, z)p = 0 \qquad (2.103)$$

We look for solutions valid in directions close to the horizontal, such as:

$$p(r, z) = F(r, z)H_0^{(1)}(k_0 r) \qquad (2.104)$$

where $H_0^1(k_0 r)$ represents the "rapid fluctuations" in r (corresponding to the term in r for the eigenmodes) and $F(r, z)$ represents the slow fluctuations in (r, z). If we use the asymptotic form of $H_0^1(k_0 r)$, there follows:

$$\frac{\partial^2 F}{\partial r^2} + 2jk_0 \frac{\partial F}{\partial r} + \frac{\partial^2 F}{\partial z^2} + k_0^2(n^2 - 1)F = 0 \qquad (2.105)$$

Neglecting $\partial^2 F/\partial r^2$ (because of its slow variation in r), there remains a parabolic equation:

$$2jk_0 \frac{\partial F}{\partial r} + \frac{\partial^2 F}{\partial z^2} + k_0^2(n^2 - 1)F = 0 \qquad (2.106)$$

This equation can only be solved numerically. The usual way is:

- to calculate the field $F(r, z)$ at an initial range r_0 by another technique (modes, rays, Gaussian field, etc.);
- to continue with r beyond r_0 with a numerical technique (e.g. split-step algorithm) or finite differences.

More detailed presentations of the parabolic approximation method are to be found in, for example, Tappert, 1977 or Jensen *et al.*, 1994. This method is currently the most used to investigate propagation in non-stratified media. Initially limited to very small angular apertures and slowly varying media, this method is now of more general validity. One will remark that, because of the accuracy needed for a gridding step comparable with the wavelength, the computation load can be very heavy.

3

Reflection, backscattering and target strength

An acoustic wave propagating in the ocean will often "collide" with obstacles either in the water column itself (fish, plankton, bubbles, submarines) or at the limits of the medium (seabed and sea surface). These obstacles will send back to the sonar system some echoes of the signal transmitted, a portion of which will be perceived by the sonar system. These echoes will then be either desirable (if the obstacle is the target wanted) or undesirable (if they are jamming the useful signal). In all cases, understanding their properties is essential to the good functioning of the sonar system, either because the echoes need to be received in the best possible conditions, or because they need to be reduced or filtered out as much as possible.

Several distinct physical processes contribute to the formation of underwater acoustic echoes. The multiple paths studied in the previous chapter are associated with the well-known and rather intuitive *reflection* by plane interfaces: the incident wave is reflected in a direction symmetrical to its direction of arrival (as with a mirror, hence the name of *specular* reflection), with a loss of amplitude. Reflection studies are therefore more particularly useful for underwater acoustic systems where the transmitter and receiver are distinct (data transmission or multistatic sonars), or for configurations where the propagation between the sonar and the target is prone to generate multiple echoes (when the propagation directions are close to the horizontal). Contrary to these reflections from the "average" bottom and surface, different types of echo are generated by local obstacles present in the water column or at the interfaces. Because of their shape, these irregularities scatter acoustic energy in all directions, and are therefore more likely to affect the signals received in any configuration. The scattering of acoustic energy back towards the sonar is called *backscattering*. Most sonar systems working in monostatic configuration (i.e., transmitter and receiver located at the same point) use backscattered echoes.[1]

---

[1] Sediment profilers (see Section 7.4.3) are the only exception: the echoes consist in *specular reflections* generated perpendicularly to the ground.

The convention of underwater acoustics is to call *reverberation* any return of energy towards the sonar system, coming from something other than the echo of the desired target.[2] Surface and bottom reverberation (boundary effects due to back-scattering of sound by the relief of interfaces) are traditionally distinguished from volume reverberation (volume effects due to backscattering from fish, plankton, suspended particles or bubble clouds). The distinction between target echoes and reverberation is of course arbitrary and depends only on the type of system con-sidered. For active detection military sonars, reverberation includes any echo that is not coming from a submarine; for example, echoes from the sea surface and the seabed. Conversely, mapping sidescan sonars will only use signals backscattered from bottom topography and obstacles; they will be perturbed by surface reverbera-tion or volume reverberation from fish schools or plankton. And fisheries sonars will instead be designed specifically to target these ...

The first section will study the reflection and transmission of sound waves at the boundaries of the propagation medium, influencing the transmission of signals in all multipath propagation configurations. We shall consider the interface between two homogeneous media, and then the case of a *layered medium*, assuming the interfaces are ideal planes. The expression of the *target index* will then be formalised, quantify-ing the propensity of a target to scatter an echo. We shall successively investigate the case of "point" targets (insonified at once, and with effective dimensions independent of the sonar characteristics), and extended targets (with effective dimensions dependent on sonar), treated as scattering volumes or surfaces. Processes due to interface roughness will then be described, as they degrade the signal reflected and, mostly, create backscattering of incident energy. Finally, we shall investigate more closely echoes from the seabed and the sea surface.

3.1 WAVE REFLECTION ON A PLANE INTERFACE

3.1.1 Interface between two fluid homogeneous media

3.1.1.1 *Reflection and transmission coefficients*

Let us now consider the case of a plane acoustic wave (assumed locally plane), incident upon an interface between two media with differing acoustic impedance. The wave propagating inside the target medium exhibits characteristics different from the incident wave, since the impedance has changed. The reflection process may be expressed as a balance between the incident wave, the transmitted wave and a wave reflected back into the first medium (Figure 3.1).

Expressing the continuity of the acoustic field at the interface (acoustic pressure and normal derivative of the sound velocity, see Appendix A.3.1 for details), one finds that:

---

[2] Perturbation of data transmission systems by multiple paths can, however, be assimilated to reverberation.

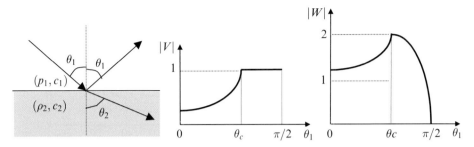

Figure 3.1. (*Left*) Reflection and transmission of an incident acoustic wave at the interface between two media. (*Middle*) Modulus of the reflection coefficient at the interface, as a function of the incidence angle θ_1. (*Right*) modulus of the transmission coefficient, as a function of the incidence angle θ_1.

- The reflected wave is propagating in a direction symmetrical to the incident wave, relative to the normal to the interface (Figure 3.1, *left*). This is called "specular reflection".
- The transmitted wave propagates in a different direction, proportional to the change in sound velocity. This is often called a "refracted wave", and it follows the Snell–Descartes law:

$$\frac{\sin \theta_1}{c_1} = \frac{\sin \theta_2}{c_2} \tag{3.1}$$

- The amplitudes of the reflected and transmitted waves are given by the reflection coefficient V and the transmission coefficient W, linked by $W = 1 + V$.

The density and sound velocity of the two media are, respectively, noted (ρ_1, c_1) and (ρ_2, c_2). The reflection and transmission coefficients[3] for an incidence angle θ_1 will be given by:

$$\begin{cases} V(\theta_1) = \dfrac{\rho_2 c_2 \cos \theta_1 - \rho_1 c_1 \cos \theta_2}{\rho_2 c_2 \cos \theta_1 + \rho_1 c_1 \cos \theta_2} \\[3mm] W(\theta_1) = 1 + V(\theta_1) = \dfrac{2\rho_2 c_2 \cos \theta_1}{\rho_2 c_2 \cos \theta_1 + \rho_1 c_1 \cos \theta_2} \end{cases} \tag{3.2}$$

If $c_2 > c_1$, there exists a critical angle $\theta_c = \arcsin(c_1/c_2)$, beyond which transmission is impossible. If $\theta_1 \geq \theta_c$, no compressional wave can propagate inside the second medium: since $\sin \theta_2 > 1$, the $\cos \theta_2$ term becomes imaginary, meaning that the refracted wave vanishes from the boundary. The reflection coefficient becomes complex, with a unit modulus, independent of θ_1. Such total reflection occurs, for example, at grazing incidence on water–sediment interfaces.

When θ_1 decreases from grazing incidence (high values of θ_1, see Figure 3.1) and crosses the critical value θ_c, the reflection coefficient V will suddenly decrease and

---

[3] More general expressions using wave propagation vectors are given in Appendix A.3.1.

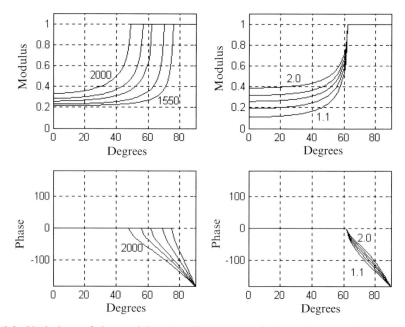

Figure 3.2. Variations of the modulus (top diagrams) and phase (bottom diagrams) of the reflection coefficient V: (*left*) constant density ratio $\rho_2/\rho_1 = 1.5$ and changing sound velocity c_2 (respectively, 1,550, 1,600, 1,700, 1,800 and 2,000 m/s); (*right*) changing density ratio ρ_2/ρ_1 (respectively, 1.1, 1.3, 1.5, 1.7, 2.0) and constant sound velocity $c_2 = 1,700$ m/s.

then vary smoothly with the angle θ_1. Close to the vertical, it will depend only on the characteristic acoustic impedances Z_1 and Z_2 of the two media:

$$V(\theta_1 = 0) = \frac{\rho_2 c_2 - \rho_1 c_1}{\rho_2 c_2 + \rho_1 c_1} = \frac{Z_2 - Z_1}{Z_2 + Z_1} \tag{3.3}$$

Figure 3.2 shows the variations of the reflection coefficient $V(\theta_1)$ as a function of the incidence angle θ_1; the sound velocity and density of the reflecting media vary independently.

3.1.1.2   Conservation of energy

The amplitude of the refracted wave is expressed by the coefficient of transmission into the second medium $W(\theta) = 1 + V(\theta)$. This implies that the pressure of the transmitted wave may be higher than the pressure of the incident wave. This apparently surprising result in fact only expresses the continuity of pressure at the interface, and does not violate the conservation of energy. If the intensity of the plane wave is $I = p^2/2\rho c$ along the propagation direction, its projection onto an interface at the angle of incidence θ will be $I(\theta) = p^2/2\rho c \times \cos \theta$. Using the expressions of the reflection and transmission coefficients given in Equation (3.2), one can easily verify that:

$$I_r(\theta_1) + I_t(\theta_2) = \frac{V^2}{2}\cos\theta_1 + \frac{W^2}{2}\cos\theta_2 = \frac{1}{2\rho_1 c_1}\cos\theta_1 = I_i(\theta_1) \qquad (3.4)$$

The sum of the reflected and transmitted intensities is indeed equal to the incident intensity, as could be expected.

3.1.1.3 Case of a strong impedance contrast

If the acoustic impedance of the reflecting medium is much higher, or much lower, than the impedance of the first medium, the reflected wave will suffer practically no energy loss. The reflection coefficient tends toward $V = 1$ (if $Z_2 \gg Z_1$) or $V = -1$ (if $Z_2 \ll Z_1$), independently of the incidence angle. In the latter case, the phase of the reflection coefficient will be shifted by π. The transmitted wave might, however, still have a noticeable magnitude.

In underwater acoustics, this ideal case of a "perfect" boundary is only met at the interface between air and water, for which the impedance contrast $Z_{air}/Z_{water} \approx 3 \times 10^{-4}$. This case is further discussed in Section 3.6.1.2.

On the other hand, typical impedance contrasts between water and "hard" reflectors are: $Z_2/Z_1 \approx 2.4$ at the interface between water and hard sediments ($\rho_2 \approx 2,000\,\text{kg/m}^3$ and $c_2 \approx 1,800\,\text{m/s}$); $Z_2/Z_1 \approx 7.5$ between water and rock ($\rho_2 \approx 2,500\,\text{kg/m}^3$ and $c_2 \approx 4,500\,\text{m/s}$); $Z_2/Z_1 \approx 26$ between water and steel ($\rho_2 \approx 7,800\,\text{kg/m}^3$ and $c_2 \approx 5,000\,\text{m/s}$). These impedance contrasts are not high enough for the approximation of a perfect boundary. Moreover, such hard elastic materials are able to transmit significant shear waves (see Section 3.1.1.5), and the model of a fluid–fluid interface is then no longer valid.

3.1.1.4 Influence of absorption inside the reflecting medium

The processes described in the previous sections (critical angle, total reflection) will need to be slightly modified if the second medium is absorbing. Absorption is expressed with a complex sound velocity c_2. Let us note α_2 the absorption coefficient in the second medium. It is most often considered to be proportional with frequency and is therefore usually expressed in dB/wavelength (acoustic wavelength λ_2 in the reflecting medium). The complex part of the wave number k_2 is given by:

$$\gamma_2 = \text{Im}\,k_2 = \frac{\alpha_2}{\lambda_2 \times 20\log e} = \frac{\alpha_2 \ln 10}{20\lambda_2} \approx \frac{\alpha_2}{8.686\lambda_2} \qquad (3.5)$$

And the complex part of the sound velocity is:

$$\text{Im}\,c_2 \approx \frac{|c_2|^2\gamma_2}{\omega} = \frac{\alpha_2|c_2|}{2\pi \times 20\log e} \approx \frac{\alpha_2|c_2|}{54.6} \qquad (3.6)$$

The simple trigonometric expressions of Equation (3.2) are no longer valid, and one has to turn to the wave vector formalism given in Appendix A.3.1. All the previous

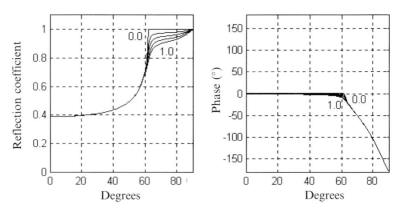

Figure 3.3. Variation of the reflection coefficient with the angle of incidence, for different absorption values of the reflecting medium. Sound velocity $c_2 = 1,700\,\text{m/s}$; density ratio $\rho_2/\rho_1 = 2$; absorption $\alpha_2 = 0.00, 0.25, 0.50, 0.75, 1.0\,\text{dB}/\lambda$, respectively).

formulae (law of Snell–Descartes, reflection coefficient) may then still be used, with the complex expression of k_2. This implies that an attenuated wave can always be transmitted and propagate through the reflecting medium, even beyond the critical angle θ_c. Compared with the perfect fluid case, the curve of $V(\theta)$ is smoothed. Total reflection does not occur beyond the critical angle; the modulus of the reflection coefficient is then slightly below unity, slowly reaching its maximum for $\theta_1 = \pi/2$ (Figure 3.3). In underwater acoustics, wave reflection on an absorbing fluid is actually occurring for fluid sedimentary seabeds. Typical values for the absorption coefficients then range from $0.1\,\text{dB}/\lambda$ to $1\,\text{dB}/\lambda$.

3.1.1.5   Case of a solid reflecting medium

When the reflecting medium is solid and elastic, shear waves may propagate along with the compressional waves. Their shear velocity c_S will be different from the sound velocity c_2. The reflecting medium being defined by its density ρ_2 and its mechanical Lamé coefficients[4] λ and μ, the two velocities are given by the respective formulae:

$$c_2 = \sqrt{\frac{\lambda + 2\mu}{\rho_2}} \tag{3.7a}$$

$$c_S = \sqrt{\frac{\mu}{\rho_2}} \tag{3.7b}$$

So, whatever the characteristics of the solid reflecting medium, $c_S < c_2/\sqrt{2}$.

The reflection of an incident wave is more complicated than in the case of a fluid,

---

[4] We shall keep the traditional notation of λ and μ here, although there is a risk of confusion with other variables of the same name.

as the incident wave can potentially excite both the pressure wave and the shear wave. The continuity conditions at the boundary are the same (continuity of pressure and normal velocity), but one needs to account for both types of wave inside the reflecting medium. The Snell–Descartes law is now expressed as:

$$\frac{\sin \theta_1}{c_1} = \frac{\sin \theta_2}{c_2} = \frac{\sin \theta_S}{c_S} \tag{3.8}$$

The reflection coefficient is now expressed as:

$$V = \frac{B-1}{B+1}, \quad \text{with} \begin{cases} B = \dfrac{\rho_2}{\rho_1} \dfrac{1}{k_S^4} \dfrac{k_{z1}}{k_{z2}} [(k_{zS}^2 - k^2 x)^2 + 4k_x^2 k_{zS} k_{z2}] \\[2mm] k_x = k_1 \sin \theta_1 \\[2mm] k_i = \dfrac{2\pi f}{c_i} \\[2mm] k_{zi} = k_i \cos \theta_i, \quad i = 1, 2 \text{ or } s \end{cases} \tag{3.9}$$

A variety of cases is then possible, depending on the relative values of c_1, c_2 and c_S (see Figure 3.4). If $c_S < c_1$, the shear wave can always be excited; hence no total reflection can ever occur, even for $\theta > \theta_c$. Conversely, if $c_S > c_1$, two critical angles

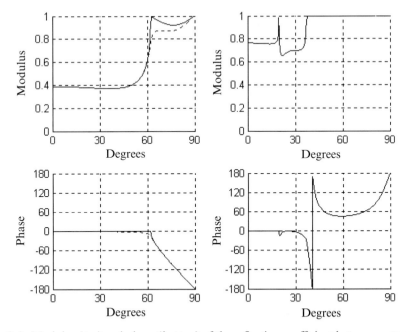

Figure 3.4. Modulus (*top*) and phase (*bottom*) of the reflection coefficient between water and a solid elastic reflecting medium, as a function of the incidence angle. *Left*: $c_S < c_1$ ($c_2 = 1{,}700\,\text{m/s}$; $c_S = 500\,\text{m/s}$; $\rho_2/\rho_1 = 2.0$). *Right*: $c_S > c_1$ ($c_2 = 4{,}500\,\text{m/s}$; $c_S = 2{,}500\,\text{m/s}$; $\rho_2/\rho_1 = 2.5$). Dotted lines correspond to absorption inside the reflecting medium.

may exist: $\theta_c = \arcsin(c_1/c_2)$ and $\theta_{cS} = \arcsin(c_1/c_S)$. There are now three angular regimes:

- at angles $\theta > \theta_{cS}$, total reflection occurs – no wave can propagate inside the reflecting medium;
- between θ_c and θ_{cS}, only the shear wave is transmitted, while the compressional wave is not;
- for $\theta < \theta_c$, both pressure and shear waves are transmitted.

As in the case of the fluid-reflecting medium, absorption smooths the processes.

3.1.2   Reflection on a layered medium

Let us first consider the case of a fluid layer of thickness d and parameters (ρ_2, c_2), sandwiched between two semi-infinite media of parameters (ρ_1, c_1) and (ρ_3, c_3) (Figure 3.5). We will investigate the reflection by Layer 2 as a function of the coefficients of reflection V_{ij} and transmission W_{ij} at the interfaces (i, j).

Taking a unit amplitude for the incident wave, the wave reflected back in Medium 1 will be decomposed into:

- the wave reflected at the interface (1–2), of amplitude V_{12};
- the wave transmitted into Medium 2, reflected from Medium 3 and retransmitted into Medium 1, of amplitude $W_{12}V_{23}W_{21}\exp(2jk_{2z}d)$, where the exponential accounts for the back and forth propagation in the layer of thickness d, with $k_{2z} = 2\pi f/c_2\cos\theta_2$;
- the wave transmitted into Medium 2, reflected from Medium 3, reflected from Medium 2, reflected from Medium 3 and transmitted into Medium 1 (i.e., with the amplitude $W_{12}V_{23}^2V_{21}W_{21}\exp(4jk_{2z}d)$;
- etc.

The total wave is the sum of all these contributions, and one can write:

$$V = V_{12} + W_{12}W_{21}V_{21}\exp(2jk_{2z}d)\sum_{n=0}^{\infty}[V_{23}V_{21}\exp(2jk_{2z}d)]^n \qquad (3.10)$$

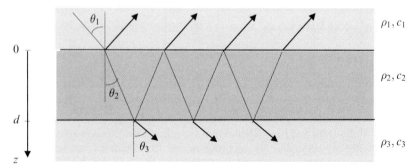

Figure 3.5. Reflection from a fluid layered medium.

The last term of Equation (3.10) corresponds to a geometric series, converging towards a finite value:

$$\sum_{n=0}^{+\infty} a^n \rightarrow \frac{1}{1-a} \qquad (a < 1) \tag{3.11}$$

Using the fact that $V_{ji} = -V_{ij}$ and $W_{ij} = 1 + V_{ij}$, we get:

$$V = \frac{V_{12} + V_{23}\exp(2jk_{2z}d)}{1 + V_{12}V_{23}\exp(2jk_{2z}d)} \tag{3.12}$$

The resulting coefficient is therefore simply an expression of the reflection coefficients at the two interfaces, and the phase shift associated with propagation inside the layer. It depends both on the angle of incidence of the wave and on its frequency. As before, the sound velocities can be complex to account for absorption inside the layered reflecting medium.

In the case where the reflecting medium consists in several stacked layers, the successive reflection/transmission processes are basically the same. But the complexity of their description grows very quickly with the number of layers, and calculations soon become intractable. It is then preferable to use the *equivalent impedance* method, as detailed in Brekhovskikh and Lysanov (1991).

Figure 3.6 presents an example, with the reflection coefficient for a one-layered medium plotted as a function of the incidence angle and the frequency of the incident acoustic wave. At low frequencies, the intermediate layer is practically transparent, and the reflection coefficient is similar to the reflection coefficient of the water–substratum interface. In particular, one can observe the critical angle (48.6°) associated with the substratum. As the frequency increases, resonance appears inside the layer, due to multiple reflections between the boundaries. The interference fringes observed are growing in intricacy as the layer thickness gets larger, compared with the acoustic wavelength. Conversely, the absorption inside the fluid layer becomes increasingly important, and the wave transmitted into the layer progressively becomes unable to reach the substratum. Finally, at higher frequencies, the reflection coefficient is given by the contrast at the upper boundary, whose critical angle (69.6°) becomes apparent.

3.2 BACKSCATTERING FROM A TARGET

3.2.1 Echo from a target

Most underwater acoustic systems are designed to receive echoes from *targets*. The latter are highly varied in nature and in structure: seabed insonified at normal or oblique angles of incidence, single fish or entire schools, submarines, objects laid on the bottom (mines, shipwrecks), buried objects (natural, like sedimentary layers, or artificial, like pipelines), etc. The incident acoustic wave will be scattered by the target in all directions of space, and a portion will be scattered back towards the transmitter. The target acts as a secondary source retransmitting the acoustic wave.

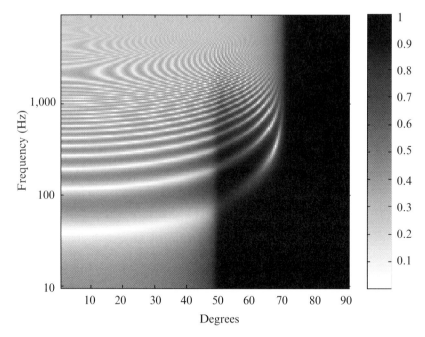

Figure 3.6. Reflection coefficient for a one-layered medium (see Figure 3.5), as a function of the incidence angle (x-axis) and the frequency (y-axis). The parameters chosen are: $c_1 = 1,500\,\mathrm{m/s}$; $\rho_1 = 1,000$; $c_2 = 1,600\,\mathrm{m/s}$; $\rho_2 = 1,500$; $\alpha_2 = 0.2\,\mathrm{dB}/\lambda$; $c_3 = 2,000\,\mathrm{m/s}$; $\rho_3 = 2,000$; $\alpha_3 = 0.5\,\mathrm{dB}/\lambda$ and $d = 10\,\mathrm{m}$. See text for interpretation.

The target strength is the ratio between the intensity sent by the target back towards the transmitter and the incident intensity. The intensity EL of the echo received by the sonar system after backscattering is:

$$EL = SL - 2TL + TS = SL - 40\log R - 2\alpha R + TS \qquad (3.13)$$

where SL is the level transmitted by the source, TL is the transmission loss (counted twice, once on the way in, once on the way back), TS is the target strength, R is the propagation range and α the absorption coefficient.

   Two types of target will be envisaged. First are the targets with dimensions small enough to be completely insonified by the sonar beam and signal (Figure 3.7, A, B). They behave as "points": their strength is an intrinsic strength, independent of the distance to the sonar or its characteristics (beam width, signal duration). Conversely, other targets will be too large to be ensonified completely at once by the same beam (e.g., large fish schools, seabed or sea surface) (Figure 3.7, C, D, E). The strengths of these targets will depend on their geometric intersection with the sound beam. The target strength is no longer a point value, but uses the insonified space (surface or volume), associated with a surface or volume backscatter coefficient. The latter is therefore an expression of the amount of energy scattered by a "unit scattering element". It is therefore expressed in dB re $1\,\mathrm{m}^2$ or dB re $1\,\mathrm{m}^3$, respectively. The

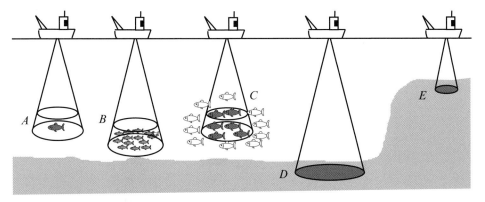

Figure 3.7. Examples of point targets (*A*, *B*) and extended targets (*C*, *D*, *E*).

backscattering strength depends on the incidence angle and the frequency, and will be specific to the processes observed.

3.2.2 Target strength

The target strength *TS* is defined as the ratio (in dB) of the intensities of the back-scattered and incident waves:

$$TS = 10 \log \left(\frac{I_{bs}}{I_i} \right) \tag{3.14}$$

It is therefore the relative amount of energy sent back by the target towards the sonar. It depends on the physical nature of the target, its external (and possibly internal) structure, and the characteristics of the incident signal (angle and frequency).

The wave incident on the target, with intensity I_i, is assumed locally plane (Figure 3.8). This assumption is valid if the sonar and the target are far enough from each other. The scattered wave is assumed spherical, radiating from the acoustic centre of the target towards the sonar, and its intensity I_{bs} is taken at 1 m from its centre. This is why the echo level must be corrected twice from propagation losses.

To be more accurate, the ratio between the intensity I_i, incident at angle θ_i, on the target and the scattered intensity I_s includes:

- The *apparent cross-section* $A(\theta_i)$ of the target as "seen" from the source. This geometrical cross-section defines the acoustic power intercepted by the target, equal to the product of this cross-section by the intensity of the incident wave: $P_i(\theta_i) = A(\theta_i)I_i$.
- The *scattering function* $G(\theta_i, \theta_s)$ of the target. This function describes the radiation of the scattered field (i.e., the spatial distribution of the energy

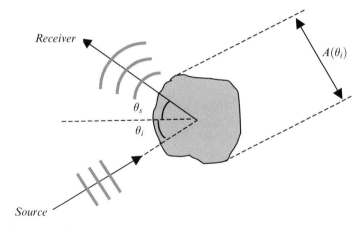

Figure 3.8. Geometry of target scattering. The incident wave is assumed locally plane, and the scattered wave is assumed spherical (see text for details). All reasoning in the text uses the 2-D representation of this figure. But one should remember that the processes should be considered in the full 3-D space.

retransmitted by the target into the neighbouring space. The intensity scattered in direction θ_s at the unit distance R_{1m} is:

$$I_s(\theta_s, R_{1m}) = \frac{P_i(\theta_i)}{R_{1m}^2} G(\theta_i, \theta_s) \tag{3.15}$$

The scattering function depends on the target structure, the signal frequency and the directions of incidence and reception.

The *effective scattering cross-section* has the dimensions of a surface and is defined as:

$$\sigma_s(\theta_i, \theta_s) = \frac{I_s(\theta_s) R_{1m}^2}{I_i(\theta_i)} = A(\theta_i) G(\theta_i, \theta_s) \tag{3.16}$$

The *effective backscattering cross-section* of the target is calculated in the particular direction back towards the source:

$$\sigma_{bs}(\theta_i) = \frac{R_{1m}^2 I_s(R, \theta_i)}{I_i(\theta_i)} = A(\theta_i) G(\theta_i, \theta_s) \tag{3.17}$$

Finally, the target strength is the value in dB of the effective backscattering cross-section:

$$TS = 10 \log \left[\frac{\sigma_{bs}(\theta_i)}{A_1} \right] \tag{3.18}$$

where A_1 is the value of the unit section $(1\,m^2)$.

One must differentiate a target strength or backscattering strength (the plane wave is reradiated from the obstacle as if it were a new point source with a directivity function), and a reflection coefficient (the reflected wave is plane, but has undergone amplitude loss and phase shift). A target strength can have a positive value in dB, which is never the case for a reflection coefficient.

Although the backscattering cross-section (or its logarithmic expression, the target strength) is the parameter most commonly used in system analysis, other derived values are sometimes encountered. The *total acoustic cross-section* is obtained from the angular integral,[5] over the entire space, of the scattered power:

$$\sigma_t(\theta_i) = A(\theta_i) \int G(\theta_i, \theta_s) \, d\theta_s \qquad (3.19)$$

If the target is absorbing, the ratio of the absorbed power P_a to the incident intensity I_i defines the *absorption cross-section*:

$$\sigma_a(\theta_i) = \frac{P_a(\theta_i)}{I_i(\theta_i)} \qquad (3.20)$$

Finally, the *extinction cross-section* sums up the power losses from scattering and absorption:

$$\sigma_e(\theta_i) = \sigma_s(\theta_i) + \sigma_a(\theta_i) \qquad (3.21)$$

These last formulae are useful when describing the effects of acoustic propagation inside a field of distributed scatterers.

3.3 POINT TARGETS

3.3.1 The ideal sphere

Let us consider an ideal rigid sphere of radius a insonified with a plane wave. We assume that the energy received by the apparent cross-section ($A = \pi a^2$, $\forall \theta_i$) is entirely and isotropically retransmitted into the entire space. The scattering function is then everywhere equal to:

$$G(\theta_i, \theta_s) = \frac{1}{4\pi} \qquad \forall (\theta_i, \theta_s) \qquad (3.22)$$

The effective scattering or backscattering cross-section then becomes simply:

$$\sigma_{bs} = \sigma_s = \frac{\pi a^2}{4\pi} = \frac{a^2}{4} \qquad (3.23)$$

This very simple result, independent of frequency, immediately yields some interesting orders of magnitude. A rigid sphere 1 cm in radius will have an effective backscattering cross-section of $2.5 \times 10^{-5}\,\text{m}^2$ (i.e., a target strength of $-46\,\text{dB}$). Conversely, a 0-dB target strength would correspond to a rigid sphere of radius 2 m.

[5] Note the integral is written here as a simple integral, function of θ_s, for the sake of convenience. As explained in the legend to Figure 3.8, all spatial directions should in fact be considered, and this integral should be a double integral.

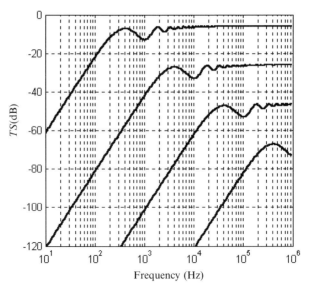

Figure 3.9. Target strengths for a perfect rigid sphere, as a function of frequency, for respective sphere radii of 1 m, 0.1 m, 0.01 m and 0.001 m (oscillations are approximate).

This simple approach is, however, meaningful only for high frequencies, where wave effects can be averaged and a mean intensity deduced therefrom. But at lower frequencies (i.e., when the wavelength is comparable with or larger than the sphere's radius), wave effects need to be taken into account. Rayleigh showed that for a fixed, rigid sphere of radius a, small compared with the acoustic wavelength, the back-scattering cross-section is given by:

$$\sigma_{bs} = \frac{25}{36}k^4a^6 \tag{3.24}$$

This is known as *Rayleigh scattering*. In this regime, the backscattering cross-section increases very quickly with frequency (in f^4, as appears from Equation 3.24). In other words, the target is hardly detectable when its size is much smaller than the acoustic wavelength.

When frequency increases, the validity limit of the Rayleigh scattering regime is roughly given by:

$$ka = \frac{2\pi a}{\lambda} = 1 \tag{3.25}$$

This occurs when the acoustic wavelength equals the circumference of the sphere. At higher frequencies, when ka ranges between 1 and 10, the target circumference reaches a few wavelengths. The target response is then dominated by interference between the reflected wave and "creeping waves" refracted around the surface of the sphere; it is therefore oscillating (see an approximate illustration in Figure 3.9).

Although this ideal case does not correspond to any actual target echo configuration, it is particularly interesting because it shows the different scattering regimes:

- *Rayleigh regime* – when the dimension of the target is small relative to the acoustic wavelength;
- *geometric regime* – when the acoustic wavelength is small relative to the target;
- *interferential regime* of circumference waves at intermediate frequencies.

Targets with more complex structures can often be interpreted using these three basic processes.

3.3.2   Fluid spheres

In the case when the spherical target is no longer ideally rigid, but fluid, its back-scattering cross-section in the Rayleigh regime is written as:

$$\sigma_{bs} = k^4 a^6 \left[\frac{1}{3} - \frac{\rho_1 c_1^2}{3\rho_2 c_2^2} + \frac{\rho_2 - \rho_1}{2\rho_2 + \rho_1} \right]^2 \tag{3.26}$$

where (ρ_1, c_1) and (ρ_2, c_2), respectively, are the densities and sound velocities of water and the target.

When ρ_2 and c_2 are large compared with the water parameters, this expression tends back toward the case of the ideal rigid sphere. But when they are comparatively small, the backscattering cross-section is dominated by the compressibility of the sphere, and tends toward the value:

$$\sigma_{bs} = k^4 a^6 \left[\frac{\rho_1 c_1^2}{3\rho_2 c_2^2} \right]^2 \tag{3.27}$$

This value is much higher than for a rigid sphere of identical radius (Equation 3.24). The target strength of an air bubble is, for example, 75 dB higher than the target strength of the ideal rigid sphere of the same radius.

3.3.3   Scattering by gas bubbles

The backscattering behaviour of gas bubbles present in sea water has been widely studied, because of its importance for many underwater acoustic applications. For instance, air bubble clouds can create undesirable reverberation from the sea surface. Gas bubbles present in sediment are also known to be an essential component of seafloor backscattering. The acoustic behaviour of gas bubbles has been the subject of many theoretical and experimental studies. However, although the individual behaviour of bubbles is well understood, the effects of random populations on acoustic propagation and backscattering are difficult, if not impossible, to predict accurately other than statistically. The interested reader is referred, for more detailed theory and results, to the books of Medwin and Clay (1998) and Leighton (1996).

The acoustic behaviour of gas bubbles is mainly marked by resonance. Around the resonant frequency (dependent on bubble size and dimensions), the effects of

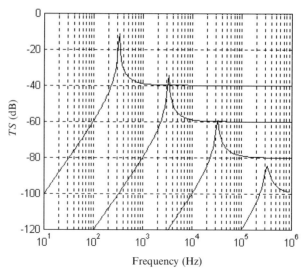

Figure 3.10. Target strength of an air bubble at atmospheric pressure, for respective radii of (from *left* to *right*) 10^{-2}, 10^{-3}, 10^{-4} and 10^{-5} m.

backscattering and absorption are enhanced. The backscattering cross-section of a gas bubble with radius a is given by the relation:

$$\sigma_{bs} = \frac{a^2}{\left(\left(\frac{f_0}{f}\right)^2 - 1\right)^2 + \delta^2} \tag{3.28}$$

where f_0 is the resonant frequency of the bubble and δ is the corresponding damping term. The cross-section thus shows a strong maximum at resonance, limited by the losses from damping. The resonant frequency can be approximated as:

$$f_0 = \frac{1}{2\pi a}\sqrt{\frac{3\gamma P_w}{\rho_w}} \approx \frac{3.25}{a}\sqrt{1 + 0.1z} \tag{3.29}$$

where ρ_w is the water density ($\approx 10^3\,\mathrm{kg\,m^{-3}}$), P_w is the hydrostatic pressure in Pa ($\approx 10^5(1 + z/10)$, z being the depth, expressed in metres) and γ is the adiabatic constant for air (≈ 1.4). The damping effect is due to the combined effects of radiation, shear viscosity and thermal conductivity. Its complete expression is rather complicated, but a good approximation may be obtained by taking $\delta \approx 0.03 f_k^{0.3}$ over the frequency range 1–100 kHz at sea level pressure, f_k being the frequency in kHz. Figure 3.10 presents target strength values computed for air bubbles with radii comprised between 10^{-2} m and 10^{-5} m, over the entire spectrum of underwater acoustic frequencies.

It should be stressed that the resonance wavelength (given by (3.29)) and the bubble dimension have completely different orders of magnitude. This is clearly illustrated in Figure 3.10.

3.3.4   Target strength of fish

In the particular case of acoustic scattering by fish, the main contribution comes from the swimbladder, a gas pocket of variable dimensions, possessed by many species and used by fish to adjust their buoyancy. This gas-filled organ shows a very strong impedance contrast with water and the fish tissues. It behaves either as a resonator (with an intrinsic frequency resonance typically between 500 Hz and 2 kHz, depending on fish size and depth), or as a geometric reflector (at higher frequencies). Its behaviour is predictably very similar to the behaviour of gas bubbles. The other main contributions to the echo come from the spine and from the muscle mass; but their impedances are very close to that of water, and their influence is not always visible. The difference in target strengths TS between species with or without swimbladders can reach 10 dB to 15 dB.

At frequencies commonly used by echo sounders (30 kHz to 400 kHz), the target strengths of individual fish are quite small: typically -30 dB to -60 dB, increasing with length. The levels observed are very variable, and can only be modelled statistically. In practice, semi-empirical models are the most often used. The following formula, derived from Love (1978), shows the order of magnitude of the average target strength TS at high frequencies:

$$TS_{\text{fish}} = 19.1 \log L + 0.9 \log f_k - 24.9 \qquad (3.30)$$

This formula is valid for dorsal echoes at wavelengths smaller than fish length L, f_k being the frequency in kHz. McLennan and Simmonds (1992) propose a more detailed synthesis, using a similar model:

$$TS_{\text{fish}} = 20 \log L - TS_{\text{spec}} \qquad (3.31)$$

where TS_{spec} at a given frequency depends on the considered species, according to Table 3.1. One should note the low values associated with mackerel, a species without a swimbladder.

At frequencies close to the resonance frequency of the swimbladder (around 1 kHz), the target strength increases, and can reach -25 dB to -20 dB. Below

Table 3.1. Target strength constant for various species of fish. Derived from MacLennan and Simmonds (1992)

Species	TS_{spec} @ 30 kHz
Cod	-28.9
Saithe	-25.8
Redfish	-27.1
Haddock	-27.9
Blue whiting	$-25.3/-26.6$
Herring	$-32.1/-33.2$
Sprat/Herring	-31.3 to -33.4
Mackerel	-39.3 to -42.8

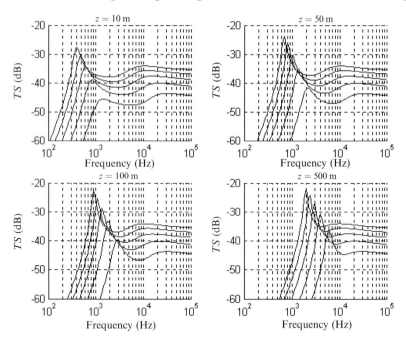

Figure 3.11. Average fish target strengths (*TS*) calculated as a function of frequency, using the model in Appendix A.3.2. The four figures correspond to fish depths of 10, 50, 100 and 500 m, respectively; *TS* decreases as pressure increases. The curves were computed each time for fish lengths *L* of 10 (*lower*), 15, 20, 25 and 30 cm (*upper*).

resonance, there is a Rayleigh regime comparable with an air bubble a few milli-metres wide, with elasticity and damping factors dictated by the wall and surround-ings of the swimbladder. Figure 3.11 shows average target strengths of fish, calculated using the model presented in Appendix A.3.2. The results show Rayleigh scattering at low frequencies, geometric scattering at high frequencies and the swimbladder resonance peak at intermediate frequencies.

3.3.5   Arbitrarily shaped target

Backscattering of a plane wave by a generic obstacle is a very difficult problem to tackle. An interesting limit case is that of an infinitely rigid body, with characteristic dimensions (radii of curvature) much larger than the acoustic wavelength. The problem can then be approached geometrically. This process is equivalently used in electromagnetic remote sensing, in the study of radar echoes from perfectly conductive targets. Its extension to underwater acoustic waves must be approached with caution, as the actual target cannot usually be assumed rigid, but is enough for a first approximation. Various elementary target shapes have been studied: the results can be extended to more complex shapes by decomposing the external

Table 3.2. Backscattering cross-sections of simple-shaped targets (calculated by the geometrical approach, valid for wavelengths much smaller than the target dimensions). Derived from Urick, 1983.

Sphere of radius a	Cylinder, with radius a and length L	Rectangular plate, with side lengths a and b	Circular plate of radius a
$\dfrac{a^2}{4}$	$\dfrac{aL^2}{2\lambda}\left(\dfrac{\sin\zeta}{\zeta}\right)^2\cos^2\theta$	$\left(\dfrac{ab}{\lambda}\right)^2\left(\dfrac{\sin\zeta}{\zeta}\right)^2\cos^2\theta$	$\left(\dfrac{\pi a^2}{\lambda}\right)^2\left(\dfrac{2J_1(\zeta)}{\zeta}\right)^2\cos^2\theta$
	$\zeta = kL\sin\theta$	$\zeta = ka\sin\theta$	$\zeta = 2ka\sin\theta$
		(in the plane including a)	

geometry of the target into its elementary components. A few classical formulae are given in Table 3.2.

3.3.6   Submarine echoes

Submarines give off sonar echoes that emanate from several types of interaction between the wave and the structure:

- specular echoes on quasi-plane surfaces (hull flanks, conning tower);
- echoes from obstacles inside the external hull;
- diffraction on angular portions of the target (fins, rudder, propulsion, etc.);
- excitation of resonance modes characteristic of the structure.

These different effects can be reduced by optimising the shape and structure of the submersible, and by covering the hull with anechoic coating. The latter is less efficient as frequency decreases, hence the development of VLF (Very Low Frequency) active sonar techniques. Urick (1983, p. 310) shows a few experimental results obtained with World War II submarines. They present a very characteristic "butterfly"-shaped directivity pattern: the echo amplitude is maximal at lateral incidence (specular echo from the hull flanks) and minimal in the axis of the submarine. According to Urick, the global target strengths of these submarines range from 0 dB to +40 dB.

The various components of a submarine sonar echo can be identified experimentally as "highlights", provided that the signals used have an adequate time resolution. The structure of these "highlights" can then be used to characterise the target.

3.4   EXTENDED TARGETS

3.4.1   Computation principle

Extended targets are delimited by their sonar characteristics (beam width, signal length) rather than by their own dimensions. This is the case for the seabed, the

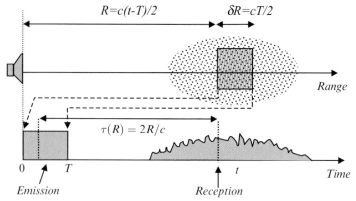

Figure 3.12. Spatial limits $(R, \delta R)$ of an extended target, for a signal of duration T and at a given observation time t.

sea surface, large fish schools, and the Deep Scattering Layer (DSL) (see Section 3.4.2.4). The backscattering cross-section can then be decomposed into two contributors: (1) the size of the target portion (surface or volume) effectively insonified by the sonar; (2) the corresponding backscattering strength (per unit surface or volume).

The backscattering cross-section then becomes:

$$\sigma_{bs} = A_{s,v}\sigma_{bs}^{s,v} \tag{3.32}$$

where $A_{s,v}$ is the effective extent of the scattering surface (or volume), and $\sigma_{bs}^{s,v}$ is the unit surface (or volume) backscattering cross-section.

The backscattered echo level is expressed in dB as:

$$EL = SL - 2TL + TS = SL - 2TL + 10 \log A_{s,v} + BS_{s,v} \tag{3.33}$$

To determine A, we must consider the fact that the sonar transmits a signal over the time interval $[0, T]$. At the observation time $t > T$, the backscattered signal comes from all parts of the target at ranges R verifying (Figure 3.12):

$$t - T < \tau(R) < t \tag{3.34}$$

where $\tau(R)$ is the two-way travel time between the sonar and a point at range R.

The propagation model used links τ and R. Assuming the propagation is spherical, we get $\tau = 2R/c$, and the elements scattering simultaneously at distances R verify:

$$\frac{c(t - T)}{2} < R < \frac{ct}{2} \tag{3.35}$$

The backscatter level can then be represented as a function of:

- Time – this is the most correct representation of the physical process actually observed by the sonar.
- Distance – if we can assume a one-to-one relation between τ and R. This is not always the case in practice (because of multiple paths). But this representation is convenient to assess reverberation as a function of detection ranges, and to

compare it with echoes from a target at a given range. It is therefore commonly used in sonar performance evaluation.

3.4.2   Volume backscattering

3.4.2.1   *Echo level*

From the preceding section one can see that the scattering volume at a time t is included between two spheres of radii $c(t - T)/2$ and $ct/2$. Let us use ψ to represent (Figure 3.13) the equivalent aperture (solid angle, in steradians) of the source/receiver system (cf. Section 5.3.1.3). Approximating the portion of the cone by a cylinder, the scattering volume becomes:

$$A_r = \psi R^2 \frac{cT}{2} \quad (T \ll t) \tag{3.36}$$

Introducing the volume backscatter strength BS_v, the backscattered level becomes:

$$EL = SL - 40 \log R - 2\alpha R + 10 \log\left(\psi R^2 \frac{cT}{2}\right) + BS_v$$

$$= SL - 20 \log R - 2\alpha R + 10 \log\left(\psi \frac{cT}{2}\right) + BS_v \tag{3.37}$$

3.4.2.2   *Volume backscattering strength*

The volume backscattering of a distribution of targets can be analysed as the incoherent sum of the contributions from each target present in an average cubic metre of water. But, for a given frequency, each contribution will depend on the size and shape of each member (the shape being less important if its dimension is much smaller than the signal wavelength) and on the nature of each member (structure, density, sound velocity of the constitutive material).

 If the scatterers are of similar nature, their backscattering cross-sections can be expressed, for a given frequency, as a function $\sigma(a)$ of their characteristic dimension a (e.g., the radius for spherical targets). If N_1 is the average number of targets per cubic metre, and $q(a)$ is the probability density function associated with the radius distribution of the individual targets, then the average volume backscatter

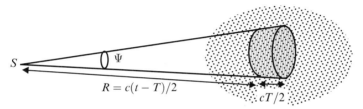

Figure 3.13. Geometrical model of volume backscattering.

cross-section (relative to $1\,m^3$) equals the weighted sum of the individual contributions of the different size classes:

$$\sigma_1 = N_1 \int_0^{+\infty} \sigma(a)q(a)\,da = N_1\bar{\sigma} \tag{3.38}$$

This is equivalent to the product of the density of scatterers and the effective, average, individual backscattering cross-section (weighted by the size distribution).

For further simplification, one can assume that all targets are identical. The backscattering strength will then equal the individual backscattering cross-section, corrected by the number of targets per cubic metre. In logarithmic notation, this writes:

$$BS_v = TS + 10\log N_1 \tag{3.39}$$

where TS is the individual target strength.

This model assumes that the intensities scattered by each target can be summed directly. This implies that, first, the contribution of the energy backscattered by multiple scattering[6] between scatterers is ignored as is, second, the masking of scatterers by each other. These assumptions are obviously more valid as the scatterers become less densely distributed.

3.4.2.3   Fish schools

Fish schools are very variable in shape and size, with typical values of a few metres vertically and several tens of metres horizontally. A school is usually composed of only one species, with individuals of nearly similar ages, and therefore similar sizes (hence similar target strengths). Their backscattering strength can consequently be modelled using the individual target strengths and the number of fish per unit volume. Fish density depends on the dimension of individuals: a good order of magnitude for the number of fish per cubic metre is $N_{1\,m^3} \approx 1/L^3$. Under these assumptions, and simplifying the TS model of Equation (3.30) into $TS_{fish} \approx 20\log L - 25$, one gets:

$$BS_{school} = TS_{fish} + 10\log N_{1\,m^3} = -25 + 20\log L - 10\log(L^3) = -25 - 10\log L \tag{3.40}$$

This shows that the volume backscatter strength decreases when fish size increases, since the number density effect primes upon individual size influence.

These simple models and assumptions, using the proportionality of volume backscattering strength and individual target strength, form the basis of the *echo-integration* technique used to assess fish school populations. The number of fish present in a given area is estimated by the total energy backscattered towards the

---

[6] A simplified criterion to neglect multiple scattering is $\sigma n/\beta \ll 1$, for a scatterer distribution with the same scattering cross-section σ, where n is the average number of scatterers per unit volume and β is the absorption coefficient for the sound pressure (in Neper/m). A more complete analysis of the case where the distribution is not homogeneous can be found *inter alia* in Boyle and Chotiros (1995a, b).

sounder, corrected for the average individual target strength (see Section 7.3.3 for more details).

Let us finally remark that backscattering from a fish school can be considered as the effect either of an extended volume target or of a local target if its dimension is small enough relative to the sonar resolution (cf. Figure 3.7B). In this case, the resulting target strength can then be expressed as a function of the fish school volume V_{school}:

$$TS_{school} = BS_{school} + 10 \log V_{school} \tag{3.41}$$

3.4.2.4 Deep Scattering Layer

The Deep Scattering Layer (DSL) is a thin layer of the ocean (tens to hundreds of metres), crowded with living organisms (e.g., phytoplankton, zooplankton, small fish). This layer has very significant effects on acoustic propagation (absorption and scattering), and is an oceanographic feature in its own right. It can be found in all the oceans, but the amount of biomass will depend on latitude: maximum close to the equator, minimum in the polar seas. A very characteristic feature of the DSL is its time dependence: in daytime, the layer stays at depths between 200 m and 600 m, whereas at night it migrates to shallower depths, typically 100 m.

The main acoustic effect from the DSL is caused by the resonance of fish swimbladders (in the 1–20-kHz range, because of fish size variations). The corresponding BS_v values are highly variable and unpredictable in this resonance-dominated regime. A striking feature is the changing of the frequency dependence with day and night, due to the effect of hydrostatic pressure changes during the daily migration of the DSL (see Figure 3.11, showing the effects of pressure on swimbladder resonance frequencies).

At higher frequencies (above 20 kHz), the acoustic effect of the DSL comes from scattering around smaller, non-resonant animals (zooplankton). Urick (1983) reports typical average BS_v values around −70 dB re 1 m³. It should be noted that this average order of magnitude is much smaller than the volume BS values observed for locally-packed fish schools (Equation 3.40).

3.4.3 Surface backscattering

3.4.3.1 Grazing incidence backscatter levels

Let us consider a sonar with an equivalent beam aperture ϕ (see Section 5.3.1.3) in the horizontal plane, insonifying the interface (sea surface or seabed) at grazing incidence (Figure 3.14). Assuming spherical propagation between the source and the interface, the surface insonified at time t is included between the two concentric circles of respective radii $c(t-T)/2\sin\theta$ and $ct/2\sin\theta$. Its projection on the interface is:

$$A_r = \phi R \frac{cT}{2} \frac{1}{\sin\theta} \tag{3.42}$$

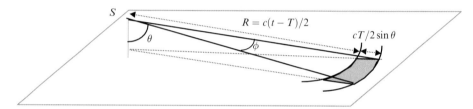

Figure 3.14. Interface backscatter at grazing incidence.

From this we can deduce the backscattered echo level:

$$EL = SL - 40\log R - 2\alpha R + 10\log\left(\phi R \frac{cT}{2\sin\theta}\right) + BS_s(\theta)$$

$$= SL - 30\log R - 2\alpha R + 10\log\left(\phi \frac{cT}{2\sin\theta}\right) + BS_s(\theta) \qquad (3.43)$$

where $BS_s(\theta)$ is the interface backscattering strength at grazing incidence.

3.4.3.2   *Normal incidence backscatter level*

Let us now consider a sound beam with a narrow aperture, insonifying a scattering surface at normal incidence. The insonified surface is the intersection of the directivity lobe (assumed vertical) and this surface (Figure 3.15, *left*):

$$A = \psi H^2 \qquad (3.44)$$

where H is the height from the source to the target and ψ is the equivalent solid aperture, in steradians. If the beam is conical, the surface can be expressed as a function of a conventional angular half-aperture φ:

$$A = \pi (H \tan \varphi)^2 \qquad (3.45)$$

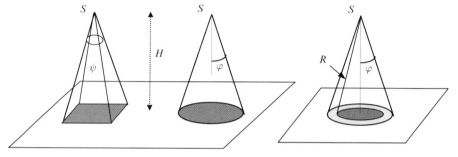

Figure 3.15. Interface backscattering at normal incidence: the dark patches show the extent of the areas ensonified for a long pulse (*left*) and a short pulse (*right*). See text for details.

The maximum echo intensity will then equal:

$$EL = SL - 40 \log H - 2\alpha H + 10 \log(\psi H^2) + BS_s(0)$$
$$= SL - 20 \log H - 2\alpha H + 10 \log \psi + BS_s(0)$$
$$= SL - 20 \log H - 2\alpha H + 10 \log(\pi \tan^2 \varphi) + BS_s(0) \qquad (3.46)$$

where $BS_s(0)$ is the bottom backscattering strength at normal incidence.

This model implies that the signal is transmitted for a sufficiently long time for the beam footprint to be insonified at once.

For a short pulse, the patch of seafloor insonified is not determined by the beam aperture, but by the pulse duration. Its projection onto the seafloor is a disk (Figure 3.15, *right*), whose radius is given by the delay between the edge and the centre. The range between the sonar and the disk edge is $R = H + cT/2$. The disk radius equals approximately $\sqrt{HcT}$ and its area equals $A = \pi HcT$. In this case, the maximum echo level becomes:

$$EL = SL - 30 \log H - 2\alpha H + 10 \log(\pi cT) + BS_s(0) \qquad (3.47)$$

The transition between the long-pulse and the short-pulse regime occurs when the signal projection onto the interface intersects the edges of the beam directivity footprint. Accordingly, it depends directly on the beam shape. For a conical beam of half-aperture φ, the limit between the two regimes occurs at water depth H:

$$H = \frac{cT}{\tan^2 \varphi} \qquad (3.48)$$

3.5   REFLECTION AND SCATTERING BY A ROUGH SURFACE

3.5.1   Roughness and scattering

The boundaries of the propagation medium (sea surface and seafloor) are usually far from ideal plane surfaces. The acoustic processes will therefore be much more complex than the reflection and transmission cases discussed in Section 3.1. These interfaces may be considered as locally plane, on average, with a microscale roughness whose influence will be significant if its characteristic dimensions are at least comparable in magnitude with the acoustic wavelength.

The effect of relief on the incident acoustic wave will be a compound of the frequency, the angle of incidence and the local characteristics of the relief. Because of their geometry, interface irregularities will scatter the incident wave in all directions (scattering process) (Figure 3.16). Part of the incident wave will be reflected with no deformation other than an amplitude loss in the specular direction (coherent part of the signal). The remainder of the energy is scattered in the entire space, included back towards the source (backscattered signal). The relative importance of the specular and scattered components will depend on surface roughness (i.e., the

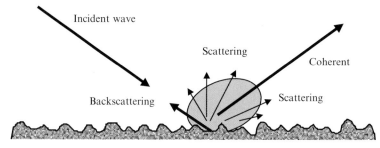

Figure 3.16. Coherent reflection and scattering of an incident wave by a rough surface.

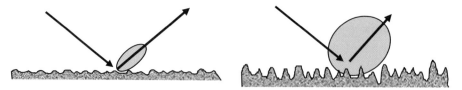

Figure 3.17. Specular reflection and scattered field for low interface roughness (*left*) and high interface roughness (*right*).

ratio between the characteristic amplitude of the relief and the wavelength of the acoustic signal).

Low interface roughness will induce a relatively larger specular component and low scattering distributed around the specular direction (Figure 3.17, *left*). Conversely, high interface roughness will strongly attenuate the specular component and spread the scattering in all directions (Figure 3.17, *right*).

Microscale roughness has various possible origins. At the sea surface, the interface relief is due to gravity or capillary waves caused by the wind. On the seafloor, the roughness depends on the geology (strong relief on rocks, tide and current ripples, and living organisms on or in the sand or mud). The microscale roughness presents a wide scale of amplitudes. For processes of interest at sonar frequencies, these amplitudes range between a millimetre and a few metres. Usually, several micro-roughness scales coexist on the same surface. For example, at the sea surface, capillary waves (centimetre scale) are superimposed on the swell (metre scale). On a sandy seafloor, the sand waves have a centimetre-scale roughness, superimposed on the existing topography. In this case, the two roughness scales might correspond to different physical processes, depending on their size relative to the acoustic wavelength (see Section 3.5.3).

The spatial spectrum of the relief is used to quantify the amplitude distribution of a rough surface. This is the Fourier transform of the relief. It shows the energy distribution of the different harmonic components of the relief (in the same way that a time signal can be decomposed into its frequency components). Every spatial spectrum component is defined by a wave number $\kappa = 2\pi/\Lambda$, Λ being the spatial wavelength. The spectrum can have specific, strong components if the relief is

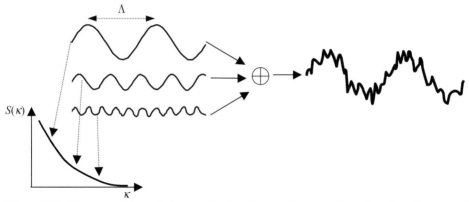

Figure 3.18. Decomposition of the amplitude of a random rough surface into its spectral components. The power spectrum $S(\kappa)$ decreases with wave number, as larger amplitudes are associated with longer wavelengths.

periodic (e.g., regular sand ripples). It can also be continuous if the relief is completely random (Figure 3.18). The latter case is the most representative: common models often use a spectrum of the type $S(\kappa) = S_0 \kappa^{-\gamma}$, with a γ factor typically between 3 and 3.5. This exponential form arises because the lowest spatial frequencies correspond to the largest amplitudes, and hence the maximum energy, whereas the highest frequencies correspond to small microscale relief. To be rigorous, the spatial spectrum should be considered along one particular direction of the average rough plane. However, in practice, the relief is often assumed to be isotropic, with a spectrum independent of the direction considered.[7] The expression of the spatial spectrum is normalised so that its integral over all components in a given area equals the variance of relief amplitudes on this area:

$$\int S(\kappa)\, d\kappa = h^2 \qquad (3.49)$$

(although the latter is not valid for the above-mentioned power-law spectrum).

3.5.2 Coherent reflection

From an acoustic standpoint, interface roughness is quantified by the *Rayleigh parameter*:

$$P = 2kh\cos\theta \qquad (3.50)$$

This equation uses the acoustic wavenumber $k = 2\pi/\lambda$, the standard deviation h of the relief amplitude and the angle of incidence θ (relative to the normal to the average surface). Surface roughness expresses the ratio between the mean amplitude of the relief and the wavelength, accounting for the angle of incidence.

---

[7] This assumption is obviously not valid for a relief strongly organised along a preferential direction (e.g., sedimentary ripples).

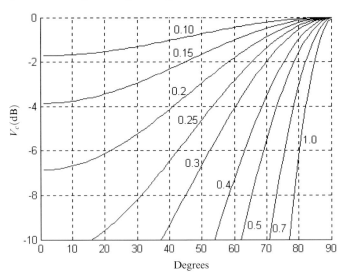

Figure 3.19. Coherent reflection loss as a function of the incident angle. The different curves correspond to respective values of $zh/\lambda = 0.1, 0.15, 0.2, 0.25, 0.3, 0.4, 0.5, 0.7$ and 1.0. The computation is limited to a maximum loss of $10\,\mathrm{dB}$, to remain in the validity domain of the model.

The coherent (specular) component of the reflected signal can be described by the reflection coefficient V_c (see Appendix A.3.3 for details):

$$V_c = V \exp\left(-\frac{P^2}{2}\right) = V \exp(-2k^2h^2 \cos^2 \theta) \qquad (3.51)$$

This reflection coefficient is a function of the Rayleigh parameter P and the reflection coefficient V on the interface without relief.

This model is valid when the Rayleigh parameter is small (i.e., for small values of the frequency and the relief amplitude) and at grazing incidence. The conventional limit of validity is $P = \pi/2$. This corresponds to $h = \lambda/(8 \cos \theta)$, and a coherent loss of $10.7\,\mathrm{dB}$ (Figure 3.19). For larger values of the Rayleigh coefficient, the coherent signal becomes negligible and the scattered field is predominant.

One should keep in mind that the coherent reflection coefficient expresses *specular reflection*, instead of scattering or backscattering. Except in particular cases (i.e., at normal incidence), it should not be assimilated to a target strength, and is not relevant when modelling the echoes sent back to a sonar directly from the seabed or sea surface.

3.5.3   Backscattered field

The scattered part of the signal shows very different orders of magnitude, depending on the characteristic dimensions of the configuration used. In practice, the

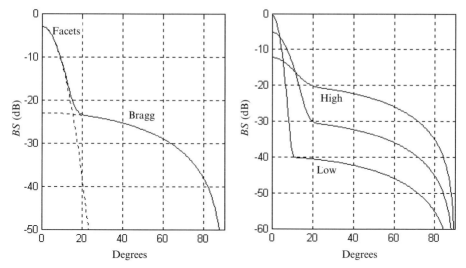

Figure 3.20. Idealised backscattering strength of a rough surface. *Left*: facet and Bragg regime. *Right*: influence of roughness. With increasing roughness, the level of the Bragg regime increases and the level of the specular component decreases.

backscattered field is the most interesting to study, because of its use by sonar systems analysing target echoes. It will strongly depend on the angle or roughness considered. Close to the normal to the surface, the incident wave is reflected by the facets oriented to reflect the specular echo; this corresponds to the maximum back-scattered energy. At oblique incidences, the backscattered wave comes from a continuum of sources distributed as a function of microscale roughness. For a given direction of observation, the spectral component of the relief corresponding to the reception in phase of the contribution from scatterers is predominant; this domain is called the *Bragg scattering domain*. Theoretical models currently available are based on these two fundamental regimes. The balance between them will depend on the microscale roughness (Figure 3.20). For example, a smoother seabed will increase the contribution from facets and yield a low backscattering strength at oblique angles; the opposite will be true for high interface roughness. However, in practice, things are not quite so well contrasted, even more so because the back-scattered levels will also depend on the impedance contrast at the interface, and on the contribution to backscattering by the underlying volume.

3.5.3.1   Rough-interface scattering strength

Let us consider an incident acoustic wave, assumed plane, insonifying a small area A of a rough interface with the incident angle θ_i. The incident wave intensity is $I_i = p_i^2/2\rho c$ (i.e., the available power on the surface A is $P_i = AI_i \cos \theta_i$). A portion η of this power (neither transmitted nor coherently reflected) is then scattered into the entire half-space above the interface. This scattering is characterised by an angular intensity distribution $G(\theta_i, \theta_s)$ typical of the rough surface, and normalised so that $\int G(\theta_i, \theta_s)\, d\theta_s = \eta(\theta_i)$.

Consequently, in a given direction of observation, the intensity of the scattered wave equals the power scattered, corrected for the scattering function in this direction, and the transmission loss (expressed here as spherical loss, without absorption):

$$I_s(R, \theta_s) = P_i(\theta_i)\frac{G(\theta_i, \theta_s)}{R^2} \tag{3.52}$$

This result can be generalised by considering the intensity scattered by a unit area A_1 (equal to $1\,\mathrm{m}^2$) at a unit distance $R_{1\,\mathrm{m}}$:

$$I_s(R_{1\mathrm{m}}, \theta_s) = A_1 I_i(\theta_i)\cos(\theta_i)\frac{G(\theta_i, \theta_s)}{R_{1\,\mathrm{m}}^2} \tag{3.53}$$

The ratio of the scattered and incident intensities is the interface scattering cross-section:

$$\sigma_s(\theta_i, \theta_s) = \frac{I_s}{I_i}R_{1\mathrm{m}}^2 = A_1 \cos\theta_i G(\theta_i, \theta_s) \tag{3.54}$$

Its value for $\theta_i = \theta_s$ is the backscattering cross-section:

$$\sigma_{bs}(\theta_i) = \frac{I_s}{I_i}R_{1\mathrm{m}}^2 = A_1 \cos\theta_i G(\theta_i, \theta_i) \tag{3.55}$$

Its value in dB is the interface backscattering strength, expressed in dB re $1\,\mathrm{m}^2$:

$$BS_s = 10\log\left(\frac{\sigma_{bs}}{A_1}\right) \tag{3.56}$$

3.5.3.2 Lambert's law

Let us now investigate the simple case where the rough surface is a perfect scatterer, so that neither η nor G depend on the angles. Then $\eta(\theta_i) = \eta_0$, and $G = \eta_0/2\pi$ (the scattered intensity is zero in the lower medium, and is equally distributed over the upper half-space, i.e., on half the solid angle of a sphere, 4π). This yields:

$$\sigma_s(\theta_i, \theta_s) = A_i\frac{\eta_0}{2\pi}\cos\theta_i \tag{3.57}$$

This is the same expression for backscattering.

The interface scattering process is now no longer isotropic, and depends on roughness. The scattered intensity will then depend on the interface slope distribution: it will basically be maximum normally to the average interface, and minimum along it. For instance, the resultant, average, radiated intensity can be taken as proportional to the cosine of the scattering angle:

$$G(\theta_i, \theta_s) = \frac{\cos\theta_s}{\pi} \tag{3.58}$$

This leads to a scattering cross-section:

$$\sigma_s(\theta_i, \theta_s) = A_1 \frac{\eta_0}{\pi} \cos\theta_i \cos\theta_s \qquad (3.59)$$

And the resulting backscattering cross-section ($\theta_i = \theta_s$) is:

$$\sigma_{hs}(\theta_i) = A_1 \frac{\eta_0}{\pi} \cos^2\theta_i \qquad (3.60)$$

This result is widely known as Lambert's law. It can be conveniently expressed in dB as:

$$BS(\theta_i) = BS_0 + 20\log\cos\theta_i \qquad (3.61)$$

Note that an upper limit of BS_0 can easily be defined. Considering a perfectly reflecting interface leads to $\eta_0 = 1$ and $BS_0 = 10\log(\eta_0/\pi)$ (i.e., around $-5\,\text{dB re } 1\,\text{m}^2$). Practically observed values of BS_0 in fact range between $-10\,\text{dB}$ and $-40\,\text{dB re } 1\,\text{m}^2$.

Despite its simplicity, Lambert's law is a good first approximation. In many practical cases, it is in acceptable agreement with the physical observations. For backscattering by a slightly rough surface (calm sea surface, soft sediments), its validity is restricted to oblique and grazing incidences. On very rough surfaces (e.g., rocky seabed) it may be employed over the entire angular domain.

It is interesting to remark that:

- Heuristic descriptions of the backscattering strength are often derived from Lambert's law. It is quite common to search for a fit between experimental results and a law in $BS(\theta) = BS_0 + 10\gamma\log\cos\theta$.
- Conversely, more sophisticated theoretical models for oblique incidence backscattering often practically lead to angular dependence akin to a Lambert-like law.

3.5.3.3   Facet reflection

Let us now consider a relatively smooth horizontal surface (e.g., the sea surface during moderate sea states, or a sedimentary seafloor). For a wave insonifying the surface at angles close to the vertical, one can consider the interface to be a mosaic of facets with random tilt angles, distributed around the average surface plane. Each facet reflects the incident wave mainly around its specular direction, depending upon its tilt (Figure 3.21). The major part of the surface is made up of facets close to the horizontal, and the backscattering strength will be maximum close to the vertical, as many facets contribute to it. If the direction of the incident wave moves away from the vertical, there will be fewer and fewer facets with the correct orientations, and the backscattering strength will decrease drastically. Since the average intensity is proportional to the total surface of the contributing facets, the result is expected to follow the facet slope distribution away from the horizontal.

Let χ represent the angle of a facet with the horizontal. Assuming that the facet slopes ($\tan\chi$) follow a Gaussian distribution in $1/(\sqrt{2\pi}\delta)\exp(-\tan^2\chi/2\delta^2)$, where δ^2 is the slope variance, the backscattering cross-section is expected to vary

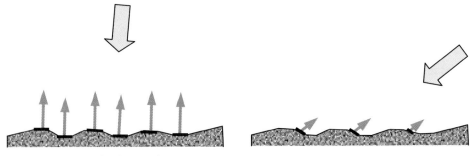

Figure 3.21. Facet backscattering: there are many facets with the right orientation at incident angles close to the vertical (*left*), and fewer at grazing angles (*right*).

proportionally with the plane wave reflection coefficient $V(0)$ at the vertical of the average surface (see Section 3.1.1.1):

$$\sigma_{bs}(\theta) \propto |V(0)|^2 \exp\left(-\frac{\tan^2 \theta}{2\delta^2} \right) \tag{3.62}$$

The complete theoretical approach can be found in the books by Ishimaru (1978a, b) or Brekhovskikh and Lysanov (1992). It gives the limit case for high-frequency waves:

$$\sigma_{bs}(\theta) = \frac{|V(0)|^2}{8\pi\delta^2 \cos^4 \theta} \exp\left(-\frac{\tan^2 \theta}{2\delta^2} \right) \tag{3.63}$$

This model is valid when the local curvature of the facets is negligible compared with the wavelength – in other words when the condition of reflection on locally plane facets (Kirchhoff approximation) is fulfilled. This approach is commonly used to model backscattering close to the perpendicular, on smooth sediment seafloors at sufficiently high frequencies.

One limitation of this tangent plane model is that it does not depend on the frequency of the incident acoustic wave. Jackson *et al.* (1994) proposed an analytical solution to the Helmholtz–Kirchhoff integral, using the roughness spectrum of the interface and keeping the frequency dependence in the final results.

3.5.3.4   *Bragg scattering*

When microscale roughness is small relative to the acoustic wavelength, the back-scattered field is made of a continuum of contributions from the points along the interface. For a given angle of observation θ, the signal is dominated by the contributions coming from scatterers where the scattered signals are in phase. These scatterers are distributed at distances d from each other, with:

$$2d \sin \theta = \lambda \tag{3.64}$$

In other words, the main contributor to scattering along this angle of observation is the roughness spectrum component given by:

$$\kappa = \frac{2\pi}{d} = \frac{4\pi \sin \theta}{\lambda} = 2k \sin \theta \tag{3.65}$$

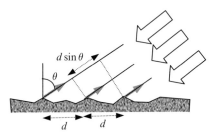

Figure 3.22. Bragg backscattering: at incidence θ, the main contribution to backscatter comes from the scatterers with spacing d such as $2d\sin\theta = \lambda$, building a constructive interference pattern.

A practical formula for the backscattering cross-section of a rough interface between two fluids in the Bragg regime is:

$$\sigma_{bs}(\theta) = 4k^4\cos^4\theta|U(\theta)|^2 S(2k\sin\theta) \tag{3.66}$$

In this equation, $U(\theta)$ is an intensity reflection coefficient:

$$U(\theta_1) = \frac{(\rho_2 c_2 \sin\theta_1 - \rho_1 c_1 \sin\theta_2)^2 + \rho_2^2 c_2^2 - \rho_1^2 c_1^2}{(\rho_2 c_2 \cos\theta_1 + \rho_1 c_1 \cos\theta_2)^2} \tag{3.67}$$

where $\theta = \theta_1$ and θ_2 are the incidence angles in the two fluid media.
  Note that the expression σ_{bs} in Equation (3.65) features a frequency dependence in f^4, bringing back to mind the considerations of Section 3.3.1 about Rayleigh backscattering by small targets.
  The actual backscattering cross-section strongly depends on the term $S(\kappa)$ (cf. Equation 3.66). The roughness spectrum of the sea surface or seabed is often approximated by the expression $S(\kappa) \propto \kappa^{-\nu}$, with the exponent ν between 2.5 and 3.5. Taking $\nu = 3$ yields the proportionality relation:

$$\sigma_{bs}(\theta) \propto k\frac{\cos^4\theta}{\sin^3\theta}|U(\theta)|^2 \tag{3.68}$$

This last expression makes it clear that the backscattering cross-section decreases at lower frequencies, albeit slower (in f) than for Rayleigh scattering, and that it decreases with angles at grazing incidences; the $\cos^4\theta$ is only partly compensated by the increase in θ of the term $|U(\theta)|^2$.

3.6   REFLECTION AND SCATTERING AT OCEAN BOUNDARIES

3.6.1   Reflection and scattering at the sea surface

3.6.1.1   Sea surface characteristics

The relief of the sea surface depends mainly on wind. It has been widely studied for many different types of application. The swell spectrum is often quantified by the classical model of Pierson and Moskowitz[8] (1964):

$$S(\Omega) = 8.1 \times 10^{-3} \frac{g^2}{\Omega^5} \exp\left(-0.74 \left(\frac{g}{\Omega w} \right)^4 \right) \tag{3.69}$$

This expresses the energy content of frequencies Ω as a function of wind speed w (in m/s); g is the acceleration constant.

The spatial wave number κ is related to the frequency $\kappa = \Omega^2/g$. It is also related to the wavelength $\Lambda = 2\pi(g/\Omega^2)$. The relief standard deviation h (in metres) for this spectrum equals $h = 5.33 \times 10^{-3} w^2$. The model of Cox and Munk gives the relation between wind speed and the slopes of the swell-induced relief, using its variance $\delta^2 = (3 + 5.12w) \times 10^{-3}$ (δ in radians). The correlation length of the relief is $L = \sqrt{2}h/\delta$.

It should be noted that the bubble layer immediately below the surface, and reaching several metres deep, strongly perturbs the observation of acoustic processes at the air–water interface. Being very absorbent, the bubble layer can completely mask the surface, and it has its own backscattering. Its existence, and its influence, are of course strongly correlated with agitation of the sea surface.

3.6.1.2   Acoustic transmission across the sea surface

At the water–air interface, the acoustic characteristics of the reflecting medium are $\rho_2 \approx 1.3\,\mathrm{kg/m^3}$ and $c_2 \approx 340\,\mathrm{m/s}$, meaning that $Z_2 \ll Z_1$. Hence, for underwater acoustic waves hitting the sea surface, there is no total reflection as $c_2 < c_1$, and the reflection coefficient is always close to $V \approx -1$, whatever the incident angle. The pressure transmission coefficient from water into air at normal incidence then reads:

$$W(\theta_1 = 0) = \frac{2Z_2}{Z_1 + Z_2} \approx \frac{2Z_2}{Z_1} \approx 5.7 \times 10^{-4} \tag{3.70}$$

---

[8] The Neuman–Pierson model (Neumann and Pierson, 1966) is also often used:

$$S(\Omega) = \frac{2.4}{\Omega^6} \exp\left(-2\left(\frac{g}{\Omega w} \right)^2 \right)$$

with a relief standard deviation of $h = 0.18 \times 10^{-2} w^{2.5}$.

This corresponds to a transmission loss of $-64.9\,\mathrm{dB}$. The power transmission coefficient corresponds to a loss of $-29.3\,\mathrm{dB}$, since:

$$\frac{P_2}{P_1} = W^2 \frac{Z_1}{Z_2} \approx \frac{4Z_2}{Z_1} \approx 1.2 \times 10^{-3} \tag{3.71}$$

At the air–water interface, the impedance contrast is the inverse, and $Z_2 \gg Z_1$. The critical angle is $\theta_1 \approx 13°$. But whatever the angle value, the reflection coefficient remains very close to $V = 1$. The pressure transmission coefficient at normal incidence from air into water is now given by:

$$W(\theta_1 = 0) = \frac{2Z_2}{Z_1 + Z_2} \approx 2 \tag{3.72}$$

This corresponds to a transmission coefficient of $+6\,\mathrm{dB}$. The power transmission loss becomes $-32.4\,\mathrm{dB}$, since:

$$\frac{P_2}{P_1} = W^2 \frac{Z_1}{Z_2} \approx \frac{2Z_2}{Z_1} \approx 5.7 \times 10^{-4} \tag{3.73}$$

In summary, although the acoustic powers are orders of magnitude different in the two media, acoustic transmission from water into air, and vice versa, is perfectly observable. Close to their source, low-frequency sonar transmissions are easily audible in the air. And underwater receivers may detect sources above the sea surface (e.g., helicopters or planes).

3.6.1.3   Reflection and scattering by the sea surface

We saw that the interface between water and air is a quasi-perfect reflector, because of the high impedance contrast between the two environments. To estimate reflection and scattering processes, we can directly apply the models of coherent reflection and scattering (facet and Bragg regimes) seen in the previous sections, replacing the local energy reflection coefficient $|V|$ by 1.

To be complete, these approaches should be complemented by two effects caused by the surface bubble layer: (1) extra attenuation for backscattered echoes reflected on the surface; (2) backscattering by bubble plumes. Explanations of these effects may be found in Chapter 2.

3.6.2   The seafloor

3.6.2.1   Interaction of acoustic waves with the seafloor

The acoustic effects of the seafloor are much more complex than for the sea surface. Many different processes may be observable and their relative importance will depend on the signal frequency:

- The seafloor basically acts as a rough interface, thus scattering the incident sound waves. The backscattered wave is the signal that is used for all seafloor mapping sonars.

- A noticeable part of the incident energy may penetrate the seabed, because of the small impedance contrast between water and sediment. Absorption inside sediments is much higher than in water (typically 0.1 dB to 1 dB/wavelength). But low acoustic frequencies can propagate with reasonable levels.
- Processes similar to those in water propagation (e.g. internal refraction and reflections) may occur inside the sediment. The sediment may also show profiles of sound speed and density, with gradients and discontinuities due to geological layering processes.
- Various scatterers lie at the interface or are buried inside the sediment (e.g., shells and living organisms, minerals, weeds and algae, gas bubbles). They may generate specific additional scattering.

The physical processes, their interpretation and their modelling will differ greatly with the frequencies considered. At high frequencies, where bottom penetration is small, the interaction is usually limited to the surface (reflection from a non-plane interface), with possible complications related to the stratified or heterogeneous nature of the sediment. If the interaction remains very shallow (at frequencies beyond a few tens of kHz), and if the seabed structure remains relatively simple, the processes can be described correctly by combining reflection at the plane interface with a model of the relief influence.

At low frequencies (below a few kHz), one must model the interaction of the wave with the sediment layer at a deeper level as the frequency is lower. The significant processes are then no longer the relief irregularities and the small heterogeneities of the environment, but the sound velocity/density profiles and mostly their discontinuities. The latter's acoustic behaviour is widely used in marine geology and geophysics (e.g., seismics, sediment profiling). In practice, acoustic energy can be reflected at the interfaces between layers. It may also undergo refraction; because of the sound velocity profile in the sediment.[9]

At highly grazing incidences and low frequencies, one can observe interface waves, propagating at the boundary between layers of distinct characteristics. These waves (Rayleigh, Love, Scholte, ...) will differ according to the nature of the interfaces. This fact is exploited in refraction seismics techniques used in geology and geophysics.

3.6.2.2   Acoustic parameters of the seafloor

Typical values for a variety of sediment types are presented in Table 3.3. They follow the classification of Shepard (1954), and the mean grain size (in Phi units[10]) is given as a link to geological characterisation. The acoustic parameters listed here are:

- the mean grain size;

---

[9] It should be noted that velocity and density gradients inside sediments are much stronger than in water: respectively, 1 m/s/m and 1 kg/m^3/m as orders of magnitude.

[10] The mean grain size M in Phi units is defined as $M\phi = -\log_2(a)$, where a is the average grain diameter in mm.

Table 3.3. Typical acoustic parameters for sediments. All these values have been simplified and rounded on purpose, to emphasise that they are just orders of magnitude (see text for details).

Sediment type	M (ϕ)	n (%)	ρ (kg m^{-3})	c_r	c (m/s)	$V(0°)$ (dB)	α (dB/λ)	c_s (m/s)	Ω_0 (cm^4)	h (cm)	δ (°)
Clay	9	80	1,200	0.98	1,470	−21.8	0.08	–	5×10^{-4}	0.5	1.2
Silty clay	8	75	1,300	0.99	1,485	−18.0	0.10	–	5×10^{-4}	0.5	1.5
Clayey silt	7	70	1,500	1.01	1,515	−13.8	0.15	125	5×10^{-4}	0.6	1.7
Sand–silt–clay	6	65	1,600	1.04	1,560	−12.1	0.20	290	5×10^{-4}	0.6	2
Sand–silt	5	60	1,700	1.07	1,605	−10.7	1.00	340	5×10^{-4}	0.7	2.5
Silty sand	4	55	1,800	1.10	1,650	−9.7	1.10	390	1×10^{-3}	0.7	3
Very fine sand	3	50	1,900	1.12	1,680	−8.9	1.00	410	2×10^{-3}	1.0	4
Fine sand	2	45	1,950	1.15	1,725	−8.3	0.80	430	3×10^{-3}	1.2	5
Coarse sand	1	40	2,000	1.20	1,800	−7.7	0.90	470	7×10^{-3}	1.8	6

- the porosity (volume percentage of water in the sediment);
- the density[11] (in kg/m^3);
- the relative compressional velocity c_r (referenced to the speed of sound in water);
- the absolute compressional velocity c for a seawater value of 1,500 m/s;
- the reflection coefficient at normal incidence in dB ($20 \log|V(0°)|$);
- the compressional wave absorption coefficient α, in dB/wavelength;
- the shear wave velocity c_s, in m/s;
- the roughness spectral strength Ω_0, in cm^4;
- the standard deviation of the roughness along a unit distance h, in cm;
- the roughness slope standard deviation δ, in degrees.

The first eight parameters in Table 3.3 (from M to c_s) describe the sediment bulk properties. They are taken from compilations (see Hamilton, 1972, 1976; Hamilton and Bachman, 1982) and provide good expectations of what the average properties should be. The three parameters linked to sediment roughness are interpolated from data compilations (APL, 1994) and do not pretend to be more than orders of magnitude. While it is commonly admitted, and experimentally verified, that there exists a strong correlation between the sediment type (grain size, porosity) and its density, velocity and absorption, things are less clear as far as roughness character-istics are concerned. It seems unreasonable to give a particular sediment type a roughness or slope variance automatically, even as an average value. There is obviously a correlation, however, between the sediment "hardness" and its roughness: fluid sediment (clays) are unlikely to present any noticeable interface roughness, whereas coarse sands may present local slopes of up to 20–30°. In this respect, the average roughness parameters of Table 3.3 increase globally with sediment impedance, without being individually associated with any precise

[11] The density is linked to porosity by $\rho = n\rho_w + (1 - n)\rho_b$, with $0 < n < 1$, ρ_w and ρ_b are, respectively, the seawater density (around 1,030 kg m^{-3}) and the bulk grain density (around 2,700 kg m^{-3}).

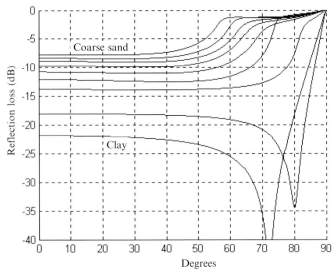

Figure 3.23. Reflection coefficients at the water–sediment interface, for the sediment types listed in Table 3.2. The reflection loss due to microscale roughness is not included (see Figure 3.19).

sediment type. Finally, the volume heterogeneity parameters are not given here any value (cf. Section 3.6.2.3). Very roughly, values of -15 to -40 dB re 1 m^3 can be encountered, depending on the sediment characteristics, the environment and the signal frequency.

   Surface sediments are saturated with sea water, and their acoustic velocities are then expected to be proportional to the sound velocity in sea water. It is therefore more relevant to characterise the sediments with their sound velocity relative to sea water. This accounts conveniently for the effects of pressure and temperature.[12]

3.6.2.3   *Reflection at the water–sediment interface*

Sound reflection at the water–sediment interface is most conveniently described by the fluid interface reflection coefficient. It may include the effect of shear waves, important for sandy sediments. To account for microscale roughness, the coherent reflection loss may be added to the reflection coefficient (Equation 3.51 and Figure 3.19). The result is then frequency-dependent.

3.6.2.4   *Seafloor backscattering*

At frequencies used by seafloor mapping sonars (tens and hundreds of kHz), the backscattering from the seabed can generally be separated into two contributions (Figure 3.24). Part of the energy is scattered by the interface relief: either by the

---

[11] Note that saturated sediment density is not proportional to seawater density. Relative density refers to the conventional value of 1,000 kg m^{-3} (freshwater density), which may be different from the value for the actual saturating fluid.

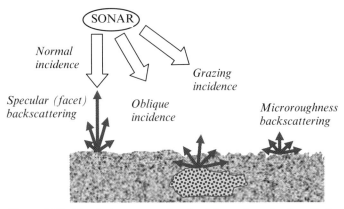

Figure 3.24. Physical processes involved in seafloor backscattering.

subhorizontal facets, at incidence close to the vertical or by microscale roughness at grazing incidence. Another part of the energy penetrates the sediment and is reflected back by volume heterogeneities. This process can become predominant at oblique incidences. Interface backscattering can be described using a statistical geometrical model of scattering at a rough interface, coupled with the reflection coefficient characteristic of the two media. Volume backscattering (see Section 3.6.2.5) depends on the characteristics of water–sediment transmission and absorption inside the bottom, together with the volume backscattering strength from the hetero-geneities inside the sediment.

3.6.2.5   *Volume backscattering strength of the sediments*

A significant part of the acoustic energy scattered at the water–sediment interface comes from the heterogeneities included in the sediment volume. These may have various origins (e.g., buried stones, shells, crustaceans), but the most visible ones are gas bubbles trapped inside the sediment. They are best described as a random distribution of scatterers, whose contributions are added incoherently.

This particular contribution to global backscattering strength is once again modelled using a geometrical description of the volume instantaneously insonified. The computation principle is thus exactly the same as in water volume backscatter-ing. It accounts for refraction at the interface (angle change and transmission loss across the boundary). The effective target considered is the isochronous volume inside the sediment, generated below a given interface area and limited in its extent by sediment absorption. For the sake of simplicity, one can approximate this isochronous surface with the plane orthogonal to the direction of propagation inside the sediment (Figure 3.25). The effective volume reads:

$$A_v = A_s \sin \theta_2 \int_0^{+\infty} \exp(-4\beta\zeta \tan \theta_2)\, d\zeta$$

$$= \frac{A_s \cos \theta_2}{4\beta} \tag{3.74}$$

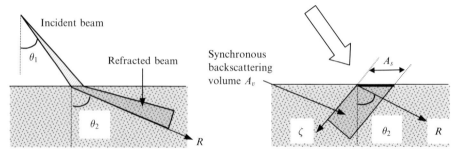

Figure 3.25. Insonified volume inside a sediment layer. *Left*: incident and refracted beam. *Right*: instantaneously insonified volume A_v, below the area A_s.

where A_s is the area of the insonified interface, θ_2 is the refraction angle of β is the pressure sediment attenuation (in Neper/m) and the factor 4 comes from the fact that we consider intensity, and not pressure, on a two-way propagation path.

The volume backscattering contribution seen from the interface (in addition to the rough interface backscattering cross-section) is then:

$$\sigma_{bs_v} = |1 - V_{12}^2|^2 \left(\frac{c_2}{c_1}\right)^2 \frac{\cos^2 \theta_1}{\cos \theta_2} \frac{\sigma_v}{4\beta} \qquad (3.75)$$

This expression features $A_s = 1\,\text{m}^2$, and the volume backscattering strength σ_v (defined for $1\,\text{m}^3$). It also accounts for the product of the intensity transmission coefficients, $|W_{12}|^2 |W_{21}|^2 = |1 - V_{12}^2|^2$, in both directions across the interface, and for the change of geometrical divergence due to the refraction: $(c_2 \cos \theta_1 / c_1 \cos \theta_2)^2$.

This expression is valid in subcritical conditions. It can be improved by accounting for the grazing incidences across the critical angle:

$$\sigma_{bs_v} = |W_{12}|^2 |W_{21}|^2 \left(\frac{c_2}{c_1}\right)^2 \frac{\cos^2 \theta_1}{4\Im(k_{z2})} \left|\frac{k_2}{k_{z2}}\right|^2 \sigma_v \qquad (3.76)$$

At this point, the main difficulty lies in estimating the elementary backscattering strength σ_v. No general approach can be proposed, because of the sheer variety of possible configurations inducing volume backscattering: gas bubbles, mineral inclusions, thin sediment layers with impedance discontinuities, bioturbation by crustaceans, shells or worms, etc. Each case justifies a specific acoustic model, and many examples are available in the literature. An arbitrary value of $-30\,\text{dB}$ re $1\,\text{m}^3$ was used for all sediment types presented in Figure 3.26.

The importance of volume backscattering at a given frequency, relative to interface roughness backscattering, depends on the type of sediment. It dominates in soft sediments at all oblique angles of incidence, whereas its influence decreases in sandy seafloor because of absorption. Frequency dependence makes things even more complex. At low frequencies, absorption decreases and the volume contribution increases, combined with a diminution of the interface roughness contribution.

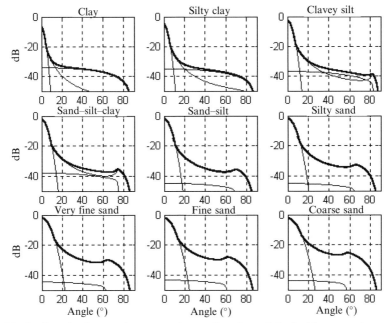

Figure 3.26. Seafloor backscattering strength as a function of the incidence angle, computed for the seafloor types listed in Table 3.3, and for a frequency of 100 kHz. This was obtained using the tangent plane model near vertical incidence, combined with a Bragg scattering model at grazing angles and volume backscattering ($10 \log \sigma_v = -30$ dB re 1 m^3). These three different components are plotted with thin lines, and the global backscattering strength is plotted with a thick line. From clay to clayey silt, volume backscattering dominates at grazing angles. From sand-silt to coarse sand, the global backscattering strength merges with the Bragg scattering curve.

This is, however, partly compensated by the frequency dependence of the volume backscattering strength (in f^4 for the bubble target strength in Rayleigh scattering).

3.6.2.6   Contrast of seafloor sonar images

Sidescan sonars and multibeam sounders record signals backscattered from the seabed at grazing angles of incidence (see Chapter 8 for more details). Mapped onto a plane, the intensities of these signals represent the variations of acoustic seabed reflectivity. These variations come from the nature of the targets, or the geometric variations inducing changes in the angles of incidence (and thus the back-scattered levels).

The quality of a sonar image is partly related to its contrast (i.e., the difference in image levels associated with a given change in the interface characteristics). To make things simpler, we shall assume that the reflectivity strength varies following Lambert's law: $BS(\theta) = BS_0 + 20 \log \cos \theta$. Figure 3.27 shows the difference in the geometry of ensonification of two facets with different slopes. One facet is horizontal

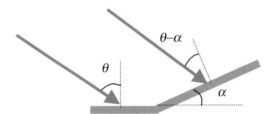

Figure 3.27. Geometry of wave incidence on two facets with different slopes.

(the angle of incidence is θ), and the neighbouring facet is inclined at an angle α (the angle of incidence is then $\theta - \alpha$).

The difference in acoustic levels backscattered from the two facets (i.e., the acoustic contrast of this portion of the sonar image) can be expressed in dB as a function of the slope α and the angle of incidence θ:

$$\Delta BS(\theta) = \left| 20 \log \frac{\cos(\theta - \alpha)}{\cos \theta} \right| \tag{3.77}$$

For $\alpha > 0$, the derivative in θ of $\Delta\sigma = \cos(\theta - \alpha)/\cos\theta$ reads:

$$\frac{d(\Delta\sigma)}{d\theta} = \frac{d}{d\theta}\left(\frac{\cos(\theta - \alpha)}{\cos \theta} \right) = \frac{\sin \alpha}{\cos^2 \theta} \tag{3.78}$$

This expression is always positive. For a given slope α, the contrast ΔBS increases therefore with θ, and is maximal at grazing incidence ($\theta \rightarrow \pi/2$); the same conclusion would be reached for $\alpha < 0$. It can be concluded that the quality of a sonar image[13] is better as the angle of incidence is closer to grazing angles (i.e., at lower altitudes above the seabed). Indeed, the sidescan sonars providing the best quality images are towed close to the bottom (see Chapter 8 for more details, and chapter 2 of Blondel and Murton, 1997).

---

[12] Although contrast should not be the only factor to consider, the resolution of the image and the noise level are important factors too.

4

Noise and signal fluctuations

Noise is an important component of underwater acoustics, covering many different processes; all of which add to the expected signals and decrease the performance of underwater acoustic systems. The causes of noise can be grouped in four categories (Figure 4.1):

- *Ambient noise.* This type of noise originates from outside the system, and stems from natural (e.g., wind, waves, rain, animals) or man-made (e.g., shipping, industrial activity) causes. It is independent of the sonar system or the conditions of its deployment.
- *Self-noise.* This is the noise suffered by an underwater acoustic system itself, when caused by the supporting platform (e.g., radiated noise, flow noise, electrical interference) or the system's own electronics (thermal noise).
- *Reverberation.* This type of noise affects only active sonar systems, as it is caused by parasite echoes (generated by the sonar's own signals). It can be so loud as to mask the detection of the expected target echoes.
- *Acoustic interference.* This type of noise is generated by other acoustic systems working in the vicinity, usually onboard the same ship or underwater platforms, sometimes from sources farther away.

From these definitions, we can see that the concept of "noise" is more related to its nature than to a specific acoustic content. Structurally, noise can correspond to very different waveforms: completely random in the case of ambient noise, diffuse echoes from the transmitted signal in the case of reverberation, or clearly recognisable signals for interference caused by other instruments.

In the large majority of cases, however, we shall consider the additive noise degrading the nominal sonar performance as a *random process*. Its description will use the classical statistical models used in signal processing theory.

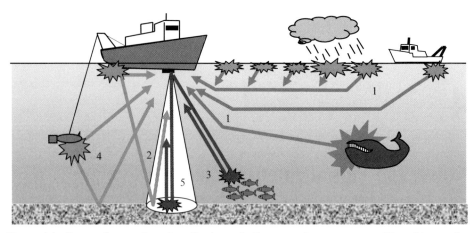

Figure 4.1. Types of acoustic noise affecting a hull-mounted sonar system: (1) ambient noise; (2) self-noise (from the ship); (3) reverberation; (4) acoustic interference (signals from other systems); (5) expected target echo.

Deformed by propagation, the useful acoustic signal can also present fluctuations, more or less random, coming from the environment. These fluctuations also degrade sonar system performance. From their result, if not from their origin, they are also classified as "noise".

In this chapter, we shall first introduce the notions of narrowband and wideband noise. These will be followed by a description of the causes of additive noise, as encountered in underwater acoustics. Specific sections will be devoted to the noise radiated by ships, and the self-noise of a sonar system. All these presentations will be accompanied by simple models quantifying these effects. We shall then present two approaches of noise modelling: the first, related to the space–time coherence of the acoustic field; the second, an analysis of the spatial distribution of the acoustic field generated by transmitters at the surface. Their physical description will include a discussion of the propagation conditions. The last part of the chapter will deal with the effects of variations in the propagation medium on the fluctuations of the signals transmitted.

4.1   NARROWBAND AND WIDEBAND NOISE

A *narrowband noise* is a monochromatic signal with amplitudes randomly varying with time (Figure 4.2, *top left*). In the frequency domain, its power is concentrated around the carrier frequency (Figure 4.2, *top right*). Conversely, a wideband noise will exhibit a frequency spectrum spread over a broader interval (Figure 4.2, *bottom*). The variation of this spectral power with frequency may take various shapes. If it is constant within a particular frequency range, the noise is called "white noise" (Figure 4.2, *bottom left*). If it varies with frequency, the noise is called "coloured

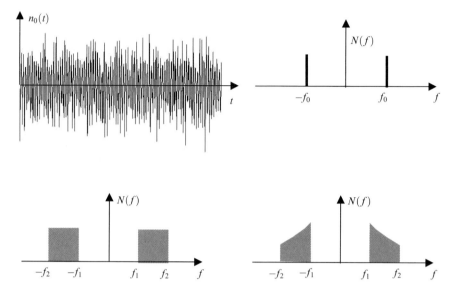

Figure 4.2. (*Top left*) Example of a narrowband time signal $n_0(t)$, as a function of time t. (*Top right*) Its frequency spectrum $S(f)$ as a function of frequency f. (*Bottom left*) Wideband spectrum for white noise. (*Bottom right*) wideband spectrum for a typical coloured noise.

noise" (Figure 4.2, *bottom right*). In this case, the spectral power density (i.e., the power in a 1-Hz band) usually decreases with frequency, most often as a negative power of the frequency (dependence as f^{-n}).

The concept of narrowband noise is convenient when modelling the additive perturbation to a sinusoidal signal, or as an elementary component of a broadband noise. A Gaussian narrowband signal n_0 is represented by the time domain expression:

$$n_0(t) = a_0(t) \exp(j(\omega_0 t + \varphi_0)) \tag{4.1}$$

where $a_0(t)$ is the amplitude of a random variable following a Rayleigh distribution (see Section 4.6.3.2 for more details), modulating a carrier sinusoidal wave with pulsation $\omega_0 = 2\pi f_0$ (as in Figure 4.2).

The monochromatic description of signals has a limited interest, as underwater acoustics makes a wide use of broadband signals. A *broadband noise* $n(t)$ can be decomposed into a sum of N elementary spectral contributions $n_i(t)$ with narrow bands:

$$n(t) = \sum_{i=1}^{N} n_i(t) = \sum_{i=1}^{N} a_i(t) \exp(j(\omega_i t + \varphi_i)) \tag{4.2}$$

The mean amplitude of the resulting signal reads:

$$\langle a \rangle = \langle |n(t)^2| \rangle^{1/2} = \left\langle \sum_{i=1}^{N} n_i(t) \sum_{k=1}^{N} n_k^*(t) \right\rangle^{1/2}$$

$$= \left\langle \sum_{i=1}^{N} a_i^2 + \sum_{i=1}^{N} \sum_{k \neq i} a_i a_k \exp(j(\omega_i t - \omega_k t + \varphi_i - \varphi_k)) \right\rangle^{1/2} \quad (4.3)$$

The mean of the alternate phase terms equals zero (because the phases of the elementary signals are assumed random and independent of each other). And as the amplitudes are also independent of each other, Equation (4.3) becomes:

$$\langle a \rangle = \left(\sum_{i=1}^{N} \langle a_i^2 \rangle \right)^{1/2} \quad (4.4)$$

The mean amplitude of the resulting signal is therefore the quadratic sum of the amplitudes of its components. The mean energy (in power or intensity) is given by the sum of the mean elementary energy contributions:

$$\langle P \rangle = \sum_{i=1}^{N} \langle P_i \rangle \quad (4.5)$$

Let us extend this discrete sum to a continuous distribution of narrowband components. The mean power of a broadband signal, over the frequency band Δf, of spectral power density $N(f)$, reads:

$$\langle P \rangle = \int_{\Delta f} N(f) \, df \quad (4.6)$$

The spectral power density is the power corresponding to the elementary spectral band of 1 Hz. It is therefore expressed in W/Hz, or, with logarithmic notation, in dB re 1 W/Hz. To remain closer to the physical measurement of acoustic pressure, one often prefers expressing it in dB re 1 μPa/$\sqrt{\text{Hz}}$. This notation is indeed consistent with the dB re 1 W/Hz, as the power is proportional to the pressure squared.[1]

   In the case of noise with a white power spectrum $N(f) = N_0$, the total power P is simply related to the power spectral density and the bandwidth Δf by the relation:

$$\langle P \rangle = \int_{\Delta f} N_0 \, df = N_0 \Delta f \quad (4.7)$$

This can be expressed in dB:

$$10 \log(P/1 \, \text{W}) = 10 \log(N_0 \Delta f / 1 \, \text{W})$$

$$= 10 \log(N_0 / (1 \, \text{W/Hz})) + 10 \log(\Delta f / 1 \, \text{Hz}) \quad (4.8)$$

For a coloured noise, with a power spectral density varying with frequency, the power in a bandwidth Δf will require the calculation of the integral in Equation

---

[1] For typographic convenience spectral density levels are often found in dB re 1 μPa/Hz (instead of $\sqrt{\text{Hz}}$),

(4.6). For example, if the frequency variation is $N(f) = N_0(f_0/f)^n$ (i.e., in dB, a spectrum decreasing in $-10n \log f$ from a level N_0 at frequency f_0), the power in a frequency band $[f_1, f_2]$ becomes:

$$\langle P \rangle = N_0 \int_{f_1}^{f_2} \left(\frac{f_0}{f} \right)^n df = N_0(n-1)f_0^n \left(\frac{1}{f_1^{n-1}} - \frac{1}{f_2^{n-1}} \right) \tag{4.9}$$

A case commonly encountered is $n = 2$ (spectrum decreasing in $-20 \log f$):

$$\langle P \rangle = N_0 \int_1^{f_2} \left(\frac{f_0}{f} \right)^2 df = N_0 f_0^2 \left(\frac{1}{f_1} - \frac{1}{f_2} \right)$$

$$= N_0 \left(\frac{f_0}{\sqrt{f_1 f_2}} \right)^2 (f_2 - f_1) = N_0 \left(\frac{f_0}{f_C} \right)^2 \Delta f \tag{4.10}$$

In this equation $\Delta f = f_1 - f_2$ and $f_C = \sqrt{f_2 f_1}$. It may be remarked that the power thus obtained would be the same for a white noise with a level at the "average" frequency f_C, different from the central frequency of band Δf. Similar expressions may be derived for other spectrum slopes.

4.2 CAUSES OF UNDERWATER ACOUSTIC NOISE

4.2.1 Ambient noise

4.2.1.1 Permanent noise

By definition, ambient noise is the noise received by the sonar in the absence of any signal and self-noise from the system. Ambient noise has several very distinct physical origins, each corresponding to particular frequency ranges:

- Very low-frequency noise generated by remote *seismic and volcanic activity* is limited to a frequency range of a few tens of hertz. Therefore they are only of marginal concern for underwater acoustic systems (except, possibly, some passive sonars).
- Between 10 Hz and 1 kHz, *shipping* is often the main cause of underwater acoustic noise, along with noise from industrial activity alongshore. The noise levels obviously depend on the regions studied, and current models using a "shipping density" parameter are only indications of the amount of noise one can expect. The proximity of harbours and "shipping lanes" concentrating the traffic are, of course, predominant factors in local variations of the noise level. They also induce spatial anisotropy (dependence of noise on azimuthal angle). The noise radiated by individual ships inside the working area of a sonar system cannot be modelled as ambient shipping noise; the ships instead act as local jammers of the signal. Noise radiation by ships is presented in more detail in Section 4.2.2.
- *Surface agitation* depends on sea state and wind speed (Table 4.1). It generates noise over a frequency range of a few hertz to a few tens of kHz. It is dominant

in two frequency ranges, associated with very distinct physical processes. At very low frequencies (below 10 Hz), wind-generated turbulence induces pressure variations akin to acoustic pressure variations. Wind effect is also dominant above a few hundred hertz: the microbubbles of air dissolved in the shallow water layer are undergoing dilation and bursting, induced by pressure changes associated with the movements of the surface under the influence of the wind (as can be appreciated when looking at the numerical values in Table 4.1). When the sea state worsens, there is additional noise from the clouds of bubbles bursting in the foam crests and from the impacts of breaking waves.

Table 4.1. Equivalent Beaufort[2] scale, sea state, descriptive terminology and wind speed values (from the *Handbook of Oceanographic Tables*, US Naval Oceanographic Office, 1966).

Beaufort scale	Sea state	Description	Wind speed (knots)	Wind speed (m/s)
0	0	Calm	<1	0–0.2
1	$\frac{1}{2}$	Light air	1–3	0.3–1.5
2	1	Light breeze	4–6	1.6–3.3
3	2	Gentle breeze	7–10	3.4–5.4
4	3	Moderate breeze	11–16	5.5–7.9
5	4	Fresh breeze	17–21	8.0–10.7
6	5–6	Strong breeze	22–27	10.8–13.8
7	7	Near gale	28–33	13.9–17.1
8	8	Gale	34–40	17.2–20.7
9	9	Strong gale	41–47	20.8–24.4
10	9	Storm	48–55	24.5–28.4
11	9	Violent storm	56–63	28.5–32.6
12	9	Hurricane	>64	>32.7

Thermal noise is created by molecular agitation, and can only be heard beyond 100 kHz. It is the dominant source of noise in this high-frequency range.

4.2.1.2 Modelling permanent ambient noise levels

Modelling the levels of ambient noise with enough confidence and accuracy is obviously a rather difficult and daunting task. Several noise models have been proposed since the 1940s, mainly for use in naval sonar performance prediction. They are heuristically defined from compilations of field observations; the most famous and commonly met is the so-called Wenz model (Wenz, 1962). The following will be limited to a few ambient noise models, which are relevant to the purpose of system performance quantification. These models are presented as power spectral densities, in 1-Hz bands, as a function of frequency (in Hz).

---

[2] The wind speed v (in knots) is related to the Beaufort scale by $v \approx 3/1.852$ Beaufort$^{3/2}$. Expressing v in m/s, $v \approx 1/1.2$ Beaufort$^{3/2}$.

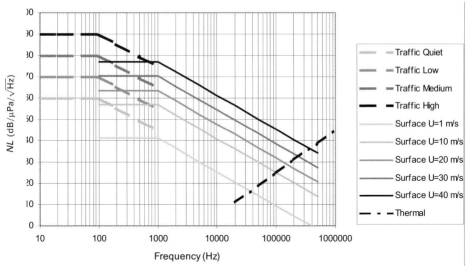

Figure 4.3. Synthesis of ambient noise models, with levels represented as a function of frequency.

- *Frequency band 10–100 Hz.* The noise in this band depends heavily on shipping density and industrial activities. The levels reported in the literature are typically comprised between 60 and 90 dB re 1 μPa/$\sqrt{\text{Hz}}$. The frequency dependence is hardly noticeable.
- *Frequency band 100–1,000 Hz.* The noise in this band is dominated by shipping contributions (decreasing with frequency, with the typical noise spectrum radiated by ships). But a significant part may come from the agitation of the sea surface. Indicative orders of magnitude can be obtained with the simple empirical model, derived from Urick (1986):

$$NL_{\text{shipping}} = NL_{100} - 20\log\left(\frac{f}{100}\right) \qquad (4.11)$$

where NL_{100} varies from 60 to 90 dB re 1 μPa/$\sqrt{\text{Hz}}$, depending on the average shipping density (see Figure 4.3).

- *Frequency band 1–100 kHz.* The contribution from the sea surface is now predominant, unless it is superseded by intermittent sources such as marine animals or rain. The best documented noise models have been presented for this particular frequency band, as it is the working range of most sonars. A simple model was developed by Knudsen *et al.* (1948), and has been widely used since:[3]

$$NL_{\text{surf}} = \begin{cases} NL_{1K} & \text{if } f < 1{,}000 \text{ Hz} \\ NL_{1K} - 17\log\left(\dfrac{f}{1{,}000}\right) & \text{if } f > 1{,}000 \text{ Hz} \end{cases} \qquad (4.12)$$

[3] The Knudsen model is traditionally used as a standardised reference level for sonar dimensioning and performance evaluation, regardless of its actual physical accuracy.

Table 4.2. Values of the coefficient NL_{1K} of the Knudsen model (in dB re $1\,\mu\text{Pa}/\sqrt{\text{Hz}}$), as a function of sea state (spectral noise level referenced at $1\,\text{kHz}$).

Sea state	0	0.5	1	2	3	4	5	6
NL_{1K}	44.5	50	55	61.5	64.5	66.5	68.5	70

where NL_{surf} is the noise level at the sea surface, in dB re $1\,\mu\text{Pa}/\sqrt{\text{Hz}}$, and NL_{1K} is a parameter depending on the sea state as given in Table 4.2. A more recent model of sea surface noise was proposed in a report from the Applied Physics Laboratory, University of Washington (APL, 1994). It is quite similar, with a slightly different frequency dependence and a more precise parameterisation of NL_{1K}, to:

$$NL_{\text{surf}} = NL_{1K} - 15.9\log\left(\frac{f}{1{,}000}\right) \tag{4.13}$$

$$\begin{cases} NL_{1K} = 41.2 + 22.4\log v & \text{for } \delta T < 1°\text{C and } v \geq 1\,\text{m/s} \\ NL_{1K} = 41.2 + 22.4\log v - 0.26(\delta T - 1)^2 & \text{for } \delta T \geq 1°\text{C and } v \geq 1\,\text{m/s} \end{cases}$$

v is the wind speed (in m/s), and δT is the temperature difference between sea water and the atmosphere at the sea surface.

- *Frequencies above 100 kHz.* Noise is then dominated by *electronic thermal noise*, following the equation:

$$NL_{\text{th}} = -75 + 20\log f \tag{4.14}$$

This is not rigorously ambient noise, as it depends on the sonar receiver itself, and is discussed in more detail in Section 4.2.3.1.

The relative contributions of the different sources of ambient noise are presented in Figure 4.3. It should be noted that:

- the noise level (power spectral density, in 1-Hz elementary bands) generally decreases when the frequency increases (except for thermal noise);
- the noise level decreases at great depths, as most noise sources are at the surface (see Section 4.3.2);
- ambient noise is higher in shallower water for identical environmental conditions (confinement effect, where the noise field is trapped between the seabed and sea surface) (see Section 4.3.2).

4.2.1.3   *Intermittent ambient noise levels*

An important source of intermittent ambient noise is biological in origin. Each species of marine mammal transmits a rich variety of acoustic signals, for communication (between individuals) and echolocation (of prey and obstacles). The first type of signal is common to all Cetaceans. The signal characteristics depend, of course, upon the species. Whales use their larynx to vocalise low-frequency signals that are capable of propagating for hundreds of kilometres. Their typical frequency range is between 12 Hz and a few thousands of hertz. These "whale songs" may last for hours, and they possess a quasi-musical quality, obviously due to their similarity to human voices. Conversely, smaller marine mammals transmit modulated sounds (cries and whistles) at higher frequencies (often in the upper part of the human auditory range, above 1 kHz). Only the Odontocetes (Cetaceans with teeth: e.g., dolphins, porpoises or killer whales) use echolocation. These mammals transmit a series of high-frequency "clicks", typically in the 50–200-kHz band, in fast bursts (up to 10 times a second). The purposes and signals are very similar to active sonars: detection of prey and obstacles, localisation in range and direction and, maybe the most impressive, target identification and characterisation.

The acoustic intensity levels transmitted by marine mammals are not negligible when compared with man-made sonars. Sounds vocalised by whales are reported to reach 188 dB re 1 µPa at 1 m in the 100–200-Hz band. Echolocation clicks transmitted by Odontocetes can reach levels as high as 180–200 dB re 1 µPa at 1 m.

As an aside, one should note that this acoustic interference between marine mammals and sonars is felt both ways. Cetaceans are known to be very sensitive to man-made acoustic signals. Whales and dolphins are routinely reported to gather around transmitting sonar systems, and even to "talk back" to the sonars. It is reasonable to assume that sonar signals can also have a repellent – and even dangerous – effect on marine mammals. Their hearing and orientation faculties may be affected by high-intensity sonar signals; in particular, naval manoeuvres involving an intensive use of active sonars are increasingly suspected to be the indirect cause of the beaching of whales. Environmental activists have seized on the issue, and acoustical oceanography experiments were accused in the 1990s of damaging marine life.[4] Sea trials are now designed in such a way as to minimise the potential risk to marine life, if any. Although fish have a less developed sense of hearing than mammals (their frequency range is limited to low frequencies, below a few kHz), they appear to be sensitive to noise radiated from fishing vessels, and avoidance reactions are frequently observed.

Other marine organisms are also generating substantial acoustic noise. The best known are the "snapping shrimps". Their presence in large colonies in warm, shallow water is a major factor of acoustic perturbation for sonars. These invertebrates have huge asymmetrical pincers, the snapping of which creates cavitation bubbles. The implosion of the bubbles generates an intense broadband noise, maximum in the 1–10 kHz frequency range. The average spectral level reached by a "snapping shrimp" colony can reach 60–90 dB re 1 µPa/$\sqrt{\text{Hz}}$. The upper limit of 90 dB re 1 µPa/$\sqrt{\text{Hz}}$ is significantly higher than commonly encountered sea surface noise levels (cf. Figure 4.3).

Another significant source of intermittent ambient noise is rain. It causes very high noise, due to the impact of raindrops on the sea surface, and the implosion of air bubbles dragged into water by the raindrops. The frequency range (1–100 kHz) is

---

[4] One particular experiment was even cancelled because of the supposed risk to marine mammals (although it seems there was some misunderstanding about the sound levels actually transmitted).

basically the same as sea surface noise, because the underlying processes are quite similar. But the spectrum of rain-induced noise does not regularly decrease with frequency. Its power spectral density shows a maximum at 20 kHz, quickly decreasing beyond that. It depends both on the amount of rain and the sea state. The increase above the sea surface noise may reach up to 30 dB for heavy rain over a calm sea surface. A detailed model of rain noise is available in the APL report mentioned previously (APL, 1994).

4.2.2 Ship-radiated noise

4.2.2.1 *Causes of ship-radiated noise*

This is another important factor, as the noise radiated by a ship is received by all sonar systems on board (self-noise) and imposes limits to the performance achievable when it exceeds ambient noise levels. In anti-submarine warfare (ASW), passive sonar applications use the noise radiated by a potential target (surface vessel or submersible) to detect it in the background of ambient noise. And, finally, ship-radiated noise (mainly from remote shipping) is the main cause of ambient noise in the 10–1,000-Hz frequency band. The self-noise level from the sonar-carrying platform can be considerably larger than the ambient noise, especially for civilian ships (old oceanographic vessels, fishing vessels, offshore drilling ships), whose noise specifications are less severe than military vessels (which must be as quiet as possible).

Several sources intervene in the overall noise radiated from a ship:

- *Propeller noise.* Rotating propellers generate spectral lines at very low frequencies, in the 0.1–10-Hz range. The frequencies of these lines depend on the speed of rotation of the propeller and its geometry. The depressions induced by the movements of the blades create some cavitation, causing a characteristic wide-band noise at higher frequencies.[5] The cavitation noise level depends on the speed of rotation of the propeller, the mechanical character of the propulsion system (fixed- or variable-step propeller) and the depth (cavitation bubbles can form only below a hydrostatic pressure threshold: submerged submarines do not have this problem). The average frequency decrease is in $-20 \log f$.

- *Flow noise.* Turbulence is generated by the flow of water on the ship's hull or on the active face of the acoustic transducer or its protection. This type of noise depends, of course, on the speed of the ship, the frequency, and the shape and emplacement of the transducer's protection.

- *Machine noise.* Many noisy machines are installed on ships: engines, reduction gear, generators and alternators, hydraulic machinery, winches, etc. They induce vibrations in the hull by solid transmission in the inner structures or through the air; the hull's vibrations are then passed on to the water. Machine noise is generally independent of the ship's speed; it is therefore felt more keenly when

---

[5] Examples of time-frequency analysis of ship propulsion noise are presented in Chapter 6, Figure 6.10.

the ship is at low speed and it is masked at high speed by flow noise and cavitation noise. The main frequencies of machine noise are usually accompanied by their harmonics.

- *Submarine transient noise*. The acoustic stealth of a submarine can be degraded by the exceptional transmission of short, transient noise, lasting from a few milliseconds to a few seconds. The causes are often impossible to avoid: opening of torpedo launch tubes, steering manoeuvres, starting mechanical or hydraulic machinery, etc. These noises are very characteristic and are used by passive sonar operators to detect and identify submarines.
- *Activity noise*. Some activities of civilian ships are very noisy (e.g., seismic surveys, drilling, trawling, towing, deployment of submersibles). They should not, however, mask the necessary acoustic operations (e.g., ship or platform positioning, data transmission, fish sounding).

4.2.2.2   Ship-radiated noise model

The spectrum of the noise radiated by a ship presents a continuous broadband background, the level of which increases with ship speed (Figure 4.4). Its maximum is around 100 Hz, and it typically decreases by 6 dB per octave (frequency dependence in $-20 \log f$) above a few hundreds of Hertz. At low frequencies, this wideband noise is superseded by narrow-band components (spectral lines), usually associated with rotating devices such as engines or gears. These lines form the ship's *acoustic signature*. They are widely used for passive sonar classification. The radiated noise level can be roughly summarised using two parameters:

- the level RNL_{1K} at 1 kHz, from which the level at other frequencies is derived with the equation $RNL(f) = RNL_{1K} - 20 \log(f/1,000)$;
- the radiated noise of spectral lines RNL_{SL} – depending on the application, it is best described by either its average or its maximum level (the latter being used for the acoustic detectability of submarines).

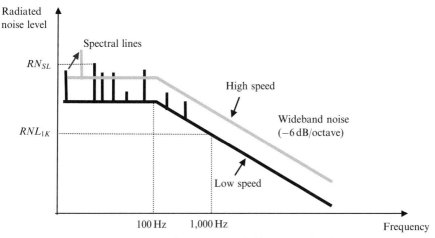

Figure 4.4. Generic spectrum of ship-radiated noise.

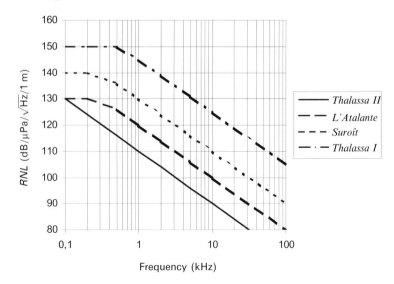

Figure 4.5. Typical ship-radiated noise spectra for four vessels from the French oceanographic fleet. These are typical orders of magnitude, corresponding to low-to-medium speeds (below 8 knots). The actual noise levels might increase at higher speeds or during specific operations (e.g., when trawling or on station).

Figure 4.5 shows typical values, obtained experimentally, of radiated noise levels for four vessels[6] of the French oceanographic fleet. These results clearly show the progress accomplished in noise reduction over the last 30 years.

The noise radiated by military vessels is paramount to passive sonar detection applications. The noise level associated with a particular vessel will define how to detect and identify it, which explains why most of these data are classified (especially for submarines). Urick (1983) presents some detailed measurements, but they are limited to World War II vessels, whose acoustic performance is, of course, not representative of today's navies. More recent elements can be found in Miasnikov (1998). Table 4.3 shows typical values for different generations and types of submarine. Except for the classical/diesel configuration (snorkelling at the surface), all these figures correspond to low-speed configurations (typically 4 knots), aimed at keeping the vessel particularly quiet. In this configuration, only

---

[6] R/V *Thalassa I* was an oceanographic vessel from the early 1960s, rigged for trawling and universally recognised as very noisy (at least for a research vessel); her radiated noise level is, however, still representative of many current merchant vessels. R/V *Suroît* was built in 1975 for general oceanographic research. R/V *L'Atalante* is a recent (1990) research vessel, equipped for seafloor mapping (multibeam sonar and seismics) and for deploying submersibles and towed systems. Finally, R/V *Thalassa II* was built in 1995 for halieutic (fisheries) research and specially designed for high acoustic performance.

Table 4.3. Orders of magnitude of the noise radiated by different types of submarine at low speed (4 knots). These values have to be increased by 1.5–2 dB per additional knot.

	RNL_{SL}	RNL_{1K}
World War II submarine (electric)	140	120
Modern submarine (electric)	100	80
Modern submarine (diesel)	140	120
Recent SSN	110	90
Recent SSBN	120	100

the internal machinery noise can be detected (so nuclear-powered submarines are noisier than traditional submarines, because of the nuclear power plant added to the electrical engine). At patrolling and transit speeds (10–20 knots), the contribution from flow noise becomes dominant and significantly increases with speed (roughly 1.5 dB to 2 dB per knot). Beyond the speed that causes the propeller to cavitate (20–30 knots), the submarine completely loses its stealth.

4.2.3   Self-noise

4.2.3.1   *Thermal noise*

In electronic circuits, any resistor creates some electric noise due to electronic agitation. The voltage u (in volts) thus generated in a resistor R is given by the Nyquist formula:

$$u = \sqrt{4KRT\Delta f} \qquad (4.15)$$

where K is the Boltzmann constant ($K = 1.38 \times 10^{-23}$ J/K), R is expressed in ohms, T is the absolute temperature (in kelvins) and Δf is the frequency band considered.

For underwater acoustic transducers, the resistivity R is not limited to the resistivity *sensu stricto* of the electronic circuits. It is in fact the resistive part of the equivalent impedance, including the electrical impedance and the motion impedance (which translates to the coupling of the mechanical part of the transducer with its surroundings).

For an ideal hydrophone, the only resistivity to consider will be associated to radiation:

$$R_r = \frac{\pi \rho c}{\lambda^2} S_H^2 \qquad (4.16)$$

where S_H is the hydrophone sensitivity (see Section 5.2.2). The electronic noise voltage for a frequency bandwidth of 1 Hz will then be:

$$u = \sqrt{4K\pi \rho c T}\, \frac{S_H}{\lambda} \qquad (4.17)$$

For the equivalent acoustic pressure:

$$p = \frac{\sqrt{4K\pi \rho c T}}{\lambda} \qquad (4.18)$$

And finally, replacing the quantities in (4.18) by their numerical values, for the power spectral density in dB re 1 μPa/$\sqrt{\text{Hz}}$:

$$NIS_{\text{therm}} \approx -75 + 20 \log f \qquad (4.19)$$

The value of the level thus obtained does not directly depend on the hydrophone's characteristics; it can be assimilated to some ambient noise. It is often considered as such, although this is not really rigorous, and correct accounting of thermal noise should use the exact characteristics of the receiving hydrophone.

For a real hydrophone with an efficiency $\eta < 1$, the noise level to consider becomes:

$$NIS_{\text{therm}} \approx -75 + 20 \log f - 10 \log \eta \qquad (4.20)$$

Finally, the thermal noise voltage generated in the transducer is increased by the noise factor of the receiver (typically 3 dB).

4.2.3.2   Platform self-noise

The self-noise level of a transducer installed on a supporting platform (surface vessel or submersible) is very difficult to assess. It results from the combination of several components:

- the acoustic noise radiated in water by the platform, and received by the transducer after transmission through water, either directly or reflected from the seabed;
- the mechanical vibrations passed on to the transducer by the platform's structure;
- the hydrodynamic noise generated around the transducer itself or its surroundings on the platform (e.g., protecting dome);
- the electronic noise radiated by other high-power electrical devices towards the sonar's circuits or its cables, if inadequately shielded.

So, self-noise depends on many different parameters, interacting in complex and unpredictable ways. In consequence, to estimate the performance of a given system, one should use systematic measurements made in normal use conditions. Without these measurements, one can consider the noise radiating from the ship as due to the propulsion system, and correct for the corresponding propagation loss and the directivity pattern of the transducer affected by noise.

Increasingly often, sonar manufacturers include in their systems some functionalities for the self-assessment of self-noise. This enables a sonar user to judge the acoustic environmental conditions of its system immediately.

One may consider that the self-noise levels observed can be described by the equation:

$$SNL = SNL_{1K} - 20 \log \left(\frac{f}{1,000} \right) \qquad (4.21)$$

SNL_{1K} is the self-noise spectral level received by the transducer at 1 kHz. Following various self-noise measurements made on the transducers of echo sounders aboard several oceanographic vessels, it was found to lie in the range 60–80 dB re 1 μPa/$\sqrt{\text{Hz}}$.

In shallow water a noticeable (if not prevalent) part of the self-noise due to the ship's radiated noise may be received after a reflection or scattering from the seafloor. This is especially true for echo-sounding sonars, whose directivity pattern is steered downwards. A model for this currently observed phenomenon is proposed in Appendix A.4.1.

4.2.4 Interference and acoustic compatibility

Acoustic systems rarely function on their own on a ship, and jamming between distinct systems is likely to occur. Unfortunately, the vessels with the most acoustic systems – and therefore the most prone to acoustic interference – are also the ones whose users are most demanding.

The main problem is that several transmitting and/or receiving acoustic transducers must coexist over a rather restricted space (a few tens of square metres at most). The level transmitted by each is enough to affect the others completely, if they are in receiving mode at this time (which is most probable). The direct signal received a few metres away reaches a much stronger level than the echo from any remote target.

Let us, for example, consider two sonar transducers mounted on a ship's hull (Figure 4.6) and separated by a distance d. The jamming level JL_i received by one sonar after the other has transmitted at level SL_j is given by the equation:

$$JL_i = SL_j - 20 \log d + D_1\left(\frac{\pi}{2}\right) + D_2\left(\frac{\pi}{2}\right) \qquad (4.22)$$

where $D_i(\pi/2)$ are the directivity pattern levels in the horizontal direction, along the horizontal line joining the two transducers. If the lowest sidelobe level is taken to be -30 dB (a typical value), and if the transducer spacing is $d = 10$ m (another typical value), the signal level received is $JL = SL - 80$. The acoustic interference level is much higher than the expected echo level of any target. The latter may be as low as $EL = SL - 180$; this is a difference of 100 dB! The masking of echoes by jamming

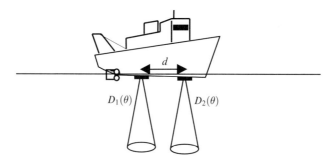

Figure 4.6. Geometry of the acoustic interference between two sonar systems.

signals will be reduced by improving the rejection of each transducer's acoustical radiation and bandpass filtering of the signals received, if the two sonars work at different frequency bands. Fortunately, this is often the case, although filters with sufficiently steep slopes might be difficult to achieve. If the combination of these two solutions does not work efficiently enough, the two sonars must be synchronised in time in order to minimise interference, which decreases their respective efficiency.

4.3   TWO APPROACHES TO NOISE MODELLING

4.3.1   Noise coherence

Although it is mostly considered random, noise is related to physical processes that are individually deterministic at certain scales of observation. Some "self-similarity" of the noise measured at close points or instants should therefore be expected. It can be quantified statistically between two points or two instants. This estimate is called *noise coherence*. It may be considered:

- between two distinct receivers at a given measurement time (spatial coherence);
- between two measurement times at a given receiver (time coherence);
- between two receivers and at two different times (spatio-temporal coherence, the general case).

The classical model given below is explained in more detail in Burdic (1984a, b). It uses two noise signals $n_1(t)$ and $n_2(t)$, measured at two points 1 and 2 distant from d, with a time delay τ. The coherence is expressed with the intercorrelation function:

$$c_{12}(d, \tau) = \langle n_1(t) n_2^*(t - \tau) \rangle \qquad (4.23)$$

It can also be normalised relative to the mean power of each of the two components:

$$\rho_{12}(d, \tau) = \frac{\langle n_1(t) n_2^*(t - \tau) \rangle}{\langle |n_1(t)|^2 \rangle^{1/2} \langle |n_2(t)|^2 \rangle^{1/2}} \qquad (4.24)$$

where $\rho_{12}(d, \tau)$ is the intercorrelation coefficient in the interval $[-1, +1]$.

Let us now consider the case of a monochromatic plane wave, whose direction of arrival is at an angle ψ to the plane of the receivers located along axis x (Figure 4.7). The two signals to consider are given by:

$$\begin{cases} n_1(t) = a_0 \exp(j(\omega t + \varphi)) \\ n_2(t - \tau) = a_0 \exp(j(\omega(t - \tau) + \varphi - kd \sin \psi)) \end{cases} \qquad (4.25)$$

The intercorrelation function now equals:

$$c_{12}(d, \tau, \psi, \theta) = |a_0(\psi, \theta)|^2 \exp(j(\omega \tau + kd \sin \psi)) \qquad (4.26)$$

Considering now that the intensity of the plane wave represents the energy contribution of the elementary solid angle $d\Omega$ around the direction (ψ, θ), the intercorrelation function for a diffuse noise composed of contributions coming from all directions

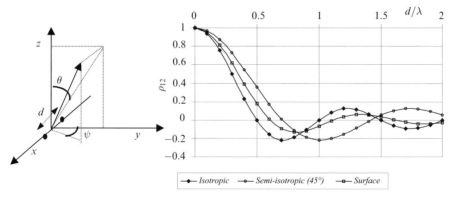

Figure 4.7. Geometry of the noise coherence model (*left*); and spatial correlation function for different types of noise (*right*).

becomes, by integration throughout the entire space:

$$C_{12}(d,\tau) = \int_{4\pi} |a_0(\psi,\theta)|^2 \exp(j(\omega\tau + kd\sin\psi))\, d\Omega \qquad (4.27)$$

This can be expressed in spherical coordinates:

$$C_{12}(d,\tau) = \exp(j\omega\tau) \int_{-\pi}^{+\pi}\int_{-\pi/2}^{+\pi/2} |a_0(\psi,\theta)|^2 \exp(jkd\sin\psi)\cos\psi\, d\psi\, d\theta$$

$$= \exp(j\omega\tau)\eta_{12}(d) \qquad (4.28)$$

The term $\eta_{12}(d)$ represents the cross-spectral density of noise at Points 1 and 2. Let us first suppose that the noise is isotropic in the entire space:

$$|a_0(\psi,\theta)|^2 = n_0(\psi,\theta) = |a_0|^2 \qquad (4.29)$$

This constant can be taken out of the integrand of Equation (4.28):

$$\eta_{12}(d) = |a_0|^2 \int_{-\pi}^{+\pi}\int_{-\pi/2}^{+\pi/2} \exp(jkd\sin\psi)\cos\psi\, d\psi\, d\theta = 4\pi n_0 \frac{\sin(kd)}{kd} \qquad (4.30)$$

The (normalised) intercorrelation coefficient can now be written with the time delay τ, and we can only consider its real part:

$$\rho_{12}(d,\tau) = \frac{\sin(kd)}{kd}\cos(\omega\tau) \qquad (4.31)$$

This coefficient equals zero (and the isotropic noise is totally incoherent) when $\sin(kd) = 0$ (i.e., for receivers spaced at multiples of the half-wavelength). This is (in addition to ensuring correct spatial sampling) one reason for the $\lambda/2$ spacing applied in multireceiver antennae.

This model of isotropic noise (energy density constant in all directions) is very simple, but it is still realistic for some practical configurations (e.g., at high frequencies, typically above 50 kHz). But other models of the spatial structure of noise are possible, and can yield more realistic expressions of the interspectral density (Burdic, 1984a, b). For example, for an isotropic noise in a conical space limited to $\theta \leq \theta_0$, the horizontal intercorrelation coefficient becomes:

$$\rho_{12}(d,\tau) = \frac{\sin(kd \sin \theta_0)}{kd \sin \theta_0} \cos(\omega\tau) \tag{4.32}$$

When the noise is generated at the surface (see Section 4.3.2), it is characterised by:

$$|a_0(\psi,\theta)^2| = n_0(\psi,\theta) = |a_0^2| \cos \psi \cos \theta \tag{4.33}$$

The horizontal intercorrelation coefficient becomes:

$$\rho_{12}(d,\tau) = \frac{2J_1(kd)}{kd} \cos(\omega\tau) \tag{4.34}$$

These results can be generalised to other spatial structures of noise (Burdic, 1984a, b).

4.3.2   Spatial structure of noise intensity

4.3.2.1   Surface-generated noise field

Underwater ambient noise can be considered as caused by a random distribution, over the entire surface of the sea, of similar noise sources. This assumption is reasonable if we consider, for example, the noise from surface agitation or from remote shipping. We will therefore use it as a starting point for a general model of ambient noise.

Each elementary noise source is located close to the sea surface, which acts as a reflecting plane and introduces an image source with a phase shifted by π. The radiation from each elementary source can therefore be assimilated to the radiation from a *dipole* whose intensity is a function of the angle of observation (Figure 4.8):

$$I_d(\theta) = I_{d0} \cos^2 \theta \tag{4.35}$$

Let us consider the average intensity created by a distribution of N identical dipoles per unit area (with $I_0 = N_{d0}$):

$$I(\theta) = I_0 \cos^2 \theta \tag{4.36}$$

The intensity component coming from angle θ is given by:

$$dI = 2\pi r\, dr \frac{I(\theta)}{R^2} \exp(-\beta r)$$

$$= 2\pi I_0 \sin \theta \cos \theta \exp\left(-\beta \frac{h}{\cos \theta}\right) d\theta \tag{4.37}$$

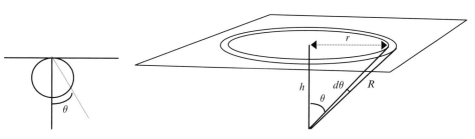

Figure 4.8. (*Left*) Directivity pattern of a radiating dipole at the sea surface. (*Right*) Model geometry for the ambient noise generated at the surface.

where β is the absorption coefficient for intensity, given by $\beta = \alpha/4.343$ (α is in dB/ m). The global intensity level therefore becomes:

$$I = \int_0^{+\pi/2} dI(\theta) = 2\pi I_0 \int_0^{+\pi/2} \sin\theta \cos\theta \exp\left(-\beta\frac{h}{\cos\theta}\right) d\theta \qquad (4.38)$$

If the water absorption coefficient β is negligible, the noise level can be simplified into the following expression, independent of the frequency or the water depth:

$$I = \int_0^{+\pi/2} dI(\theta) = 2\pi I_0 \int_0^{+\pi/2} I_0 \sin\theta \cos\theta \, d\theta = \pi I_0 \qquad (4.39)$$

But if the water absorption is not negligible, the integral in Equation (4.38) must be computed numerically (which is relatively easy), or expressed analytically using the approximation:

$$\frac{I(h)}{I(h=0)} = \frac{2\exp(-\beta h)}{2+\beta h} \qquad (4.40)$$

Expressed in dB, this approximation becomes:

$$10\log\left(\frac{I(h)}{I(h=0)}\right) = -\alpha h - 10\log\left(1+\frac{\beta h}{2}\right) \qquad (4.41)$$

This simple equation shows a frequency dependence (via β) and a dependence on the receiver's depth (via h). It may be used to correct the surface-generated noise level for depth effects.

4.3.2.2   Bottom-reflected noise field

The model presented above supposes that noise only comes from the surface, hence arrives at the receiver upwards. Practically, considering a confined field, for instance in a shallow water configuration, account should be taken of the intensity contribution from noise reflected from the seafloor and surface. For example, limiting ourselves to only one seafloor reflection at a given angle θ, one should add the contribution from the image sources obtained by symmetry relative to the seafloor. It is easily shown that the intensity contributions from reflected noise sources are expressed as:

$$dI_R = 2\pi I_0 \sin\theta\cos\theta V^2(\theta)\,d\theta \qquad (4.42)$$

where $V^2(\theta)$ is the intensity plane wave reflection coefficient at the water sediment interface. The total noise level is then given by:

$$I = \int_0^{\pi/2} (dI + dI_R) = 2\pi \int_0^{\pi/2} I_0 \sin\theta\cos\theta(1 + V^2(\theta))\,d\theta \qquad (4.43)$$

Admitting that the reflection coefficient varies little over the effective angular sector, and that it stays close to its value V_0 at normal incidence (still without attenuation), we have:

$$I = \pi I_0(1 + V_0^2) \qquad (4.44)$$

The model may be complicated further by accounting for multiple reflections, refraction effects, etc.

4.3.2.3   Noise field in guided propagation configurations

The high-frequency ambient noise generated by sea surface agitation can be considered as a local process, which can only be described by angular contributions close to the vertical. But this assumption does not hold for low-frequency noise coming from distant sources. In this case the noise field is dominated by contributions close to the horizontal, resulting from long-range propagation guided in the SOFAR channel, for example, or in a surface duct, or in a shallow water area, or any combination thereof. In such configurations, the noise structure can be described using the same models as for intensity propagation. We shall not detail these models here, but the following points need to be stressed. When defining array processing, the noise field and the spatial structure of the signal are assumed uncorrelated. Therefore, a given receiving array will lower the noise contribution (through its classical array directivity index in the case of an isotropic noise field). But things will be different in the case of guided propagation: the noise and the signal will have undergone similar propagation phenomena, and their spatial coherence functions will be similar. Therefore, a particular array processing will give more or less the same results for both signal and ambient noise, meaning that the nominal signal-to-noise ratio cannot be improved as expected.

4.4   REVERBERATION

4.4.1   The concept of reverberation

Reverberation is one of the most important processes affecting underwater acoustic signals, analogous to radar clutter in airborne or spatial imaging. It is mainly a limitation for sonar systems that are trying to detect targets, but it also affects communication systems. Reverberation consists in the superposition on the useful

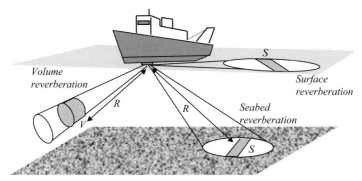

Figure 4.9. Reverberation caused by backscattering from the water column, the seabed or the sea surface. At time t, the reverberated echo can come from any one of the three areas represented, since the signal confusion volume V, and the signal footprints S on the surface or the seabed are all located at distance $R = ct/2$.

signal of contributions from parasite echoes, induced by the propagation medium, its boundaries (surface and bottom) and its inhomogeneities (bubbles, living organisms, intrinsic fluctuations). These contributions can be continuous (e.g., a signal backscattered by the interfaces and received by a sonar) or discrete (e.g., multiple paths affecting the reception by a transmission system). The physical processes creating reverberation are precisely those that create useful echoes (reflection and scattering), and the characteristics of the resulting signal are often very close to those of the useful signal. This creates many difficulties when trying to get rid of reverberation.

4.4.2 Reverberation modelling

The reverberation level at a time t is the echo intensity caused by scatterers present inside the volume or surface element delimited by the sonar (Figure 4.9), at a range R corresponding to t:

$$RL(t) = SL - 2TL(R(t)) + TSS(R(t)) \tag{4.45}$$

where SL is the level transmitted by the sonar (in dB re 1 µPa at 1 m), $TL(R(t))$ is the transmission loss at range $R(t)$ and $TSS(R(t))$ is the target strength of the scatterers active at the time of observation t. Reverberation modelling usually considers that the acoustic wave propagates spherically. The relation between the range R and the time t is therefore simply $R(t) = ct/2$, where c is, of course, the acoustic velocity. $t = 0$ at signal transmission, and the transmission loss is $TL(R) = 20 \log R + \alpha R$.

The target strength of the active surface on volume depends on its geometry and on the intrinsic backscattering strength of the reverberation medium (i.e., the target strength of a unit volume or unit area). The first term is defined by sonar characteristics (beam aperture and signal duration). The second term is characteristic of the medium (see Chapter 3 for more details). The target strength can thus be expressed in dB:

$$TSS(R(t)) = 10 \log A(R(t)) + BS \tag{4.46}$$

$A(R(t))$ is the instantaneous active volume or surface, and BS is the backscattering strength (i.e., the target strength for $1\,m^3$ or $1\,m^2$, accordingly).

The technique for computing echo levels from diffuse volume or surface targets was presented in Chapter 3. In the case of volume reverberation, volume V active at range R can be approximated to a truncated cone:

$$V(R) = \psi R^2 \frac{cT}{2} \tag{4.47}$$

where ψ is the solid angle of the equivalent beam aperture (see Chapter 5 for more details) and $cT/2$ is the equivalent signal length. Writing the backscattering strength for a unit volume of water BS_V, the volume reverberation level can now be expressed as:

$$RL_V(t) = SL - 40\log R(t) - 2\alpha R(t) + 10\log V(R(t)) + BS_V$$
$$= SL - 20\log R(t) - 2\alpha R(t) + 10\log\left(\psi\frac{cT}{2}\right) + BS_V \tag{4.48}$$

Since $R(t) = ct/2$, the volume reverberation decreases with time as $-20\log(t/t_0)$.

For seafloor reverberation at oblique incidence θ, the active surface at range R is contained between two ellipses of increasing size. The active area can be approximated to:

$$S(R) = \phi R\frac{cT}{2\sin\theta} \tag{4.49}$$

where ϕ is the beam equivalent aperture in the horizontal plane, and $1/\sin\theta$ accounts for the projection of the signal length on the seafloor. Writing $BS_B(\theta)$ for the backscattering strength for $1\,m^2$ of seafloor, the interface reverberation level may be written:

$$RL_B(t) = SL - 40\log R(t) - 2\alpha R(t) + 10\log S(R(t)) + BS_B(\theta)$$
$$= SL - 30\log R(t) - 2\alpha R(t) + 10\log\left(\phi\frac{cT}{2\sin\theta}\right) + BS_B(\theta) \tag{4.50}$$

The interface reverberation level decreases with time as $-30\log(t/t_0)$ (i.e., more quickly than the volume contribution).

4.4.3 Consequences of reverberation

As far as underwater acoustic signal detection is concerned, reverberation acts like noise: it adds an undesirable random component to the expected signal. Its main differences from ambient noise are:

- for a given transmission, the level of reverberation in reception decreases with time (but more slowly than the target's echo);
- The spectral characteristics of reverberation and the signal (target echo) are nearly identical, except, maybe, for a Doppler effect if the target is moving at sufficient speed; this makes it difficult to filter out reverberation in the spectral domain.

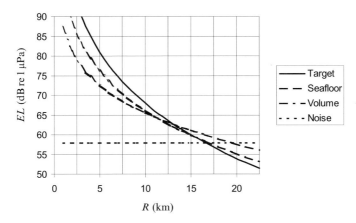

Figure 4.10. Example modelling the relative contributions of target echo, seafloor reverberation, volume reverberation and ambient noise for detection by an active sonar. The target echo is detectable up to 12 km. It is then masked by volume reverberation and seafloor reverberation around 15 km. The ambient noise becomes dominant beyond 20 km. (Input data: $SL = 230\,\text{dB re } 1\mu\text{Pa}$ at 1 m; $\alpha = 0.1\,\text{dB/km}$; $IC = 0\,\text{dB}$; $BS_F = -40\,\text{dB re } 1\,\text{m}^2$; $BS_V = -70\,\text{dB re } 1\,\text{m}^3$; $T = 0.01\,\text{s}$; $\psi = 0.0077\,\text{sr}$; $\phi = 10°$; $NL = 70\,\text{dB re } 1\,\mu\text{Pa}/\sqrt{\text{Hz}}$.)

Let us consider the detection of a point-like target, in the presence of volume and interface reverberation (and, of course, ambient noise). For a particular transmitted signal, at every instant/range of the receiving process, four components need to be taken into account: (1) the target echo, in $-40\log R$; (2) the interface reverberation, in $-30\log R$; (3) the volume reverberation, in $-20\log R$; (4) the ambient noise, constant with time/range. Figure 4.10 shows an example of the relative contributions of these four components according to range. This goes to illustrate that sonar performance in a given detection configuration depends on noise and reverberation in an intricate fashion.

4.5 UNDERWATER ACOUSTIC NOISE REDUCTION

The basic principles for reduction of noise effects in underwater acoustics systems are rather evident and based on common sense; unfortunately, most often their practical application is not! We briefly mention these points here, but reserve the practical aspects of these to Chapters 5 and 6.

External ambient noise being random by nature, the simplest way to protect against it is to maximise the signal-to-noise ratio by transmitting the signal with maximum power while constraining it, as much as possible, to the useful angular sectors. On reception, one filters out as much as possible the useful part of the spectrum of the signal expected, to minimise noise contribution. In a more elaborate way, it is possible to learn the noise structure characteristics from dedicated measurements, in order to subtract the noise from the signal as efficiently as possible. This is done by using *adaptive filters*, the properties of which follow the noise fluctuations. It is also possible to use several receivers, summing their contributions so that the coherent component of the signal stands out against incoherent noise. Finally, directional receivers aimed at the target can act as simple spatial filters, preferentially receiving the useful signal and receiving less of the noise from the other directions.

The best way to diminish an underwater acoustic *system's self-noise* is to try and identify the roots of the problem in order to cancel or reduce the perturbation cause; for example, by decreasing the noise radiated by the platform (ship or submersible). However, this is often technically difficult or impossible, especially a posteriori, and in all cases is very expensive. It is more common to optimise the location of transducers on the hulls, and to enclose them in profiled fairings that decrease turbulence. *Electronic noise* level will reflect the quality of the components chosen to build the system.

Reverberation would not be affected by increasing the transmission level as the reverberation level would increase too. It is better to decrease the signal duration (or, equivalently, to increase its spectral width); this will lower the level of undesirable, reverberated echoes. Also the directivity of antennae needs to be as spatially selective as possible, to decrease the contribution of parasite diffuse targets. These two measures should not interfere with the formation of the echo on the target. In some cases, one can use the difference in Doppler shifts between the target and the surrounding medium, through spectral processing.

Acoustic interference by neighbouring systems on the carrying platform must first be reduced or avoided by making sure all systems onboard have compatible frequencies. But this is not always possible; moreover, there is no legal regulation of underwater acoustic frequencies, as there is with electromagnetic frequencies. Furthermore, even systems with different acoustic frequencies are likely to interfere: bandpass filter rejection can be inadequate, or interference can come from the harmonics of the main signal. Spatial filtering, with optimal spacing of the transducers and use of acoustic barriers, can improve matters. The only safe solution is to synchronise the systems in time; but the downside is the lower transmission rate, which is often unacceptable.

Radiated noise must be carefully controlled for certain types of ship, especially those concerned with acoustic stealth (e.g., warships, submarines and oceanographic vessels, in particular). This is first done during the shake-down cruise and then from time to time in sea trials, because of expected degradation with age. Special measurement facilities are designed for this purpose. They are either fixed or deployable: hydrophone arrays are positioned on the seabed or in the water column, and linked by underwater cables to a recording and analysis system. The vessel studied crosses the hydrophone field, the positioning system carefully logging her navigation parameters. This is usually done at various speeds and propulsion configurations. The noise radiated is recorded as a function of range, and corrected for propagation losses. It is then analysed in the frequency domain, traditionally in one-third octaves for wideband frequency noise, and in 1-Hz bands for low-frequency spectral lines. The signals from the array must either be averaged for the global acoustic signature or processed spatially to get the directivity patterns, or even to identify local radiation sources on the hull itself. Systems for self-control of radiated noise are currently implemented on most modern submarines, enabling quick diagnostics of their acoustic stealth while they are under way.

4.6 ENVIRONMENT VARIABILITY AND SIGNAL FLUCTUATIONS

4.6.1 Variations in the propagation medium

By nature, the ocean is a heterogeneous and unstable medium, both spatially and temporally. This is a source of major perturbations to underwater acoustic transmissions. Several scales of variability can be considered in space and in time.

At small scales (relative to the wavelength or duration of the acoustic signal), the heterogeneity of the propagation medium induces acoustic scattering and fluctuations in the signal itself (amplitude or phase jitter, time spread). The instantaneous measurement quality is degraded; these signal fluctuations are akin to random degradation of the signal-to-noise ratio.

At intermediate scales (relative to the sampling rate, e.g., the ping rate of a sonar), instabilities in the propagation medium such as swell, movements of the sonar platform, or interference may induce other types of effect in signal and data processing (e.g., time delays in signal arrivals, amplitude fading).

At large scales, slow and large variations in the propagation medium (such as sound velocity profiles or water depths) intervene in a deterministic but often uncontrollable way. They cause errors in the data obtained from acoustic measurements (e.g., inducing permanent biases in target positioning).

4.6.1.1 Spatial variability

Inhomogeneities in the water column induce acoustic scattering, and therefore possible variations of the acoustic signal. Fish, plankton, suspended particulates, bubbles, ... are the most common examples of such inhomogeneities. Local sound velocity variations, created by thermal microstructures, can induce both temporal and spatial fluctuations of the signal. These processes are predominant factors in the instability of direct, high-frequency signals (echo sounders, transmitters, ...).

The underwater acoustic propagation channel is considered as stratified at a first approximation. This is often enough, and proves very attractive, for modelling studies. But this assumption is sometimes not appropriate, and one must take into account other processes such as:

● *Bottom relief.* Seafloor topography changes at very variable scales, inducing a variety of acoustic propagation effects that cannot be described in a few lines. At large scale, slow depth variations (average slopes 1° or less, on continental shelf or abyssal plains) have little effect on low-frequency propagation; strong topography features (slopes over 10°: seamounts, ridges, canyons, continental slopes) cause barrier effects, refract the sound field through reflections on inclined surfaces, and increase reverberation. Small-scale relief must be considered as a local boundary condition for reflected sound paths, causing either reflection or scattering processes, according to its roughness.

- *Bottom-type variation.* Variations in the type of seabed will add to the effects from the relief, and the acoustic field will vary even more from the ideal assumption that its propagation medium is stratified. These variations occur on generally very large scales in deep waters (seabed types are then often homogeneous for tens of kilometres), but variations can be very rapid in coastal areas (changes may occur at scales of tens of metres).
- *Sound velocity profiles.* They are often assumed to be layered, to simplify acoustic propagation models. But sound velocity profiles can change spatially, locally because of geographical or environmental constraints (currents or gyres, freshwater input near estuaries, exchanges between oceanic basins with very distinct hydrologies) or regionally (over hundreds of kilometres) because of climate change.

4.6.1.2   *Temporal variability*

Swell is a major cause of fluctuations in the acoustic field. Its dominant effect is indirect, as it causes movements of the platforms that carry the acoustic systems (ships, underwater vehicles, towing or mooring lines), and consequently small geometric perturbations of the acoustic path. Swell also directly affects the signals reflected at the surface, modifying their frequency spectrum by adding sidebands, corresponding to the movements in time of the water surface (typical frequency shifts around 0.1 Hz).

The first effect of currents is spatial, with the modification of local sound velocity profiles. They can also induce temporal variability, through three effects. The first stems from variations in the sound propagation velocity, as their intrinsic speed adds to the velocity of the acoustic signal. This effect can be safely neglected for most applications, except for some configurations in ocean acoustic tomography, where one measures the differences in propagation times of a signal back and forth between two points. The second effect is related to the variations in their speed (and not the speed itself), and is a Doppler effect. The third effect is "scintillation" (amplitude instabilities), due to scattering by water turbulence generated by currents.

Internal waves are generated by vertical oscillations due to the depth profile of water density. These waves induce sound velocity variations of several metres per second, with relatively long time periods (a few minutes to a few hours). These fluctuations induce losses in the spatial coherence of the acoustic signals. There have been many studies of internal waves and their influence on acoustic propagation since the 1970s (see Flatté, 1979).

Tides have a period of half a day, and evidently they do not influence elementary signals. But, in shallow water, they can lead to radical modifications of the sound field structure in a few hours.

Daily and seasonal variations in temperature induce modifications of the sound velocity profiles (mostly close to the surface). These can noticeably affect the sound field structure, and therefore the possible sonar cover of a particular area. Even in the same day, but at different hours, the conditions of propagation can vary considerably.

Movements of the source, receiver and target relative to each other, of course, induce Doppler effects, which are not negligible even for slow objects. Furthermore, movements inside a diffuse acoustic field or in the presence of interference can yield fluctuations in phase and in amplitude. In many cases, the movements of these objects form the main cause of signal fluctuation.

We shall remark that the propagation medium (water mass, interfaces) can most often be considered as fixed at the time scale of the observation of the signal, relative to the movements of sources, receivers and targets. Therefore, the position fluctuations inside the spatial heterogeneities of the acoustic field (due to scattering or multipath interference) will very often be the root cause of temporal fluctuations observed in transmitted signals.

4.6.2   Nature of signal fluctuations

Signal fluctuations can be analysed either as time fluctuations (source and receiver are stationary in a moving environment) or as spatial fluctuations (source and/or receiver/target moving in a stationary environment); in practice, the two causes are often mixed up.

Three main types of signal fluctuation must be remembered:

- The Doppler effect that results from the relative movements of the source, target and receiver, and the movements of the propagation medium (currents, swell, ...). These movements induce noticeable variations in the length of the propagation path during one period of the signal.
- Amplitude and phase fluctuations coming from multiple paths. They are often the most disadvantageous, particularly if the paths have comparable energies and are not far apart in time.
- Amplitude and phase fluctuations that result from scattering. During propagation, each path is accompanied by a continuum of scattered "micropaths", whose relative energies increase with distance. Similarly, after reflection on a sufficiently large target, an echo is generated by a series of scatterers whose contributions add randomly.

4.6.3   Amplitude distributions associated with signal fluctuations

After propagating in a scattering medium, or in the presence of multiple paths, the amplitudes of the resulting signals need to be added as a sum of elementary contributions. The following statistical distributions are very useful in this task, and are more generally usable in the description of underwater acoustic signals.

4.6.3.1   *Gauss distribution*

A variable a follows a normal distribution, or Gauss distribution, if its probability density function corresponds to an equation of the type:

$$f(a) = \frac{1}{\sigma\sqrt{2\pi}} \exp\left(-\frac{(a-a_0)^2}{2\sigma^2}\right) \qquad (4.51)$$

where a_0 is the average value of the distribution and σ^2 is its variance. The Gauss distribution is a fundamental statistical law, easy to employ and therefore often used. It can be used to describe the instantaneous value of a narrowband, random signal; it is therefore an important element in the theory of more elaborate distributions (see Section 4.6.3.2). Apart from that, its interest for the statistical description of under-water acoustic processes lies more in its ease of use than in its actual representative-ness in practical observations.

4.6.3.2   *Rayleigh distribution*

Let us consider the mixing of N signals with the same narrow band (sinusoidal signals with the same frequency), and with random phases equally distributed in $[-\pi; +\pi]$. The resulting signal can be written as:

$$x = \sum_{n=1}^{N} a_n \exp[j(\omega t + \varphi_n)] = a \exp[j(\omega t + \psi)] \qquad (4.52)$$

Neglecting the common term $\exp(j\omega t)$ of time dependence, one can write:

$$x = x_1 + jx_2, \quad \text{with} \begin{cases} x_1 = \sum_{n=1}^{N} a_n \cos\varphi_n \\ x_2 = \sum_{n=1}^{N} a_n \sin\varphi_n \end{cases} \qquad (4.53)$$

When N tends toward infinity, using the central limit theorem, one can see that x_1 and x_2 tend toward centred Gaussian variables with the same variance $\sigma^2 = \sum_{n=1}^{N} a_n^2$.

The variable $X = x_1^2 + x_2^2$ follows a *(chi-squared)* χ^2 law with two degrees of freedom. Its probability density is therefore expressed as:

$$f(X) = \frac{1}{2\sigma^2} \exp\left(-\frac{X}{2\sigma^2}\right) \qquad (4.54)$$

This *exponential law* describes the probability density of the energy associated with the resulting signal. Conversely, the distribution of the amplitude $a = \sqrt{X}$ reads:

$$f(a) = \frac{a}{\sigma^2} \exp\left(-\frac{a^2}{2\sigma^2}\right) \qquad (4.55)$$

This distribution is known as the *Rayleigh distribution* (see Figure 4.11).

The most probable value of a is σ, and its mean value is $\langle a \rangle = \sigma\sqrt{\pi/2}$. The

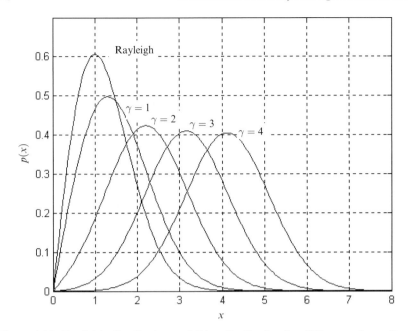

Figure 4.11. Rayleigh distribution and Rice distribution for different values of γ.

amplitude standard deviation is $\sigma\sqrt{2 - \pi/2} \approx 0.655\,\sigma$. The (constant) ratio between the standard deviation and mean is then $\sqrt{(4 - \pi)/\pi} \approx 0.522$. The intensity $X \equiv a^2$ has the average $\langle X \rangle = 2\sigma^2$ and the variance of X is $\langle X \rangle^2 = \langle (X - \langle X \rangle)^2 \rangle = 4\sigma^2$ (meaning that the entire signal power fluctuates). The phase ψ is equally distributed in the interval $[-\pi; +\pi]$.

The Rayleigh distribution is widely used in underwater acoustics because it describes the amplitude fluctuations of a signal resulting from the contribution of a large number of echoes. These echoes can be multiple paths for waveguide propagation, or echoes from a volume distribution of scatterers (in the water column), or a surface distribution (at the bottom or on the surface). This distribution corresponds therefore to extremely common physical configurations (see discussion in Section 4.6.3.4).

4.6.3.3 The Rice distribution

Let us look again at the summation of a large number of narrowband signals, by considering a dominant component with non-random characteristics:

$$x = a_0 \exp[j(\omega t + \varphi_0)] + \sum_{n=1}^{N} a_n \exp[j(\omega t + \varphi_n)] = a \exp[j(\omega t + \psi)] \qquad (4.56)$$

The two real and imaginary parts x_1 and x_2 are still tending toward Gaussian processes, for N large. But their means are now different (respectively, $m_1 = a_0 \cos\varphi_0$; $m_2 = a_0 \sin\varphi_0$), whereas their variances have the same value $\sigma^2 = \sum_{n=1}^{N} a_n^2$.

It can be demonstrated that the variable $X = x_1^2 + x_2^2$ follows the probability density:

$$f(X) = \frac{1}{2\sigma^2} \exp\left(-\frac{X + a_0^2}{2\sigma^2}\right) I_0\left(\frac{a_0}{\sigma^2}\sqrt{X}\right) \tag{4.57}$$

where $a_0^2 = m_1^2 + m_2^2$ and I_0 is the modified Bessel function of the first kind and order zero.

If one now looks at the probability density of the amplitude $a = \sqrt{X}$ of the signal, from Equation (4.57), one gets the *Rice distribution*:

$$f(a) = \frac{a}{\sigma^2} \exp\left(-\frac{a^2 + a_0^2}{2\sigma^2}\right) I_0\left(\frac{aa_0}{\sigma^2}\right) \tag{4.58}$$

The Rice distribution corresponds to the amplitude distribution of the sum of a signal and noise with narrow bands, parameterised with the ratio $\gamma = a_0^2/2\sigma^2$ between the mean coherent energy $a_0^2/2$ and the mean energy σ^2 of the random component. This ratio represents the "randomicity" of the mix.

The Rice distribution is more generally usable than the Rayleigh distribution, and is also known as *generalised Rayleigh distribution*.

There are two interesting cases:

- when the power of the coherent signal tends toward 0; γ tends toward 0, the randomness is maximum and the Rice distribution becomes the Rayleigh distribution again, which could be expected:

$$f(a) = \frac{a}{\sigma^2} \exp\left(-\frac{a^2}{2\sigma^2}\right) \tag{4.59}$$

- conversely, when the power of the noise tends toward 0, considering that $a \approx \sqrt{aa_0}$ and using the asymptotic form of the Bessel function, the probability density of x tends toward a normal distribution $N(a_0, \sigma)$ centred on a_0 and with variance σ^2:

$$f(a) = \frac{a}{\sigma^2} \frac{1}{\sqrt{2\pi}} \sqrt{\frac{\sigma^2}{aa_0}} \exp\left(-\frac{aa_0}{\sigma^2}\right) \exp\left(-\frac{a^2 + a_0^2}{2\sigma^2}\right)$$

$$\approx \frac{1}{\sigma\sqrt{2\pi}} \exp\left(-\frac{(a - a_0)^2}{2\sigma^2}\right) \tag{4.60}$$

4.6.3.4   *Physical interpretation of the Rayleigh and Rice distributions*

The Rayleigh distribution results from the mixing of signals with the same narrow bands and comparable amplitudes, observed simultaneously. In principle, it can therefore be encountered in multipath propagation conditions, as long as none of the paths is predominant. However, this condition is not always met, and the delays between paths are often larger than the duration of the signals used (at least in "active" acoustic sensing). And the amplitude distribution of the sum of a large number of paths is not always of great interest, except in a few specific cases (such as SOFAR propagation).

More interesting is the case of signals scattered in a volume, or at the interfaces of a medium, or by a large target. The fluctuations are indeed following the Rayleigh distribution when backscattering can be assimilated to the contribution of many *highlights* of relatively comparable amplitudes. The most typical cases are:

- reflection/scattering by a surface with a strong relief at sufficiently high frequencies: rocky seabed or choppy sea surface (corresponding to large values of the Rayleigh parameter $P = 2k\sigma \cos\theta$);
- volume backscattering by a number of elementary scatterers that have a sufficiently high volume insonified at a particular time (bubbles, plankton, etc.);
- backscattering from a large target, made of several important highlights, for a signal lasting long enough to insonify the entire target simultaneously.

The existence of fluctuations may be related to the presence of sonar in the "near field" of the target (below the Fresnel distance). The target is either a limited-size obstacle (e.g., submarine, mine, fish) or part of a medium (interface or volume) insonified by the system at a particular time.

The Rice distribution is encountered when noise is superposed on a coherent signal, both with the same bandwidth. A most interesting example is the scattering of a signal around its main direction. Two cases are commonly seen:

- The propagation of a signal in an inhomogeneous medium. The heterogeneities act as scatterers perturbing the geometrical path. The deterministic acoustic ray is accompanied by a continuum of random "micropaths" due to this scattering, acting as an incoherent part of the signal received.
- The reflection of a wave on a small-roughness interface (with a low Rayleigh parameter P). The signal reflected in the specular direction is then accompanied by diffuse contributions due to the relief around the geometric impact point in the insonifying patch of the acoustic signal transmitted.

4.6.3.5 Chi-squared distribution

We have just seen the expression of the Rayleigh distribution as the sum of two squared Gaussian variables, centred and with the same variance. This concept can be generalised to the sum of N variables, using the variable $X = \sum_{i=1}^{N} x_i^2$. The resulting probability density function of X is the χ^2 *distribution with N degrees of freedom*:

$$f(X) = \frac{1}{\sigma^2 \, 2^{N/2} \, \Gamma\left(\dfrac{N}{2}\right)} \left(\frac{X}{\sigma^2}\right)^{(N/2)-1} \exp\left(-\frac{X}{2\sigma^2}\right) \tag{4.61}$$

where Γ is the gamma function.[7] Applying the same reasoning used with the

---

[7] The Γ function possesses the following useful properties: $\Gamma(N+1) = N\Gamma(N)$, with $\Gamma(1) = \Gamma(2) = 1$, and $\Gamma(1/2) = \pi^{1/2} \approx 1.7725$.

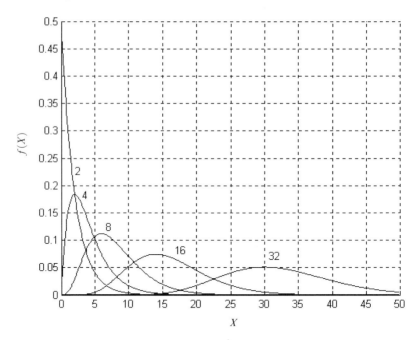

Figure 4.12. Probability density function of the χ^2 distribution, for a variance normalised to unity and different values of N.

Rayleigh distribution, we find that the mean value of X is $N\sigma^2$ and its variance is $2N\sigma^4$. When N becomes large, the χ^2 distribution tends toward a gaussian. Figure 4.12 shows the χ^2 distributions for several values of N, with the variance normalised to $\sigma^2 = 1$.

The χ^2 distribution is often useful in underwater acoustics, as it is used to describe the statistics of signals at the output of quadrature receivers, processing the squared amplitude of the signal with summations or integrations. The degree of freedom N is then the number of independent samples used in the processing.

4.6.3.6   Dynamics of underwater acoustic signals

Underwater acoustic signals received by a particular system for a given application can present values in a very large dynamic range. This must be accounted for when designing the electronics (analogue or digital) for the processing of received signals. The main causes of signal fluctuations are:

- variations in propagation loss depending on the distance between source and receiver, or sonar and target;
- variability of the mean target index;
- fluctuating character of the signals received, due to the propagation medium or the behaviour of the target.

A brief analysis of the latter is rather interesting. Let us consider a signal whose amplitudes follow a Rayleigh distribution (a good hypothesis for a first approximation). The minimal and maximal amplitudes corresponding to a given probability are easily determined. For example, the value below which 99.9% of the samples fall is given by:

$$\int_0^{A_1} \frac{a}{\sigma^2} \exp\left(-\frac{a^2}{2\sigma^2}\right) da = 0.999 \quad \Rightarrow \quad A_1 = 3.72\sigma \tag{4.62}$$

The value defining the 0.1% lowest values is similarly given by:

$$\int_0^{A_2} \frac{a}{\sigma^2} \exp\left(-\frac{a^2}{2\sigma^2}\right) da = 0.001 \quad \Rightarrow \quad A_2 = 0.045\sigma \tag{4.63}$$

The dynamic range is defined by $A_1/A_2 \approx 83$ (i.e., 38.3 dB). This is in addition to the "slow" variations of the average signal intensity.

Analogue-to-digital conversion of signals will need to take this into account. If the signal between the two limits above is to be coded without losing any information, the N-bit dynamic range of the converter will need to respect $2^{N-1} > 83$ (i.e., a minimum of $N = 8$ bits). This does not imply that an 8-bit converter is sufficient, but rather that its dynamic range is just enough to follow the instantaneous fluctuations of the signal, independently of variations in the average value, related, for example, to propagation losses or changes in target aspect. These slow variations are generally compensated by a time-varying gain. But such a system only makes approximate corrections, and a safety margin needs to be preserved. These considerations are imperative for systems that measure amplitudes. For receivers that only measure times or phases, the constraints may be less severe.

5

Transducers and array processing

5.1 UNDERWATER ELECTRO-ACOUSTIC TRANSDUCERS

5.1.1 Fundamental principles

5.1.1.1 Definitions

Electro-acoustic transducers are essential for the transmission and reception of acoustic signals underwater. In a similar way to how loudspeakers and microphones are used in the air, they convert acoustic energy into electric energy, and vice versa. Underwater acoustic sources are called *projectors* (equivalent to loudspeakers). The reception transducers are called *hydrophones* (by analogy with microphones). Extended transducers are named *antennas*, or *arrays*. The last name is usually reserved for structures made up of several elementary transducers. A *transmitter* is made up of a projector and its associated electronics. Similarly, a *receiver* is made up of a hydrophone array and its associated low-level electronics (preamplifier and filters).

5.1.1.2 Piezoelectricity

Underwater acoustic transducers can call on several physical processes to generate or receive sound waves. Most of them use the *piezoelectric* properties of some crystals, natural or artificial (*ceramics*). An electric field applied to these materials causes a deformation related to electrical excitation. These mechanical deformations in turn create acoustic waves (Figure 5.1). The opposite effects are used in reception: a piezoelectric material stressed by sound waves will generate an electric potential between its sides.

Natural piezoelectric crystals (such as quartz or Seignette salt) were used in the early days of underwater acoustics. Paul Langevin, for example, built a quartz source for his early experiments in 1917. Such crystals are now replaced by synthetic

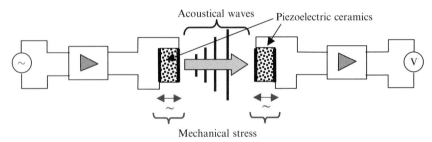

Figure 5.1. Piezoelectric effect: (*left*) transmission: applying an electric signal to a piece of piezoelectric material induces a mechanical deformation, generating an acoustic wave; (*right*) reception: the mechanical stress caused by the acoustic wave is transformed by the piezo-electric material into an electric voltage.

ceramics. They are made by mixing components under high temperature and high pressure (*sintering*). The resulting material is then machined to the dimensions required and coated with metal. The ceramics produced is not spontaneously polarised; this is created artificially by applying a very intense electric field to induce a *remanent polarisation*. The piezoelectric effect will be linear and reversible around this remanent polarisation.

The fundamental equations of piezoelectricity are presented in Appendix A.5.1; they link together the mechanical, electrical and piezoelectrical values of ceramics. We will only mention here that the thickness a of a ceramic plate, submitted to a voltage V between its sides (assumed perpendicular to the direction of polarisation, see Figure 5.2), will vary proportionally to the amplitude of the excitation:

$$\Delta a = d_{33} V \tag{5.1}$$

where d_{33} is the *piezoelectric constant* of the ceramic in the direction of polarisation. The resulting mechanical displacements are very small, due to the usual values of d_{33} (e.g., $d_{33} \approx 40\text{--}750 \times 10^{-12}$ m/V for PZT[1]). Ceramics used to transmit high powers show a typical $d_{33} \approx 300 \times 10^{-12}$ m/V; thus, a voltage of 1,000 V will yield a thickness variation $\Delta a \approx 0.30\,\mu$m. The mechanical effect can be amplified by stacking several piezoelectric ceramic plates, to which electric excitations will be applied in parallel, hence cumulating small displacements. On reception (inverse piezoelectric effect), a ceramic plate of thickness a and surface S undergoing a compression F parallel to its direction of polarisation will generate a voltage:

$$V = g_{33} a \frac{F}{S} \tag{5.2}$$

The constant g_{33} equals ca. $15\text{--}30 \times 10^{-3}$ V m/N for PZT.[2]

---

[1] Lead and titanium zirconate.
[2] For comparison, a natural quartz piezoelectric crystal has characteristic values of $d_{33} \approx 2 \times 10^{-12}$ m/V and $g_{33} \approx 50 \times 10^{-3}$ m/V.

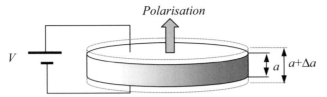

Figure 5.2. Deformation of a piezoelectric ceramic disk submitted to an electrical tension.

5.1.1.3 Other working principles

Presumably, 90–95% of underwater acoustic transducers are piezoelectric devices. However, specific applications may require transducers based on other physical processes:

- *Magnetostriction* is comparable with piezoelectricity, but the driving field is magnetic and not electric. Materials in use are *ferromagnetic* (Al–Fe alloys, Ni–Co ferrites) or rare earths (e.g., *Terfenol*). More expensive than piezoelectric devices, magnetostrictive transducers have poor efficiency and a narrow frequency bandwidth. But they withstand large vibration amplitudes, and hence are convenient for high-power applications at low frequencies.
- *Mechanics* or *hydraulics*: the active part of the transducer is driven directly by a mechanical or hydraulic power source. This is more relevant for transmissions at ultra-low frequency.
- *Electrodynamics*: a mobile coil vibrates in a permanent magnetic field, as in aerial loudspeakers. This principle allows very wide frequency bands, but at the expense of extremely low efficiency, and it is limited to very shallow depths.

5.1.1.4 Nominal frequency, and directivity

When transmitting, underwater transducers generally work around their resonant frequency, to yield the best output level achievable. But it is often possible to look for a compromise with a bandwidth broad enough to pass several close frequencies, or a wide-spectrum modulated signal. The receiving transducers used in sonars generally work around their resonance regime. However, hydrophones that are used for laboratory measurement are wideband devices.

Finally, directional transducers are often preferred, as specific directions of transmission and/or reception can be achieved and controlled. The *directivity pattern* of an antenna can be obtained either from the transducer geometry or from signal processing with an array of elementary transducers. These characteristics are paramount to the correct operation of a sonar system. They control both the signal-to-noise ratio of the measurement (via the directivity index) and the target angle estimation, essential in many sonar systems.

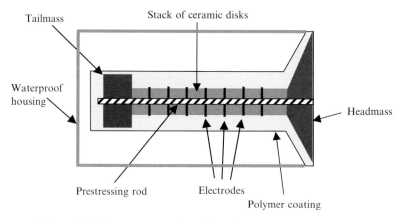

Figure 5.3. Transverse section of a typical Tonpilz transducer.

5.1.2   Underwater acoustic sources and hydrophones

5.1.2.1   *Tonpilz transducers*

Tonpilz technology is the most used in underwater acoustic transducers. Piezoelectric ceramic plates are separated by electrodes (Figures 5.3 and 5.30, p. 176) and stacked under strong static pressure imposed by a *prestressing rod*. This stack is interdependent of a radiating *headmass* (balanced by a *tailmass* at the other end). It transmits to the surrounding water the vibrations induced by a driving electric field applied along the electrodes of the stack of piezoelectric disks. The entire system, covered with a polymer coating, is packaged inside a waterproof housing, filled with air to limit backward radiation of the headmass. This air filling precludes the use of Tonpilz transducers at large depths, since the housing risks being crushed by high hydrostatic pressures. Filling it with oil increases the depths achievable, at the expense of lower efficiency.

The size of the piezoelectric ceramics that make up the transducer determines the resonance frequency, the transmission level and the electrical impedance. The diameter and the thickness of the headmass, acting as a transformer adapting the active ceramics to the propagation medium, influence both the resonance frequency and the transmission level. The use of a sufficiently light metal (e.g., aluminium or magnesium) makes it possible to broaden the bandwidth. The role of the tailmass is to limit backward acoustic radiation, and to tune the resonance frequency. To be effective, it must be made of a dense enough material (e.g., steel or bronze).

Tonpilz transducers are based on a resonance concept. They can achieve high transmission levels with good power efficiency, but they only allow limited bandwidths. Quality factors as low as 2 or 3 can be obtained.[3] Because of their simple design, Tonpilz transducers are very successful in the majority of applications at

---

[3] This means that the bandwidth is $\frac{1}{2}$ or $\frac{1}{3}$ of the resonant frequency (see Section 5.2.1 for details).

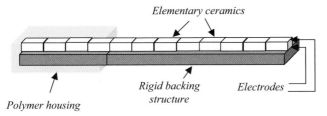

Figure 5.4. High-frequency linear transducer based on monolithic ceramics.

frequencies typically between 2 kHz and 50 kHz. At frequencies around 1 kHz and below, their size and their weight make them too cumbersome for practical applications. Conversely, at higher frequencies, their dimensions are so small that they become difficult to build, and other, simpler solutions are then preferred.

5.1.2.2   High-frequency transducers

At high frequencies, where Tonpilz technology can no longer be used, one uses blocks of piezoelectric ceramics, electrically driven directly by surface electrodes. Many different shapes can be built: rods or parallelepipeds, rectangular or round plates, and rings. They work best at the resonant frequency of the constitutive ceramics, determined by its thickness, equal to the nominal half-wavelength. This type of transducer is used at frequencies typically larger than 100 kHz, but lower frequency dimensioning is achievable down to 50 kHz. These transducers are strongly resonant, and their bandwidth (with a typical quality factor between 5 and 10) is less advantageous than that of Tonpilz transducers.

If the dimensions of an elementary ceramic block are not large enough, several of them are built into the antenna by fixing on a rigid backing structure (Figure 5.4). The most common shapes are rectangles or rings, depending on the type of directivity sought. The mechanical behaviour of the backing (material and dimensions) is very important, as it limits backward acoustic radiation, which should be as small as possible.

To ensure good acoustic radiation from the ceramics into the water, the entire set is either moulded in an elastomer matrix, or embedded in an acoustically transparent fluid-filled housing (most commonly castor oil). This type of equipressure packaging makes these transducers particularly well suited to large depths.

At high frequencies, another possibility is the *composite ceramic* technology. Piezoelectric sticks are grouped to form a given projector shape, and embedded in a polymer matrix ensuring mechanical rigidity. This makes it possible to manufacture transducers of varied shapes with a relatively good performance, both in efficiency and in bandwidth.

5.1.2.3   Low-frequency transducers

At very low frequencies (below 1 kHz), acoustic source technology encounters serious limitations (Decarpigny *et al.*, 1991). The transducer must be capable of

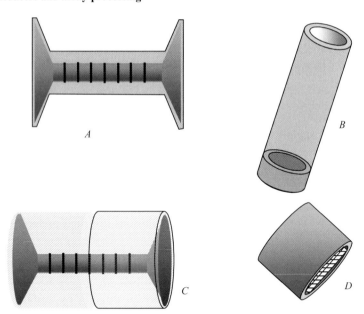

Figure 5.5. Examples of low-frequency acoustic transducers: Janus Tonpilz (*A*); Helmholtz resonator (*B*); Janus–Helmholtz (*C*); Class IV flextensional (*D*). See text for details.

withstanding the large amplitudes determined by its transmitting surface; and weight and dimension constraints are very heavy.

Several solutions have been proposed (Figure 5.5), each one being generally adapted to solve a particular problem. Among them, one can cite:

- The extension of Tonpilz technology to low frequencies with some modifications. For example, the *Janus* concept equips the Tonpilz transducer with two opposing projectors (Figures 5.5*A* and 5.30, p. 176). This type of solution is particularly well suited when high transmission levels are required.

- Sources based on the *Helmholtz resonator* technology. These are commonly used by oceanographers for acoustic tomography experiments. An open metal tube is excited at one end by a piezoelectric driver (Figure 5.5*B*). The entire structure resonates at a frequency given by $L = \lambda/4$, where L is the length of the tube. Initially designed for frequencies between 400 Hz and 250 Hz, this solution is simple, robust, low-cost and insensitive to hydrostatic pressure. Unfortunately, it shows poor efficiency, limited power and very narrow frequency bandwidths.

- A Helmholtz resonator can be coupled with a Janus transducer, leading to the *Janus–Helmholtz concept* (Figure 5.5*C*). Coupling the resonance of the transducer with that of the Helmholtz resonator yields a wide bandwidth and is efficient at the same time. This means that the elasticity inside the resonator cavity must be increased, using either compliant tubes or a compressible fluid. Initially designed for military low-frequency active sonars, and usable at great

depths, this concept has since been extended to sources used in physical oceanography and marine seismics.

- *Flextensional* transducers are also an appealing solution for high-power applications such as military sonars. They consist in an elastic shell, in which an electro-
  acoustic driver is inserted. This is the piezoelectric stack, whose longitudinal
  vibrations induce deformations in the radiating shell. The Class IV type
  (Figure 5.5D) is the most commonly used: the shell is an elliptical cylinder,
  and the ceramic bar is inserted along its main radial axis. These transducers
  have a high efficiency at low frequencies, with reasonable dimensions. However,
  they cannot withstand high pressures, as the static deformation of the shell
  uncouples it from the piezoelectric driver.
- *Hydraulic* technology has been used for acoustic thermometry experiments
  requiring large, broadband transmissions around 60 Hz. A hydraulic block,
  electrically controlled, moves radiating shells with jacks. Well suited to very
  low frequencies, this transducer concept requires high electric power and
  specific cooling devices; it is therefore ill-adapted to autonomous sources.
- *Electrodynamic* sources similar to aerial loudspeakers can be used to transmit
  broadband low-frequency signals. But the levels available are very limited
  because of mediocre efficiencies, and it is very difficult to compensate for hydrostatic pressure below a few metres.

5.1.2.4   Non-linear sources

When an acoustic wave is transmitted at a very high level, the zones of high and low
pressures in the propagation medium create local variations in sound velocity. Hence
the pressure maxima propagate slightly faster, and the pressure minima propagate
slightly slower, than the average velocity. The acoustic wave is therefore distorted
(Figure 5.6), and higher harmonics appear in the fundamental frequency.

Parametric arrays are based on this concept of non-linear generation of acoustic
waves. During simultaneous transmission, at very high levels, of two close frequencies, the non-linearity of propagation induces the apparition of a secondary wave
with a frequency equal to the difference between the two primary frequencies. This
secondary wave has the directivity pattern of the two primary waves (high frequencies, hence highly directive). It is therefore possible to get a very narrow beam
(virtually without sidelobes), at low frequency, with a high-frequency transmitting
antenna of modest size. It is also possible to transmit large bandwidths around the
secondary frequency.

The radiation of a parametric array can be represented conceptually in a relatively simple manner (a classical model was established by Westervelt, 1963; see
Appendix A.5.2). Near the antenna, the primary waves define a narrow beam of
high intensity, attenuating with distance, where non-linear generation processes
occur. Along its effective length, this beam behaves like a line of sources at the
secondary frequency. The radiation diagram of this *end-fire array* is very narrow
and without any sidelobes (see Figure 5.6).

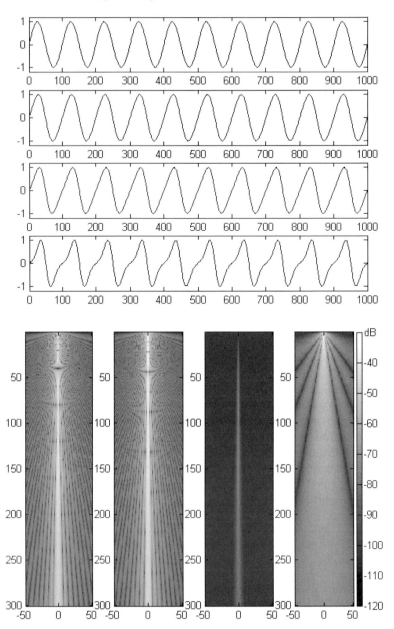

Figure 5.6. Non-linear acoustic transmission. (*Top*) Distortion of a sine wave as it propagates (along the *x*-axis). (*Bottom*) radiation from a parametric array. The source is located at (0, 0), each plot being a cross-section of the insonified space, with grey levels corresponding to dB levels; the axis scales are in metres. From *left* to *right*: primary beams, 100 kHz and 105 kHz, respectively; secondary beam at 5 kHz, −30 dB average level below the primary level, with no sidelobes; classical directivity pattern at 5 kHz if one had used a physical array of the same length.

These many advantages are compensated by very poor energy efficiency in the production of the secondary wave. The levels thus created are much smaller than those obtained with a classic antenna. After many theoretical studies and experimental tests since the 1960s, non-linear acoustics has so far only been used convincingly in a few applications. Sediment profilers based on this principle are, however, available. A few useful formulas for parametric arrays are given in Appendix A.5.2.

5.1.3   Hydrophones

Hydrophones are receiving transducers, designed to convert the acoustic pressures into electrical signals. They usually are piezoelectric devices, most often made of PZT, featuring good sensitivity and low internal noise levels. *Lithium sulphate* is reserved for high-frequency measurement hydrophones. Physically large hydrophones (e.g., submarine flank arrays) are sometimes made in PVDF (*polyvinylidene bifluoride*), a versatile material which can be tailored into very large plates, easily fitted on curved surfaces. Finally, as with projectors, *piezo-composite materials* are increasingly used: they are made of ceramic elements embedded in a polymer matrix.

Contrary to projectors, hydrophones are often capable of working over a wide frequency band. This is because they do not actually need to be tuned to a particular resonance frequency. Their efficiency (ratio of the output electric power to the input acoustic power) is usually not problematic, as the electric signal can always be amplified. But it is imperative that low acoustic signals can be detected amidst the internal noise of the receiver (combination of the internal noise of the ceramics and the self-noise of the amplifier). Measurement hydrophones are usually small compared with acoustic wavelengths, and their frequency resonance is rejected beyond the upper limit of the flat part of the frequency response (see Figure 5.8 on p. 148). They therefore show low spatial selectivity. A required directivity pattern can be obtained by combining several hydrophones into a large array.

The same transducer is often used for transmission and reception in many sonar systems (e.g., single-beam echo sounders, ADCPs [acoustic Doppler current profilers], sidescan sonars). Apart from its immediate advantage, in terms of simplicity and cost, this configuration improves directivity characteristics: it is particularly recommended when directivity requirements can be fulfilled with such a geometrical arrangement.

5.1.4   Transducer modelling and design

The design of underwater transducers with specific characteristics (e.g., frequency and bandwidth, efficiency, directivity pattern) is a difficult undertaking, as many aspects need to be considered concurrently. Theoretical dimensioning of transducers can be conducted in different ways (Wilson, 1985).

A common technique of dimensioning is based on *electro-acoustic analogies*. The principle is that mechanical (or acoustic) systems are described by equations similar to those that describe electric circuits. Assimilating an electric voltage to a mechanical force, and an electric intensity to a speed, the different mechanical components

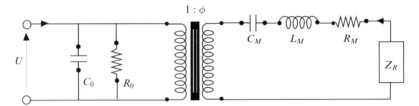

Figure 5.7. Electric circuit equivalent to an electro-acoustic transducer. See text for details.

(characterised by their mass, stiffness, damping) can be identified with electric components (characterised by their inductance, capacitance and resistivity). An electro-acoustic transducer, made up of both electric and mechanical components, can therefore be modelled with a fully electric circuit, relatively amenable to calculation (Figure 5.7). The results will, however, remain approximate, because of the necessary simplifications. The basic principles of this approach are provided in Appendix A.5.3.

Figure 5.7 represents the electric circuit equivalent to an electro-acoustic transducer. Its left part (R_0, C_0) corresponds to the dielectric characteristics of the transducer. The transformer describes the electromechanical conversion itself. On its right side, the elements of the circuit represent the mechanical characteristics (dynamic mass L_M, elasticity C_M, mechanical damping R_M). The *radiation impedance* $Z_R = R_R + jX_R$ represents the radiation of the acoustic wave into the propagation medium. The equivalent impedance of the right part of the circuit (mechanical and acoustic) is called the *motional impedance*.

If the behaviour of the mechanical elements must be studied in detail, the electromechanical analogy is not sufficient, and one must use *finite element* computations. The entire vibrating structure of the transducer must be represented by a sufficiently fine grid, with the propagation medium close to the transducer, making the best use of geometrical symmetries to lighten the computation load. The results thus obtained can be very accurate, and enable a very fine analysis of the behaviour of a transducer. But if the structures are very complex and the frequencies are high, the calculations can become extremely ponderous.

5.1.5　Transducer installation

Underwater acoustic transducers are most often installed on ship hulls or on towed bodies. Several types of installation are possible for hull-mounted transducers:

- *Flush mounting*: the radiating surface of the array follows the hull without discontinuity, and it is covered by an acoustically transparent window ensuring waterproofing and acoustic transmission into the water. This type of installation is reserved for transducers whose dimensions can be limited by the geometric constraints of the hull's shape (see Figure 5.31, p. 176).
- If the shape of the hull does not fit with the shape of the transducer, it must be installed into a special fairing, or "blister", fixed under the hull (Figure 5.33,

p. 177). One of the constraints is that it should not degrade the hydrodynamic performance of the vessel.

- The transducer can be installed inside a dome fixed below the hull or at the bow of a ship. This solution is often used for active sonars on navy ships.
- For large antennas (e.g., multibeam sounders), the "flush" or "blister" installations are not feasible, and one solution is a special housing. It consists in a profiled structure, shaped like the delta wing of a plane, fixed below the hull with stays. This "gondola"-mounting allows very large horizontal dimensions, is satisfactory regarding noise received by the transducers, and does not affect the ship's hydrodynamics too much.
- Small systems used only occasionally can be deployed through the hull in special vertical tubes. Also a flexible solution retained aboard small boats is to install the transducer at the tip of a dipped vertical pole fixed alongside the hull.

Installation on a *towed body* (Figure 5.32, p. 177) is preferred for some applications. The main advantages are low noise level (the self-noise from the vessel being reduced), more favourable propagation conditions (due to avoidance of perturbations in surface layers) and movements smaller than for a vehicle sailing at the sea surface. However, this requires special installation and deployment procedures, and in many cases determines slower speeds for the towing vehicle.[4]

Transducer installation is always a delicate operation, as its success will govern the good working of the entire sonar system. The basic principles for successful installation are now well known. The emplacement selected should fulfil several conditions:

- low acoustic perturbations by self-noise from the ship or other sonar systems;
- no masking by bubble plumes from the bow;
- shortest possible distance to the preamplifying electronics, to limit the risks of electric interference;
- compatibility with other acoustic systems;
- compatibility with the hydrodynamics of the ship and dry dock operation requirements.

However, these conditions might be mutually exclusive and the final installation will usually result in a compromise between the different objectives. Furthermore, some of the design operations are still largely empirical, since no simulation, computation or tank tests can totally guarantee the final result.

5.2 TRANSDUCER CHARACTERISTICS

5.2.1 Frequency response

As shown earlier, a projector (acoustic source) is generally designed for use at a particular frequency. This frequency is usually given by the resonant regime of the

---

[4] For instance, for a surface ship towing a vehicle close to the bottom, in deep water, the drag of the cable (several kilometres long) imposes a maximum speed of around 2 knots (1 m/s).

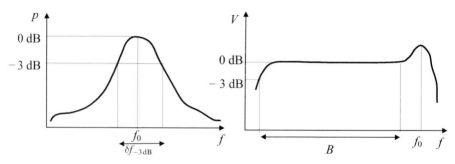

Figure 5.8. (*Left*) Frequency response curve and bandwidth $\delta f_{-3\,dB}$ of a resonant transducer. (*Right*) Frequency response curve of a hydrophone: the resonance frequency f_0 is rejected beyond the effective bandwidth B.

projector, with a narrow bandwidth. The projector may also be used for reception at the same frequency: its narrow bandwidth corresponds to optimal bandpass filtering at the input of the receiver. But some hydrophones used for measurement purposes are very poorly selective in frequency, and are designed for a wide frequency band rather than one single frequency.

The bandwidth is the width of the band of frequencies that the transducer can transmit efficiently around its nominal frequency. The bandwidth is smaller as the resonant behaviour of the transducer is more marked. Using the frequency response curve (Figure 5.8), the bandwidth is conventionally measured at $-3\,dB$ on each side of the maximum. However, this notion is meaningless for a broadband system: a receiving hydrophone often works outside its resonance mode, located beyond the useful bandwidth B determined by the flat point of the response curve (Figure 5.8).

The resonance mechanical quality factor quantifies the relative bandwidth. It reads:

$$Q_m = \frac{f_0}{\delta f_{-3\,dB}} \tag{5.3}$$

This is the ratio of the central frequency to the bandwidth. Typical values range from 2 to 10 for usual underwater acoustic transducers.

5.2.2 Sensitivity

The sensitivity of a transducer quantifies the quality of the electro-acoustic conversion. It expresses the relation between the input and output values of the transducer (acoustic pressure and electric voltage).

For transmitting transducers (projectors), it reads:

$$SV = 20\log\left(\frac{p_{1\,V}}{p_{ref}}\right) \quad \text{(in dB re } 1\,\mu\text{Pa}/1\,\text{m}/1\,\text{V)} \tag{5.4}$$

where $p_{1\,V}$ is the acoustic pressure 1 m away from the transducer, in a given direction, for an electric voltage of 1 V and p_{ref} is the reference acoustic pressure (1 μPa). The sensitivity SV is usually given in the direction of maximum radiated amplitude. It can be expressed along other directions by using the spatial directivity function of the transducer.

The transmission sensitivity of an elementary Tonpilz projector is usually between 120 dB and 150 dB re 1 µPa/1 m/1 V.

The reception sensitivity of a hydrophone reads:

$$SH = 20 \log \left(\frac{V_{1\,\mu Pa}}{V_{ref}} \right) \quad \text{(in dB re 1 V/1 µPa)} \tag{5.5}$$

where $V_{1\,\mu Pa}$ is the output electric voltage of the hydrophone, for an incident acoustic pressure of 1 µPa, and V_{ref} is the reference voltage (1 V).

For a wideband-receiving transducer, the sensitivity usually ranges between -220 dB and -190 dB re 1 V/1 µPa.

The *electro-acoustic power efficiency* of a transducer is the ratio between the output and input powers. It only has meaning for projectors, at a given frequency or for a narrow frequency band, and is then given by:

$$\beta = \frac{P_{ac}}{P_{el}} \tag{5.6}$$

where P_{ac} is the acoustic power delivered and P_{el} is the input electric power. Typical values of β range between 0.2 and 0.7 for piezoelectric sources.

5.2.3   Source level

The acoustic level delivered by a transducer evidently depends on the electric power provided, but it also depends on its own characteristics: electro-acoustic energy efficiency and directivity gain (distribution of the acoustic wave transmitted in space).

The acoustic power radiated P_{ac} is the product of the electric power P_{el} provided and the electro-acoustic efficiency β. Assuming the transmission is spherical, the corresponding acoustic intensity at a distance R reads:

$$I(R) = \frac{P_{ac}}{4\pi R^2} \tag{5.7}$$

And assuming the wave is plane at reception, the intensity is related to acoustic pressure by the relation:

$$I(R) = \frac{p^2(R)}{\rho c} \tag{5.8}$$

The transmitted acoustic pressure, brought back to $R = 1$ m, is then given by:

$$p_{1\,m}^2 = \frac{\rho c}{4\pi} \beta P_{el} G_d \tag{5.9}$$

This expression introduces the *directivity gain* G_d of the transducer (see Section 5.3.1.2). Expressing this in dB, we now have:

$$SL = 170.8 + 10 \log P_{el} + 10 \log \beta + DI \tag{5.10}$$

where SL is the level transmitted, in dB re 1 µPa/1 m; P_{el} is expressed in watts; and $DI = 10 \log G_d$ is the *directivity index*. Typical values of SL range between 170 dB and 240 dB re 1 µPa/1 m; the directivity index will depend on the antenna geometry (see Section 5.3.1.2).

For an input voltage $U = 1$ V, we have the following relation between SL, SV, β, DI and the real part R_p of the input electric impedance of the transducer (with $P_{el} = U^2/R_p$):

$$SL = SV = 170.8 - 10 \log R_p + 10 \log \beta + DI \tag{5.11}$$

5.2.4   Near field and far field

The radiating of a transducer into space can be decomposed, depending on the distance of the observation point to the transducer, into two regions:

- *Near field* (or Fresnel zone): the contributions from the different points on the transmitting face of the transducer are strongly out of phase with each other. The resulting field oscillates with distance, and its average intensity decreases more slowly than spherical spread in $1/R^2$.
- *Far field* (or Fraunhofer zone): the path differences between the signals coming from different points on the antenna are small and the interference disappears. The intensity of the field decreases monotonously with distance, and tends at large distances toward the spherical regime.

Let us now consider a transducer of characteristic dimension L transverse to the radiation direction (Figure 5.9) (L is the length of a linear or rectangular antenna, or the diameter of a disc), and a signal of wavelength λ. The transition between the near field and the far field can be determined by considering the signals coming from the centre of the antenna and a point on the edge. Provided that $x \gg L$ the difference in acoustic paths equals:

$$\delta R = R_L - x = \sqrt{x^2 + \left(\frac{L}{2}\right)^2} - x \approx \frac{L^2}{8x} \tag{5.12}$$

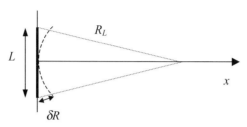

Figure 5.9. Acoustic path difference for a linear antenna.

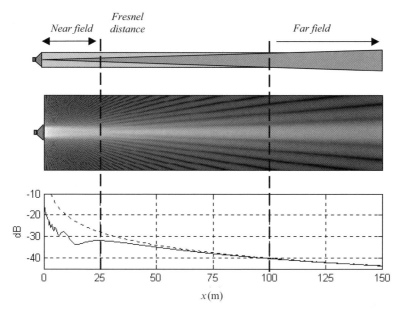

Figure 5.10. Radiation regimes of a transducer. In the near field, the directivity beam is not formed yet, the acoustic level fluctuates and the beamwidth is fairly constant. In the far field, beyond the Fresnel range (ca. 25 m, here), no oscillations are possible, and the pressure amplitude decreases monotonously; the spreading beam pattern is fully formed. However, the average intensity along the axis (full line) is not equal to the spherical spread (dotted line) until it reaches three or four times the Fresnel distance.

The field oscillations with distance along the axis are possible as long as the acoustic path difference is at least equal to half a wavelength; that is, for distances satisfying:

$$\frac{L^2}{8x} \geq \frac{\lambda}{2} \quad \Leftrightarrow \quad x \leq \frac{L^2}{4\lambda} \tag{5.13}$$

This limit is called the *Fresnel distance*:

$$D_F = \frac{L^2}{4\lambda} \tag{5.14}$$

Beyond the Fresnel distance, the field cannot oscillate any more. But the radiating regime of the antenna in the far field (with the pressure amplitude decreasing with inverse distance) is not yet reached, the distance at which this condition is fulfilled can be three or four times larger than the Fresnel distance (Figure 5.10). Geometrically, it may be considered that the far field begins when the width of the ideal diverging beam equals the transverse dimension of the transducer[5] (Figure 5.10).

---

[5] It is easily shown that this corresponds to four times the Fresnel distance, if λ/L is the beam angular aperture; the limit range for the far field condition is then L^2/λ.

These notions of near field and far field are exactly the same for a receiving transducer.

5.2.5   Maximum transmission level – cavitation

The use of transducers at very high powers is limited by two types of reason. The first is technological: too large a voltage on a transducer leads to a non-linear response of the materials, followed by degradation (breaks, etc.) due to excessive mechanical constraints, and finally to dielectric rupture of the active material used.

The second type of limit comes from the propagation medium itself. This is the *cavitation* process, due to vaporisation of water close to the active face of the projector, linked to the local pressure lows imposed by the acoustic wave. Because gas bubbles appear in front of the transducer, the electro-acoustic efficiency is limited, and can even decrease. This effect does appear when the acoustic pressure on the projection wall is larger or equal to a threshold value P_{cav} equal to the hydrostatic pressure; that is, when $p_{cav} = p_{atm} + 10^4 z$ (in Pascals, where p_{atm} is the atmospheric pressure and z is the depth in metres). This *cavitation threshold* can be related to the transmission level at 1 m from the transducer (a parameter that is practically attainable), by considering that the corresponding acoustic power is:

$$P_{cav} = S \frac{|p_{cav}|^2}{2\rho c} \qquad (5.15)$$

where S is the transmitting surface. This gives the transmission level, using:

$$p_{1m}^2 = \frac{\rho c}{4\pi} P_{cav} G_d = \frac{S}{8\pi} |p_{cav}|^2 G_d \qquad (5.16)$$

where G_d is the directivity gain. The final level, expressed in dB re 1 µPa/1 m, is:

$$SL_{cav} = 186 + 10 \log S + DI + 20 \log(10 + z) \qquad (5.17)$$

This result for the cavitation threshold value assumes that the low-frequency signals last for a long time, sufficient for the cavitation to settle. It is increased for high frequencies and short pulses (typically above 10 kHz and below 10 m/s, according to results given in Urick, 1983).

5.2.6   Underwater acoustic transducer measurements

The usual measurements performed by acoustic transducers (Robber, 1988) are electrical impedance, electro-acoustical sensitivity, frequency response and directivity pattern. In the last three cases, an auxiliary reference transducer is needed, aimed at receiving the emitted signal (when measuring a projector) or at emitting it (when measuring a receiver). According to the characteristic measured, either the signal frequency band is scanned (for a frequency response), or the measured transducer angular sector (for a directivity pattern).

Measurement of underwater acoustic transducer characteristics is a much more difficult matter than its aerial counterpart. A first family of difficulties is obviously linked to the problems associated with the use of electrical devices in an underwater environment. Specific instrumentation has to be used (and often specially developed) for this purpose: other than the transducers themselves, cables and connectors have to be carefully designed to withstand immersion in sea water, the waterproofing difficulty increasing with targeted hydrostatic pressure. Even the measurement devices themselves (amplifiers, recorders, analysers) have in some cases to be packaged inside waterproof containers. The technological solutions to these issues are not really difficult to develop, but they involve extra costs, and, therefore, they call for extreme care from the people in charge of such operations.

A second category of difficulties is related to the environment in which measurements are made, which requires the development of devoted facilities. An underwater acoustic laboratory or factory working in the field of transducers has to be equipped with a water-filled measurement tank, whose dimensions are related to the size and wavelength of the transducers that are to be measured. On one hand, to be reliable, level or directivity measurements have to be performed beyond the Fresnel distance of the transducer; in many cases, this is a very demanding condition, since this distance may be several metres or tens of metres away. This issue arises especially for high-frequency antennas with narrow directivity. On the other hand, a measurement tank must be designed so that the emitted signal used for measurement can be received directly without interference from the multipaths reflected from the walls. In aerial acoustics, this is obtained by working inside anechoic rooms, the walls of which are covered with absorbent material; unfortunately this solution is hardly transposable to underwater acoustics, because no absorbent material is efficient enough for the usual wavelengths of sonar signals. The usual way to combat the problem of a confined field is to work in the time domain with signals as short as possible, and to record/measure the direct signal before the wall-reflected signals arrive. But at a given frequency, a signal needs to last for a certain period in order to get a steady state; this makes clear the limitation of this method, which is evidently better adapted to higher frequency signals. Other concerns associated with in-tank trials are the background noise level (acoustic and electric), mechanical vibrations and the existence of air microbubbles, the presence of which may ruin the quality of measurements.

While practical solutions are currently available for small-size transducers, above say 10 kHz, in tanks of reasonable size, for low-frequency transducers (thousands of Hertz and below), accurate laboratory measurements are a real challenge, despite the emergence of methods aiming at "deconfining" the measured acoustic field. For the lowest frequency transducers (hundreds of Hertz), the best (and only?) solution may be to perform the measurements directly at sea, in sufficiently deep water; however, this implies the development of specialised instrumentation.

Note finally that direct calibration measurements of sonar transducers once installed onboard ships are almost impossible to conduct in sufficiently precisely controlled conditions; usual solutions consist in checking transducer performance indirectly by electrical measurements (impedance, noise level) or by analysis of the global performance of the sonar system.

5.3   TRANSDUCER AND ARRAY DIRECTIVITY

The directivity of a transducer expresses, in transmission, the angular distribution of the acoustic energy transmitted and, on reception, the electric response as a function of the direction of arrival of the acoustic wave. The directivity function, describing these spatial variations, depends on frequency as well as the shape and size of the transducer. The antenna is more directive as its dimensions grow larger compared with the wavelength. These important notions will be detailed in this section.

5.3.1   Notion of directivity

5.3.1.1   Directivity pattern

Transmission directivity allows concentration of the energy transmitted into a particular angular sector, thus increasing local acoustic pressure, for a given transmitting power. On reception, the use of the antenna increases the signal-to-noise ratio thanks to the coherence properties of noise (assumed decorrelated between the different receivers of the antenna) and signal (assumed perfectly coherent): combining the output of the transducers favours the signal rather than the noise. The directive antenna also allows selection of the specific directions of arrival of the signal; this can, for example, be useful for getting rid, at least partially, of multiple paths. Finally, the directive antenna allows measurement of the target angular direction.

The directivity pattern of a transducer expresses the energy response in the far field, for a given frequency, as a function of angles and normalised respective to its maximum. It expresses the integral, over the surface Σ of the transducer (Figure 5.11) of the contributions from elementary transmitters $d\Sigma(M)$ distributed over its surface and assimilated to spherical sources:

$$D(\theta, \phi) = R_0^2 \left| \int\int_\Sigma \frac{\exp(-jkR(M))}{R(M)} d\Sigma(M) \right|^2$$

$$\approx \left| \int\int_\Sigma \exp(-jkR(M)) \, d\Sigma(M) \right|^2 \tag{5.18}$$

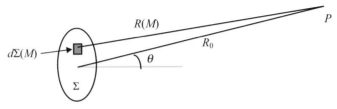

Figure 5.11. Geometry used for calculating the directivity pattern of a transducer. See text for details.

where $R(M)$ is the distance from point M (on the transducer's surface) to the point of observation P. R_0 is the average distance from the transducer to the observation point, assumed far enough and located in space with its angular coordinates (θ, ϕ).

The directivity pattern is normalised respective to its maximum, often corresponding to $(\theta, \phi) = (0, 0)$. By convention, it is therefore equal to unity in this direction.

The exact shape of the directivity pattern depends therefore on the antenna geometry and on the frequency. The simplest shapes (line, disc, rectangle: see Table 5.1 on p. 157) correspond to exact theoretical functions. Some analytical developments are available in Appendix A.5.4.

5.3.1.2   Directivity characteristics

The directivity pattern (Figure 5.12) generally shows a *main lobe*, characterised by its aperture at $-3\,\mathrm{dB}$ on each side of its maximum (called the "$2\theta_3$" value). This is the width of the corresponding beam at $D(0, \phi) = 0.5$; it expresses the *angular resolution* of the antenna. This width is a function of the ratio between the wavelength and the antenna direction. Around this main lobe, there are *sidelobes*, which are always undesirable. Their attenuation relative to the main lobe is one of the main quality

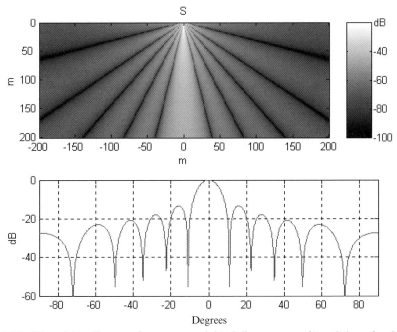

Figure 5.12. Directivity diagram for a non-weighted linear array ($L = 0.4\,\mathrm{m}$; $f = 20\,\mathrm{kHz}$). (*Top*) Cross-section of the radiated field, grey levels coding the acoustical intensity; the source is in $(0, 0)$, the axis scales are in metres. (*Bottom*) Directivity pattern as a function of radiation angle. Note that the levels of the sidelobes are quite high ($-13.5\,\mathrm{dB}$ for the first one).

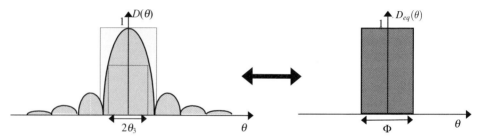

Figure 5.13. Directivity pattern $D(\theta)$, with beamwidth $(2\theta_3)$ at $-3\,\text{dB}$, and equivalent pattern $D_{eq}(\theta)$ and aperture Φ.

factors of the antenna. The level of the sidelobes depends on the antenna geometry, and weighting techniques can be applied during processing (Section 5.3.5.3).

The *directivity index* (or gain) DI expresses the "spatial gain in energy" obtained by a directive antenna (calculated by integrating the directivity pattern in the entire space), relative to the same antenna with no directivity (for which this integral equals 4π). It can be expressed in dB as:

$$DI = 10\log\frac{4\pi}{\iint D(\theta,\phi)\cos\theta\,d\theta\,d\phi} \tag{5.19}$$

At reception, the directivity gain expresses the decrease in noise perceived by the receiver, due to the directivity of the antenna (assuming the ambient noise is isotropic). At transmission, it expresses the increase in acoustic energy along the main direction.

5.3.1.3   Equivalent aperture

The *equivalent aperture* of a beam is the aperture of the ideal directivity pattern (such that $D_{eq}(\theta,\phi) = 1$ in the main lobe and 0 elsewhere, see Figure 5.13), integrating the same amount of energy as the real pattern. It can be defined in a plane (equivalent angular aperture Φ):

$$\Phi = \int_{-\pi}^{+\pi} D(\theta)\,d\theta \tag{5.20}$$

It can also be defined in space (*equivalent solid angle* Ψ):

$$\Psi = \iint D(\theta,\phi)\cos\theta\,d\theta\,d\phi \tag{5.21}$$

Note that, using this definition, the directivity index reads $DI = 10\log(4\pi/\Psi)$.

5.3.2   Theoretical results for simple geometry antennas

5.3.2.1   Plane arrays

The exact characteristics of antenna directivity patterns can be derived theoretically for some simple geometries. Appendix A.5.4 shows the details of the calculation for a

Table 5.1. Directivity characteristics for simple antenna types, in transmission *or* on reception (upper table) or in transmission *and* on reception (lower table). The angle θ is relative to the normal to the antenna. The expressions given are approximations valid for small angular apertures. The rectangular antenna is to be considered as the combination of two linear antennas; hence, the angular characteristics in the directions of the sides a and b are identical to those of linear antennas of respective lengths a and b.

Transmission *or* reception (one-way configuration)	Linear, length $L > \lambda$	Disc, diameter $D > \lambda$	Rectangle, sides $a, b > \lambda$
Directivity pattern $D(\theta)$	$(\sin A/A)^2$ $A = (\pi L/\lambda)\sin\theta$	$(2J_1(A)/A)^2$ $A = (\pi D/\lambda)\sin\theta$	
Directivity index DI (dB)	$10\log(2L/\lambda)$	$20\log(\pi D/\lambda)$	$10\log(4\pi ab/\lambda^2)$
Main lobe width $2\theta_3$ ($°$)	$50.8\lambda/L$	$58.9\lambda/D$	
First sidelobe level (dB)	-13.3	-17.7	
Equivanent aperture Φ (rad)	λ/L	$1.08\lambda/D$	
Equivalent solid angle Ψ (sr)	$2\pi\lambda/L$	$(4/\pi)(\lambda/D)^2$	$\lambda^2/(ab)$

Transmission *and* reception (two-way configuration)	Linear, length $L > \lambda$	Disc, diameter $D > \lambda$	Rectangle, sides $a, b > \lambda$
Directivity pattern $D(\theta)$	$(\sin A/A)^4$ $A = (\pi L/\lambda)\sin\theta$	$(2J_1(A)/A)^4$ $A = (\pi D/\lambda)\sin\theta$	
Main lobe width $2\theta_3$ ($°$)	$36.6\lambda/L$	$42.3\lambda/D$	
First sidelobe level (dB)	-26.5	-35.4	
Equivalent aperture Φ (rad)	$(2/3)(\lambda/L)$	$0.77\lambda/D$	
Equivalent solid angle Ψ (sr)	$(4\pi/3)(\lambda/L)$	$(1.84/\pi)(\lambda/D)^2$	$(4/9)\lambda^2/(ab)$

linear antenna. Table 5.1 presents the formulas for the three most common types of antenna: linear, disc-shaped and rectangular.

5.3.2.2 Curved arrays

The evident advantage of plane arrays is their simplicity of design and realisation. But they have a major defect: increasing their dimensions (e.g., to increase the power transmitted) leads to a narrowing of the directivity lobe, at a given frequency. To remedy this and get wide directivity diagrams, one uses curved arrays. Their directivity diagram is imposed by their geometry rather than by the dimension/wavelength ratio. The most common geometry is cylindrical (or part cylindrical, Figure 5.14).

It can be shown (cf. Appendix A.5.5) that the field radiated in a given direction by a circular antenna is mostly radiated by an angular sector of the circumference, corresponding to the first Fresnel zone. The angular extension of this Fresnel zone

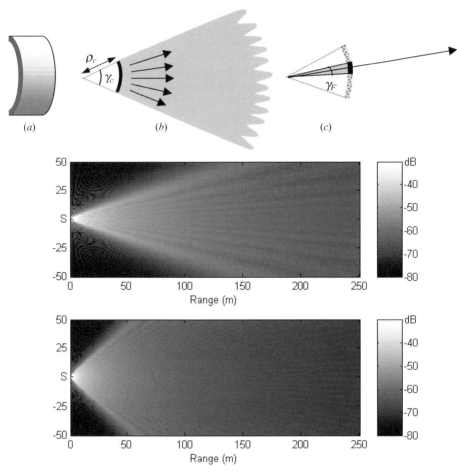

Figure 5.14. (*Upper part*) Radiating geometry of a cylindrical antenna (*a*). The main lobe width (*b*) is imposed by the emitting sector aperture γ_c. However, in a given direction (*c*) the radiated field mainly comes from a limited Fresnel sector γ_F. (*Lower part*) Field radiated by a cylindrical antenna of aperture 45° (*top*) and 90° (*bottom*), with a radius of curvature of 1 m and a frequency of 100 kHz.

reads:

$$\gamma_F = 2\sqrt{\frac{\lambda}{\rho_c}} \tag{5.22}$$

If γ_F is small relative to the angular extension γ_c of the entire antenna, the directivity pattern is very close to the geometric aperture of the physical antenna. Inside the main lobe, the acoustic field fluctuates with angles around its main value (by 1 dB to 2 dB, typically).

Figure 5.14 shows the field radiated by a cylindrical antenna with a geometric aperture of 45°, and then 90°, at a frequency of 100 kHz. The directivity pattern clearly depends on the geometric aperture. One can also see the modulation of the main lobe by the oscillations of the field, due to the integration of contributions from all points on the antenna.

5.3.3 Combining directivity patterns

5.3.3.1 Discrete arrays

In the case where several transducers are combined into an array, the resulting directivity is the combination of two components:

- the intrinsic directivity pattern D_{trans} of the elementary transducers;
- the directivity pattern D_{array} obtained by considering an array of point-like transducers (with no individual directivity).

It can easily be shown (and it is intuitive) that the resulting directivity pattern expresses the mathematical product of the two directivity functions:

$$D_{res}(\theta, \varphi) = D_{trans}(\theta, \varphi) D_{array}(\theta, \varphi) \tag{5.23}$$

The elementary transducers are much less directional than the total array. Their directivity will therefore have little effect on the resulting pattern, as long as angles close to their main lobe axis are considered. In this case, it may be enough only to consider $D_{res}(\theta, \varphi) \approx D_{array}(\theta, \varphi)$, for example, to compute the width of the main lobe. On the other hand, the elementary transducer directivity pattern may influence the angular behaviour of the array at oblique incidences. It then needs to be taken into account if the resulting directivity pattern must be accurately described in the whole angular range. An application example of directivity pattern combination is given in Figure 5.22

5.3.3.2 Combined transmission and reception

In the case of sonar systems for which the directivity patterns have to be considered for both transmission and reception antennas (noted, respectively, in the following as Tx and Rx), the resulting function is the product of the two one-way patterns:

$$D_{res}(\theta, \varphi) = D_{Tx}(\theta, \varphi) D_{Rx}(\theta, \varphi) \tag{5.24}$$

A common configuration is when the same transducer is used for transmission and for reception. In this case, since $D_{Tx}(\theta, \varphi) = D_{Rx}(\theta, \varphi)$:

$$D_{res}(\theta, \varphi) = D_{Tx}^2(\theta, \varphi) \tag{5.25}$$

If the axes of the two main lobes coincide, the 3-dB width of the resulting main lobe may be approximated using the values of the individual one-way lobe aperture:

$$\theta_{res} \approx \frac{\theta_{Tx}\theta_{Rx}}{\sqrt{\theta_{Tx}^2 + \theta_{Rx}^2}} \tag{5.26}$$

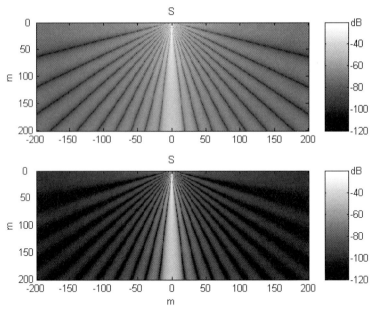

Figure 5.15. Directivity diagram for a linear array ($L = 0.3\,\mathrm{m}$; $f = 50\,\mathrm{kHz}$). Cross-cut of the radiated field, grey levels coding the acoustical intensity; the source is in $(0, 0)$, the axis scales are in metres. Directivity of a linear array, for transmission only (*top*), and transmission and reception combined (*bottom*). The main lobe narrows (by a factor $1/\sqrt{2}$ for the -3-dB beamwidth), and the sidelobes are lowered (by a factor 2 in dB).

With the same directivity at transmission and at reception, the resulting lobe is then approximately equal to:

$$\theta_{res} \approx \frac{\theta_{Tx}}{\sqrt{2}} \tag{5.27}$$

Note that, in this case, the sidelobe level expressed in dB is only one-half of its one-way value.

An example of the combination of transmission and reception effects on a directivity pattern is shown in Figure 5.15. It makes it clear that the main lobe narrows and the sidelobe decreases.

5.3.4 Directivity of wideband signals

5.3.4.1 *Wideband noise*

The directivity pattern was defined above for a specific frequency, and it is therefore not valid when the signals to process have a significant bandwidth. It is interesting to study what this means for a signal of finite bandwidth $\Delta f = [f_1; f_2]$. We will consider that the array considered is linear, without weighting, with a *sinc* directivity pattern at a fixed frequency, that the signal power spectrum is flat (taken to unity) all over the bandwidth, and that the resulting directivity pattern over the frequency band Δf is given by the integral of its energy frequency components:

$$D_{\Delta f}(\theta) = \frac{1}{\Delta f} \int_{f_1}^{f_2} D(\theta, f) \, df \qquad (5.28)$$

This model particularly suits large-band (random) noise (e.g., noise radiated by ships and detected by passive sonars).

Developing analytically the integral (5.28) for a narrow band Δf, it can be found that the resulting directivity pattern for the main lobe and the first sidelobes is:

$$D_{\Delta f}(\theta) \approx \frac{\sin^2(a_m\theta)}{(a_m\theta)^2} \cos\left(\frac{a_m\theta}{\sqrt{3}}\frac{\Delta f}{f_m}\right) + \frac{1}{12}\left(\frac{\Delta f}{f_m}\right)^2 \qquad (5.29)$$

with $a_m = \pi L/\lambda_m$, where λ_m is the wavelength at the central frequency f_m of the band Δf:

$$f_m = \frac{f_1 + f_2}{2} \qquad (5.30)$$

This expression (5.29) features the classical squared-sinc function (the average directivity pattern at the central frequency), modulated by a cosine term (decreasing the sidelobe level). and an additive term filling the troughs of the sinc oscillations.

More simply, it can be considered that, around the main average lobe given by the central frequency, the preiod of the directivity function oscillations changes with frequency, and the widening of the bandwidth therefore induces a smoothing of the monochromatic pattern. The oscillations will therefore be more attenuated, as the band considered is wider. At oblique incidences and for a wide enough bandwidth, the lobe level can be approximated by the average value of the oscillating squared-sinc function:

$$\left\langle \left(\frac{\sin x}{x}\right)^2 \right\rangle = \frac{1}{2x^2} \qquad (5.31)$$

The wideband directivity function therefore tends at oblique incidences toward:

$$D_{\Delta f}(\theta) \rightarrow \frac{1}{2}\left(\frac{\pi L}{\lambda_m}\sin\theta\right)^{-2} \qquad (5.32)$$

Figure 5.16 shows the directivity functions associated with a linear array for a pure frequency, and for bands of one-third of an octave ($f_2 = 2^{1/3} f_1$) and one octave ($f_2 = 2f_1$). It shows that the asymptotic model (5.32) presented here is a good way of representing these wideband directivity patterns.

5.3.4.2 Short-duration signals

Another common case is that of short-duration active signals, such that the equivalent length of the signal received on the array ($cT/\sin\theta$) is smaller than the array dimension (Figure 5.17). In this case, the array cannot be fully insonified at the same time, and the general array performance is degraded. The output signal is geometrically lengthened (it lasts for $L\sin\theta + T$ instead of T), and the array directivity is

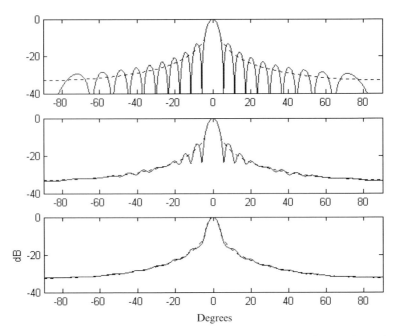

Figure 5.16. Directivity function of a linear array ($L = 10\lambda$) in monochromatic mode (*top*), a frequency band of one-third of an octave (*middle*) and a frequency band of one octave (*bottom*). The dashed lines represent the asymptotic model of wideband directivity.

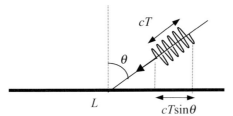

Figure 5.17. Linear array insonification by a short signal. The signal equivalent length $cT \sin \theta$ is shorter than the array length L.

the directivity of an array of length equal to the instantaneous ensonified length $cT/\sin \theta$ instead of L.

5.3.5 Beamforming

5.3.5.1 Principle

For an array made of independent discrete transducers, appropriate phase or time shifts can be imposed on the transducers to steer the main lobe of the array in the direction of choice. This process is called *beamforming*, and is usable as well on transmission and reception. It enables sweeping the space to explore without any

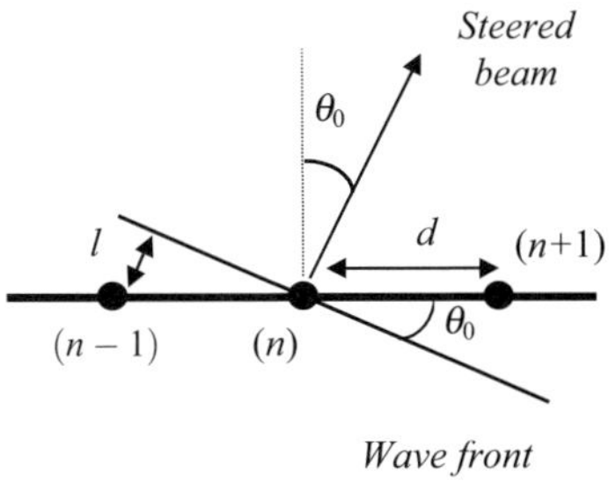

Figure 5.18. Beamforming with a linear array: geometry and notation.

mechanical movement of the array. Let us, for example, consider a linear antenna of length L, made of $N = 2k + 1$ transducers and forming a beam steered in the direction θ_0 (Figure 5.18). If l is the elementary path difference between two transducers distant from d, then:

$$l = d \sin \theta_0 \tag{5.33}$$

The delay at the transducer n, referenced to the centre of the array, equals:

$$\delta t_n = n \frac{l}{c} = n \frac{d}{c} \sin \theta_0 \qquad n = -\frac{N}{2}, \ldots, +\frac{N}{2} \tag{5.34}$$

And the corresponding phase shift reads:

$$\delta \varphi_n = 2\pi f \frac{nl}{c} = 2\pi \frac{nd}{\lambda} \sin \theta_0 \tag{5.35}$$

For all transducers in the array, the signal is a sine wave with unit amplitude along direction θ, and when combining the contributions after the phase shifts $\delta \varphi_n$ (neglecting the individual directivity patterns of the transducers), the resulting signal becomes:

$$S(\theta_0, t) = \sum_{n=-k}^{+k} s_n(t) \exp\left(2j\pi n \frac{d}{\lambda} \sin \theta_0 \right)$$

$$= \sum_{n=-k}^{+k} \exp\left(j\left(\omega t - 2\pi n \frac{d}{\lambda} \sin \theta \right) \right) \exp\left(jn \frac{d}{\lambda} \sin \theta_0 \right) \tag{5.36}$$

After some algebraic manipulations and elimination of the time dependence, the directivity pattern is finally expressed as:

$$D(\theta) = \left| \frac{\sin A}{N \sin(A/N)} \right|^2 \tag{5.37}$$

with:

$$A = \pi \frac{L}{\lambda} (\sin \theta - \sin \theta_0) = \pi \frac{Nd}{\lambda} (\sin \theta - \sin \theta_0)$$

For angles θ close to the steering angle θ_0, one gets the approximate expression:

$$A \approx \pi \frac{L}{\lambda} \cos \theta_0 \sin(\theta - \theta_0) \tag{5.38}$$

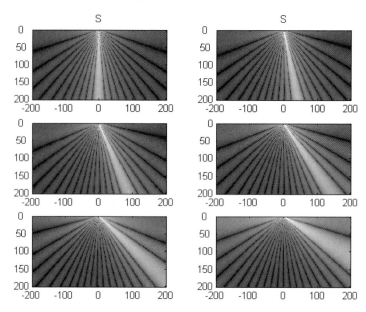

Figure 5.19. Beamforming at angles of 0° and 12° (*upper*), 24° and 36° (*middle*), 48° and 60° (*lower*), for a line array ($L = 10\lambda$, 21 sensors at $\lambda/2$). Note the lobe aperture broadening when the steering angle increases.

Setting $\nu = \theta - \theta_0$, and approximating $N \sin(A/N)$ with A using (5.38), the directivity pattern reads:

$$D(\theta) \approx \left| \frac{\sin A}{A} \right|^2 \approx \left| \frac{\sin \pi \dfrac{L_0}{\lambda} \sin \nu}{\pi \dfrac{L_0}{\lambda} \sin \nu} \right|^2 \qquad (5.39)$$

This function is the same as for a linear array with length $L_0 = L \cos \theta_0$ (i.e., the apparent array length along direction θ_0).

Finally, the directivity pattern of a beam formed with a discrete array is the pattern of a linear antenna steered in the direction θ_0 of the beam, with a length reduced in $\cos \theta_0$. This has important consequences:

- beam steering does not affect the basic shape of the directivity pattern;
- the sidelobe levels, in particular, are unchanged, with the first maximum at -13.3 dB;
- the width of the main lobe is widened by $1/\cos \theta_0$;
- the directivity index does not change, whatever the steering angle. This is strictly true only for linear arrays. For rectangular arrays, the directivity index follows the widening of the main lobe, in $10 \log(\cos \theta_0)$, in dB.

Figure 5.19 shows the field radiated by a linear array with 21 sensors at $\lambda/2$ ($L = 10\lambda$), with beams formed at 0°, 12°, 24°, 36°, 48° and 60°. Note that the width of the main lobe increases with the steering angle, while the effect of the sidelobe remains constant. One should keep in mind that the directivity effect of the elementary transducers has been completely neglected in this particular example.

5.3.5.2 Beamforming for narrowband signals

Beamforming may be considered either in the time or in the phase domains: the signals received by the array transducers must either be delayed or dephased. One approach or the other is to be preferred, depending on the type of input signal considered. For narrowband signals, the phase method may be used with good results. This is the simplest algorithm implementation, in which the signals on the array transducers need to be simply multiplied by complex correction terms with appropriate phase shifts.

An interesting property used in beamforming results from the sum of the outputs $s_i(t)$ of the N transducers, phase-delayed to point in the direction θ_0. Using the above expression for the phase delays of the transducers, the signal resulting from beamforming in the direction θ_0 is:

$$S(\theta_0, t) = \sum_{n=-k}^{+k} s_n(t) \exp\left(+ 2j\pi n \frac{d}{\lambda} \sin \theta_0\right) \qquad (5.40)$$

This expression corresponds to a discrete Fourier transform in the space of steering angles (or more exactly their sines), and the space of transducer positions, expressed relative to the wavelength. It can therefore be performed using Fast-Fourier Transform (FFT) algorithms, made all the more efficient with specialised digital signal processing boards (see, e.g., Nielsen, 1991).

This approach is very efficient numerically, but this advantage is less clear as the computers used with sonars evolve. It also has two major disadvantages:

- it is only suited to monochromatic signals, such as narrowband random noise or active signals that are sufficiently long compared with array length (see Section 5.3.5.4);
- the series of angular directions of the beams is fixed, a function of the number of points of the FFT; this can be a problem when versatile beam-steering is required (e.g., in multibeam sounders).

5.3.5.3 Array shading

By appropriately correcting the sums between transducers (applying weighting laws to amplitudes along the array, bell-shaped to decrease the role of the transducers at the extremities), one can effectively decrease the importance of the sidelobes. Unfortunately, this improvement is compensated by a widening of the main lobe of typically 40% (Figure 5.20). This comes from the fact that the "effective" length of the array is decreased by lowering contribution of the extremities' transducers, and the directivity gain is also degraded.

The weighting laws used in array shading are, of course, the ones classically used with FFTs (see Harris, 1978), as the operations are identical: Hamming, Hanning,

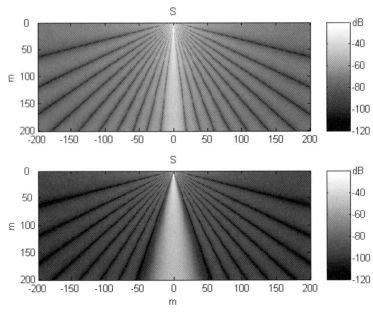

Figure 5.20. Influence of array shading on the directivity pattern, for a linear array ($L = 10\lambda$). (*Top*) No weighting. (*Bottom*) Hamming weighting. Note the decrease in sidelobe level and the widening of the main lobe.

Dolph–Chebychev, etc. The Dolph–Chebychev law is particularly interesting: it allows imposing a given constant level on the sidelobes, and minimising the aperture of the main lobe.

Appendix A.5.6 shows the main weighting laws in common use, with their characteristics. It should, however, be noted that the performance expected from theoretical considerations is often not met in practice. These laws are indeed defined to weigh transducers that are ideally identical. But in practice, the ceramics and the electronic circuits used have differing amplitude and phase sensitivities. This degrades the quality of the final results. With common transducers, the sidelobe rejection levels are at best about -25 dB to -30 dB.

5.3.5.4 *Beamforming for wideband signals*

The processing of signals in the time domain is preferable for wideband signals. Instead of investigating the phase shifts, one writes that the steering along direction θ_0 corresponds to an elementary time delay between transducers, equal to:

$$\delta t = \frac{l}{c} = \frac{d}{c} \sin \theta_0 \tag{5.41}$$

Colour plates

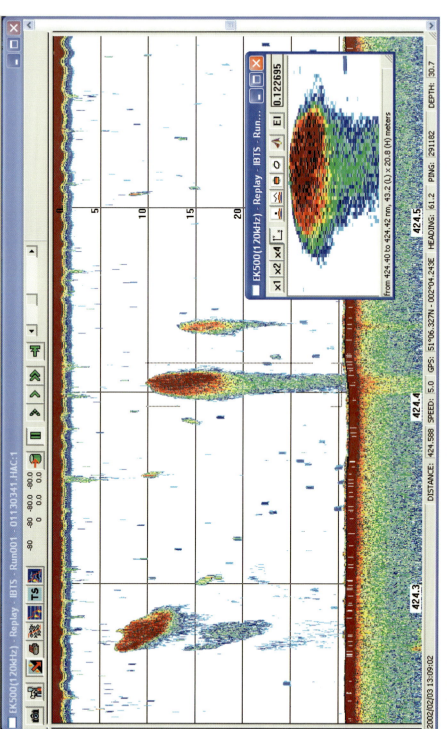

Plate 1. Display of a fisheries echo sounder (Simrad EK500, 120 kHz; processing software movies by IFREMER). The main image gives a 700-m long cross-section of the water column, in a shallow area (around 30-m deep). The first school echo on the left is accompanied by vertical replicas due to multipaths. The school in the middle has been zoomed to the secondary image, and deformed in order to retrieve its real geometrical proportions. (Data courtesy of Noël Diner, IFREMER, France).

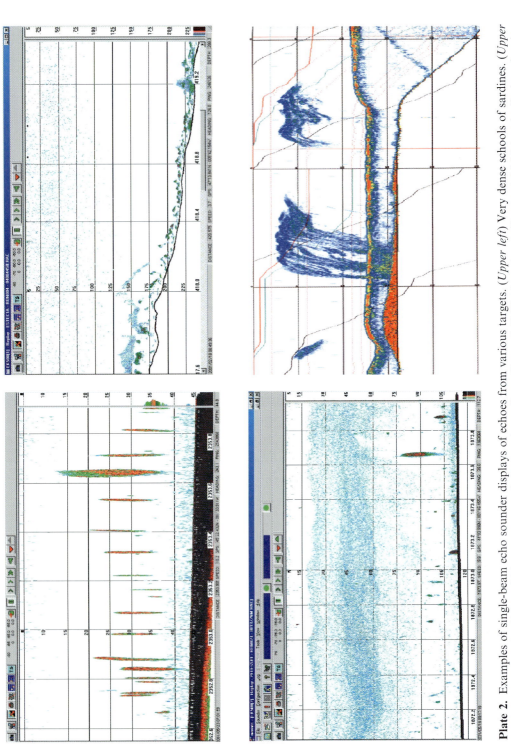

Plate 2. Examples of single-beam echo sounder displays of echoes from various targets. (*Upper left*) Very dense schools of sardines. (*Upper right*) Sparse schools of blue whiting concentrated close to the seafloor. (*Lower left*) Plankton layer. (*Lower right*) Netsonde display, showing the capture of three successive schools, the second one of exceptional size. Data courtesy of Noël Diner, IFREMER, France.

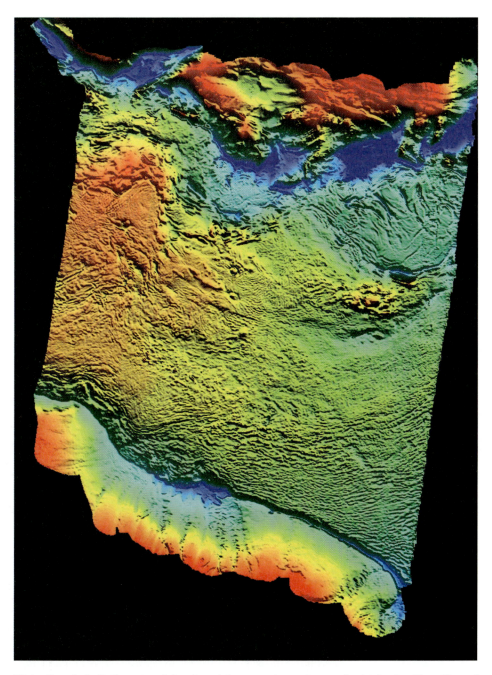

Plates 3 and 4. Bathymetry (*above*) and imagery (*opposite page*) obtained with a Simrad EM12 multibeam sounder in the Eastern Mediterranean. The area is approximately 250-km long. Bathymetry is colour-coded, from red (shallow water) to dark blue (deep troughs, down to 4,000 m); grazing-light shading has been added to enhance the small relief details. The sonar imagery represents high reflectivities as dark grey. Data from the cruise *Prismed 2*, courtesy Jean Mascle, CNRS/INSU Géosciences-Azur, France. Comments on next page.

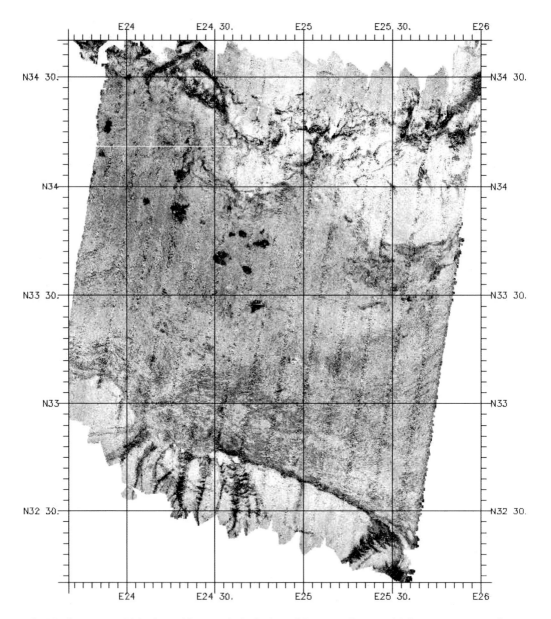

The Mediterranean Ridge is a wide morphological swell between Crete and Libya. It corresponds to a thick sedimentary mass deformed by the subduction of the African plate beneath Eurasia; the convergence rate between the two plates being of the order of 1 cm/year. The southern bathymetric scarp represents the deformation front of the Ridge, which directly overthrusts the Libyan continental margin. To the north, the sediment mass is intensively folded and deformed as a consequence of its general shortening. Near the Crete margin, an area of less deformed plateaux acts as a rigid buttress. North of this area, the Hellenic troughs separate this system from the continental margin off Crete. Several 'mud volcanoes', related to tectonic activity (compression), and delivering fluids and mud flows, can be seen on the central part of the bathymetry data, as more or less circular reliefs. The mud flows appear on the sonar image as large dark patches (see Mascle *et al.*, 1999).

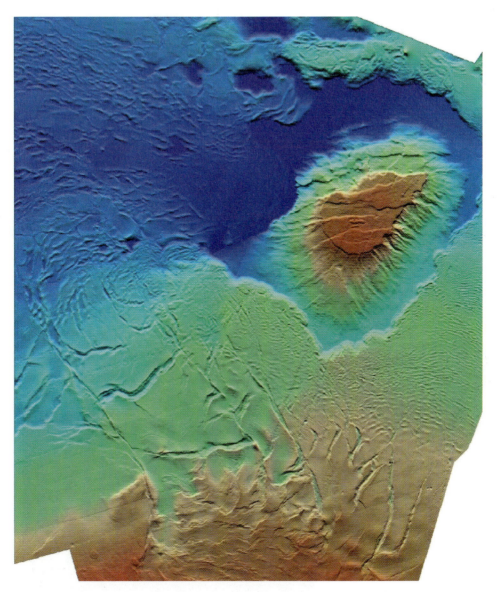

Plate 5. Bathymetric map obtained by compilation of Simrad EM12 and EM300 multibeam echo sounder data. The same bathymetric conventions are used as in Plate 3. The area (approx. 250 × 250 km) is located north-east of Egypt. This part of the deep Nile delta is crossed by a tectonic corridor, whose surface deformation is mainly due to the motion along the continental slope of underlying mobile salt-rich deposits. The relief seen in the north-eastern portion of the map (Mount Eratosthenes) corresponds to a continental fragment, once detached from Africa and presently colliding with the continental margin just south of Cyprus. This collision induces strong fracturation of this rigid block (see Mascle *et al.*, 2000). Data from *Prismed 2* and *Fanil* cruises provided courtesy of Jean Mascle, Olivier Sardou and Lies Loncke, CNRS/INSU Géosciences-Azur, France.

Plate 6. Bathymetry and imagery data obtained with a Simrad EM300 multibeam echo sounder (32 kHz). Located in the Western Mediterranean, this area (approx. 50×50 km) features the underwater canyon of the Var River, south of Nice. The water depth goes from 100 m (red) down to 2,100 m (dark blue). A partial sonar image is given for the area delimited by dashed lines; it represents high reflectivities as dark grey. Note the clear correpondence between the bathymetry features and reflectivity. Data of cruise *Ess300* provided courtesy of Jean-Paul Allenou, Genavir/IFREMER, France.

The resulting signal is therefore:

$$S(\theta_0) = \sum_{n=-k}^{+k} s_n\left(t - n\frac{d}{c}\sin\theta_0\right) \tag{5.42}$$

This corresponds to a summation of delayed time samples from the array transducers.

The most common way to implement this processing implies two steps. First, the signal time samples are shifted according to expected delays. However, the accuracy of this shift is limited by the sampling period of the digital signal, and is usually insufficient. Hence, in a second step, the signals need to be either interpolated or phase-shifted with a shift at the central frequency of the signal.

When arrays need to work over a rather wide frequency band, there comes the problem of optimising the total length and the spacing of transducers. Ideally, the array length would be given by the lowest frequency of the spectrum, and the spacing would be given by the highest frequency. As this led to sometimes unpractical configurations, the array is often irregularly sampled. The total dimension remains determined by low frequencies, but as it is over-dimensioned for high frequencies it can be finely sampled only in some parts.

5.3.5.5   Spatial sampling – grating lobes

When an array is made of separate transducers, a correct spacing between them is paramount. It must indeed be small enough to ensure the correct spatial sampling of the field. As a rule, the spacing must not be larger than $\lambda/2$, and this value is generally used in actual arrays. It is also taken as the conventional value considered by default when designing or analysing an underwater acoustic system.

If the spacing between the transducers is smaller than $\lambda/2$, each angle of the acoustic wave is related to a unique configuration of the complex field on the array transducers, excluding any angular ambiguity. But if the spacing between the transducers is larger than this limit value, the configuration is no longer unique. Different propagation angles can correspond to identical structures of the field on the transducers. Depending on the angles of the signal, there can occur a phenomenon akin to spectrum-folding in the digital Fourier transform: the directivity pattern is duplicated on itself, and parasite copies of the main lobe appear (Figure 5.21). They are called *grating lobes*.

The maxima of this series of lobes appear when the signals are in phase at the transducers:

$$2\pi\frac{d}{\lambda}\sin\theta_n = 2n\pi \tag{5.43}$$

where d is the transducer spacing. The corresponding angles θ_n are therefore given by:

$$\sin\theta_n = n\frac{\lambda}{d} \qquad n = 1, 2, 3, \ldots \tag{5.44}$$

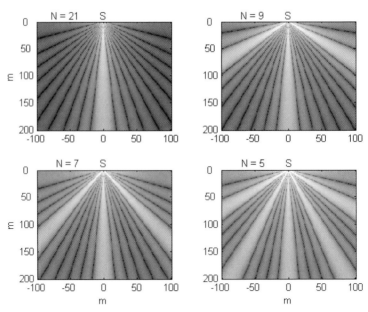

Figure 5.21. Effect of spatial undersampling, for an array of length $L = 10\lambda$, with a number N of sensors, respectively, 21, 9, 7, 5. Note the series of grating lobes, at the same intensity level as the main lobe.

In some cases, because of the specific angular structure of the field, the grating lobes are not problematic, and an undersampled array is sufficient. This option is, however, not recommended, and, generally, one will aim at sampling correctly the length of the array.

Another way to fight the presence of grating lobes is to take advantage of the directivity pattern of elementary transducers. Getting the individual directivity nulls in the directions of the grating lobes yields an acceptable directivity pattern, even for an undersampled array (Figure 5.22).

With active sonars, an alternative consists in using antennas with different gaps, at transmission and at reception. The elimination of the grating lobes is then done by combining the transmission and reception directivity functions.

5.3.5.6 Cylindrical array

Although beamforming is usually applied to linear arrays, the same processing can be applied to curved arrays. The case of a cylindrical array is very common and particularly interesting. To form a beam along a direction θ_0, one uses a part of the array, intersecting an angle sector γ_M and oriented in the direction desired (Figure 5.23). The different points of this arc circle are located with their angle γ (in the interval $[\theta_0 - \gamma_M/2; \theta_0 + \gamma_M/2]$). They are phase-shifted (or delayed, depending on

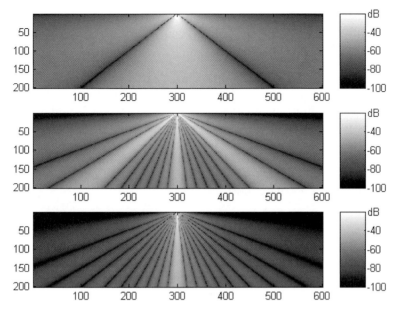

Figure 5.22. Combination of directivity patterns: elementary transducers (*top*), array (*middle*) and resultant (*bottom*). In the latter case, the grating lobes of the undersampled array are compensated by the zeroes of the elementary transducer pattern. This is only true if the elements are touching. The suppression of the image lobes by the directivity of the elementary transducers is less efficient if the array is electronically steered.

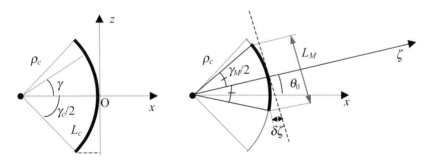

Figure 5.23. Beamforming using a cylindrical array.

the type of beamforming used), as a function of the difference in acoustic path calculated for a target assumed very far away:

$$\delta\zeta(\theta_0, \gamma) \approx \rho_c(1 - \cos(\theta_0 - \gamma)) \approx \rho_c \frac{(\theta_0 - \gamma)^2}{2} \qquad (5.45)$$

Once the acoustic path difference due to array curvature is compensated, we are back
at the configuration of a beam formed perpendicularly to a linear array of length
equal to the chord of the arc circle γ_M:

$$L_M = 2\rho_c \tan \frac{\gamma_M}{2} \qquad (5.46)$$

For another steering direction, another portion of the antenna would be considered.
This geometry is interesting because it allows formation of perfectly identical beams
for any steering angle, as long as it is between $-(\gamma_c - \gamma_M)/2$ and $+(\gamma_c - \gamma_M)/2$.
Beyond these limits, increasingly inclined beams can be formed, but with steering
relative to the axis of symmetry of the array sector used, and therefore degraded
characteristics of the directivity lobes.

5.3.6   Array focusing

When the array dimension is such that the near field condition is met (see
Section 5.2.4), the acoustic waves cannot be assumed to be plane any longer, and
beamforming is not optimal. One must then focus the array, by phase-shifting (or
delaying) the elements, to compensate for the differences in phase (or time) of a
spherical wave centred on the target. For a linear array focused on a point at
distance x_0 on the transverse axis, the acoustic path difference between the centre
of the array and a point on the abscissa y (Figure 5.24) reads:

$$\delta R = \sqrt{x_0^2 + y^2} - x_0 \approx \frac{y^2}{2x_0} \qquad (5.47)$$

The phase shift to consider along the array, to compensate signal propagation, is
therefore:

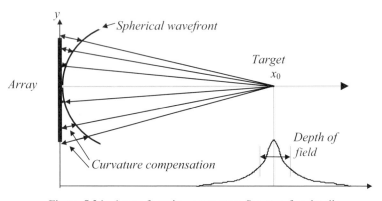

Figure 5.24. Array-focusing geometry. See text for details.

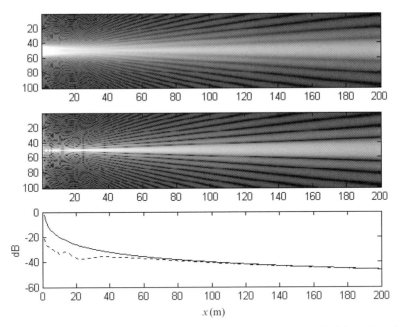

Figure 5.25. Effect of dynamic focusing for a linear array. (*Top*) Standard beamforming; the beam is formed only beyond the Fresnel range (here, around 37 m). (*Middle*) Dynamic focusing; the resolution is much sharper for short ranges, while the processing has little effect beyond the Fresnel range. (*Bottom*) Signal level along the main lobe axis, for standard beamforming (dotted) and dynamic focusing (solid).

$$\varphi(y) = k\, \delta R = \frac{ky^2}{2x_0} \tag{5.48}$$

And the corresponding time delay is:

$$\tau(y) = \frac{\delta R}{c} = \frac{y^2}{2cx_0} \tag{5.49}$$

Focusing is, of course, only valid locally, around the focal distance x_0 targeted. The amplitude of the resulting signal decreases on each side. The domain of validity of focusing (field depth) is given by the spacing between the points at $-3\,\mathrm{dB}$ around the nominal focus, and depends on the measurement geometry and the wavelength of the signal. For example, for a linear array of length L, focused at distance x_0, it can be shown that the field depth equals ca. $7\lambda(x_0/L)^2$.

In practice, for a sonar system using this kind of processing at reception, this leads to using a structure of *dynamic focusing*, the position of the focal point being made to move with the distance to the expected target. Figure 5.25 shows the result of a dynamic focusing processing.

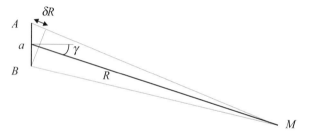

Figure 5.26. Geometry and notation for interferometric angle measurements.

5.3.7 Interferometry

A target is often spatially located by measuring the phase difference between two receivers. This fine measurement of the difference in acoustic path allows assessment of the angular direction very accurately.

Let us consider the configuration of Figure 5.26; the phase difference $\delta\varphi_{AB}$ between A and B is given by the difference in the acoustic paths issuing from the target point M:

$$\delta\varphi_{AB} = k\,\delta R = ka\sin\gamma \tag{5.50}$$

Measuring $\delta\varphi_{AB}$ yields the direction γ of the target, using:

$$\gamma = \arcsin\left(\frac{\delta\varphi_{AB}}{ka}\right) \tag{5.51}$$

This type of measurement is often used to measure bathymetry with a sidescan sonar or a multibeam sounder (see Chapter 8 for details).

The defect with this technique is that the direction is defined with some ambiguities, as the phase is measured modulo 2π, as soon as the spacing a is larger than $\lambda/2$. The phase difference ambiguities are more numerous and denser as the elements of the interferometer are further apart (Figure 5.27). This can be corrected using different techniques: several receivers can be used with different spacings; common solutions can be sought to remove the ambiguity (*vernier* method); or one can "unwrap" the phase continuously from a non-ambiguous value.

5.3.8 Synthetic aperture sonar

The most limiting performance of classical beamforming is its poor resolution parallel to the length of the array. The beam aperture being fixed (typically to λ/L), its spatial resolution is then proportional to the array target range D:

$$\delta y = D\frac{\lambda}{L} \tag{5.52}$$

This raises two problems: first, the spatial resolution is not even all along the operation range. Second, it may become unacceptable beyond a sufficient distance. Narrowing the beam pattern by increasing the array length is not a satisfying way to answer both problems.

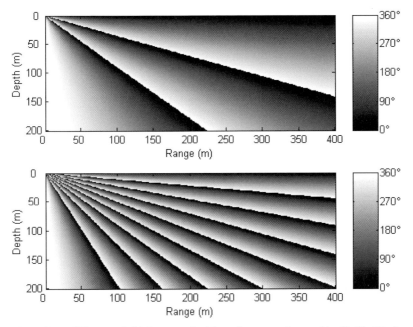

Figure 5.27. Phase difference field for a vertical interferometer located in $(0, 0)$. (*Top*) $a = 3\lambda$. (*Bottom*) $a = 9\lambda$.

In the case of an array supported by a moving platform, synthetic aperture sonars (SAS) provide a better way of addressing these limitations. This technique is widely used today in satellite radar imagery (SAR, for Synthetic Aperture Radar; see Oliver and Quegan, 1998), in which the array target ranges are enormous and preclude the use of classical beamforming. The basic principle is to record the signals received on a short ("physical") array as it proceeds along the track. The signals can then be combined to create an artificial ("synthetic") array independent of the physical length, only by post-processing the signals recorded.

In practice, the target has to be illuminated while the physical array (of length L_{Tx}) passes it by, radiating inside its physical aperture θ_{Tx} (Figure 5.28):

$$\theta_{Tx} = \frac{\lambda}{L_{Tx}} \tag{5.53}$$

The synthetic array thus needs to have a limited length, smaller than $L_{\max SAS} = D\theta_{Tx}$, where D is the distance across-track. This length imposes the ultimate minimum synthetic aperture $\theta_{\min SAS}$:

$$\theta_{\min SAS} = \frac{\lambda}{2L_{\max SAS}} \tag{5.54}$$

The factor 2 is because the process applies both in transmission and reception. In

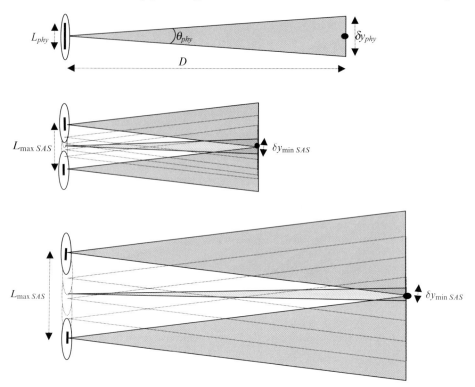

Figure 5.28. Synthetic aperture array principle. (*Top*) Classical array processing: spatial resolution is obtained by using a long physical array. However, for a fixed length physical array, the spatial resolution depends upon the range x across-track. (*Middle* and *bottom*) Recording the array signals at various instants allows making a long artificial array, with its length limited by the two extreme positions shaded in grey. The resolution obtained is independent of the range across-track: the farther the target, the longer the synthetic array length. The narrow central beam illustrates the synthetic array aperture (actually it is meaningful only close to the target, since the SAS length is defined for range D).

turn, it yields the best resolution achievable along-track $\delta y_{\min SAS}$:

$$\delta y_{\min SAS} = D\theta_{\min SAS} = D\frac{\lambda}{2L_{\max SAS}} = \frac{L_{Tx}}{2} \tag{5.55}$$

This leads to a very important property of SAS processing: along-track resolution does not depend on the range across-track but instead on the physical length of the transmit array. Furthermore, and this is quite counter-intuitive for people trained to think with classical beamforming, the shorter the physical array, the better the resolution!

However, in practice things are not so easy; and, although proposed long ago (Cutrona, 1975, 1977), SAS has long been considered a purely theoretical concept. Several constraints make SAS practically more difficult than SAR:

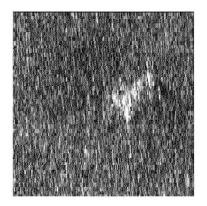 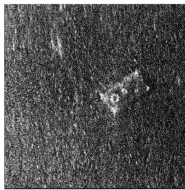

Figure 5.29. Example of SAS processing on a rectangular target of dimensions $2\,\text{m} \times 1\,\text{m}$. (*Left*) Sonar image collected by the physical array aperture. (*Right*) Sonar image after synthetic array formation and autofocusing (data provided courtesy of DCE-GESMA, France).

- the random movements of a ship or even a towfish around the average track are much more penalising than those suffered by a satellite;
- the positioning accuracy of the array transducers has to be better than typically $\lambda/8$;
- the underwater environment perturbs the propagation of acoustic waves far more than space or planetary atmospheres can perturb radar waves; the speed of sound is also much smaller than the speed of light.

Recent technological progress in attitude sensors has made it possible now to navigate the arrays with sufficient accuracy. Moreover, post-processing methods (known as autofocusing) have also allowed correcting the fine details of the array's movements, using the correlation of acoustic signals backscattered by the seafloor.

Another serious constraint is brought by the need to sample the synthetic array correctly, to avoid any gap when building it. Failure to do so would result in grating lobes in the synthetic array directivity. This implies that in the time lag between the echoes from two consecutive pings, the receiving array does not move by more than half its own length L_{Rx} (see Bruce, 1992). If $D_{\max}$ is the maximum range across-track, this imposes the following trade-off:

$$\frac{2D_{\max}}{c} = \frac{L_{Rx}}{2V} \tag{5.56}$$

where V is the speed of the sonar along-track. For a given array length, a large swathe width will be achievable if the sonar is slowed in proportion. The hourly coverage will thus remain unchanged, and it will only be limited according to the relation:

$$4VD_{\max} = cL_{Rx} \quad (\text{in m}^2/\text{s}) \tag{5.57}$$

While this limitation is not really penalising for spaceborne radars, for which satisfactory compromises can be reached, it is a serious restriction for sonars, because of the slow velocity of acoustic waves. Therefore, an SAS system with a fine alongtrack resolution and an acceptable hourly coverage should necessarily feature a long receiving array.

Figure 5.30. (*Left*) Tonpilz transducer, operating at a frequency of 25 kHz (diameter 5 cm, length 6 cm). (*Right*) Janus driver, with a response frequency of 1 kHz (diameter 40 cm, length 55 cm). Images courtesy of Yves Le Gall, IFREMER.

Figure 5.31. A sediment profiler array, installed onboard R/V *Le Suroît* (IFREMER, France), made of seven Tonpilz transducers (individual diameter 30 cm; frequency band [1.5–8 kHz]). The array is normally protected by an acoustically transparent window. Images courtesy of Yves Le Gall, IFREMER.

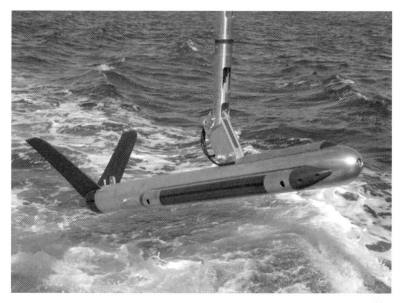

Figure 5.32. A sidescan sonar towfish being deployed at sea. This Klein-3000 sidescan sonar uses long dual-frequency arrays (130 and 445 kHz), positioned on each side of the towfish (1.2-m long). Image courtesy of Klein Associates Inc.

Figure 5.33. The arrays of a multibeam echo sounder (in this case a Simrad EM-300, nominal frequency 32 kHz, installed onboard R/V *Le Suroît*). The plane rectangular arrays are installed inside a profiled fairing. Their approximate lengths are, respectively, 1.6 and 3.0 m for reception (athwartship) and transmission (alongship) arrays. Image courtesy of Yves Le Gall, IFREMER.

6

Signal processing – principles and performance

6.1 INTRODUCTION

Underwater acoustic systems use a restricted variety of signals, chosen for their capacity to carry the information sought by the end-user in specific applications (e.g., detection, localisation, measurement, characterisation, communication). In this way, they are not much different from electromagnetic systems with the same functions, which work in the atmosphere or in space. The differences result from the physics of the surrounding underwater environment (propagation, noise, types of transducer).

There are two main aspects to the good functioning of an underwater acoustic system:

- the definition and use of signals well suited to the objective, and to the environmental conditions known a priori;
- the use in the reception chain of processing techniques combining the best performance achievable (once again taking into account the intrinsic characteristics of the signal and perturbations from the environment), and a level of complexity and cost compatible with the objectives of the system.

Active acoustic systems use controlled signals, whose characteristics are imposed during the transmission phase (duration, frequency content, level, ...). Their reception chain is therefore based on the suitable filtering of these characteristics, possibly accounting for the perturbations brought by the environment and the target.

Conversely, passive sonar systems cannot control the characteristics of the signals they need to analyse. Their processing chains are thus based on a priori assumptions on the signals received (broadband, narrowband, pulsed, ...).

The present chapter does not aim at presenting the fundamental theory of signal processing in its generality. Technological details of transmission and reception chain design will not be presented here either, as they are based on rapidly evolving digital technologies. We will instead present the basic principles of sonar signal processing, for the most common applications in underwater acoustics. This will be complemented with practical results directly usable for dimensional system analysis, the evaluation of system performance and their interpretation. The reader interested in more theory is invited to read any of the many books on signal processing in general (e.g. Bendat and Piersol, 1971; Papoulis, 1977; Van Trees, 1968; Proakis and Manolakis, 1988) or specially orientated at sonar (Nielsen, 1991). Two excellent syntheses about sonar signal processing are to be found in Winder (1975) and Knight *et al.* (1981). About principles and theories applicable to sonar, the book *Underwater Acoustic System Analysis*, written by Burdic (1984a, b), is an excellent textbook on the matter, with many complete and thorough theoretical developments.

6.2   PRELIMINARY NOTIONS

6.2.1   Signal design

6.2.1.1   *Frequency spectrum*

Any time domain signal of finite energy can be decomposed uniquely into a sum of an infinity of elementary sinusoidal signals. It makes no difference whether we look at this signal in the time domain or in the frequency domain. The spectrum $S(f)$ of a time domain signal $s(t)$ is formed of all the frequency components of the signal. It is obtained by an operation called the Fourier transform (noted here FT):

$$S(f) = \mathrm{FT}\{s(t)\} = \int_{-\infty}^{\infty} s(t) \exp(-j2\pi ft)\, dt \tag{6.1}$$

Similarly, the time domain signal can be obtained from the frequency spectrum, via the inverse Fourier transform:

$$s(t) = \mathrm{FT}^{-1}\{S(f)\} = \int_{-\infty}^{\infty} S(f) \exp(j2\pi ft)\, df \tag{6.2}$$

Results related to Fourier transforms are given in Appendix A.6.1, reminding us of the useful properties of this operation and the correspondence between the time and frequency representations of common signals.

A particularly useful case is the *rectangular function*, whose frequency spectrum is a sinc function (see Appendices A.6.1 and A.1.2):

$$\begin{cases} s(t) = A \quad \text{for } t \in \left[-\dfrac{T}{2} ; +\dfrac{T}{2} \right] \\ s(t) = 0 \quad \text{elsewhere} \end{cases} \tag{6.3}$$

$$\Rightarrow \quad S(f) = \mathrm{TF}\{s(t)\} = AT\,\frac{\sin(\pi Tf)}{\pi Tf} = AT\,\mathrm{sinc}(Tf)$$

Conversely, the inverse Fourier transform of a rectangular spectrum is a sinc function in the time domain:

$$\begin{cases} S(f) = A & \text{for } f \in \left[-\dfrac{B}{2}; +\dfrac{B}{2} \right] \\ S(f) = 0 & \text{elsewhere} \end{cases} \tag{6.4}$$

$$\Rightarrow \quad s(t) = \text{TF}^{-1}\{S(f)\} = AB\frac{\sin(\pi Bf)}{\pi Bf} = AB\,\text{sinc}(Bt)$$

The frequency spectrum is often presented as the *power spectrum* $|S(f)|^2$, expressing the amount of energy at a given frequency.

The total amount of energy in the frequency spectrum equals the energy of the time signal (Parseval theorem):

$$\int_{-\infty}^{\infty} |s(t)|^2 \, dt = \int_{-\infty}^{\infty} |S(f)|^2 \, df \tag{6.5}$$

Using the example of the rectangular signal above, its energy in the time domain is:

$$\int_{-\infty}^{\infty} |s(t)|^2 \, dt = \int_{-T/2}^{T/2} A^2 \, dt = A^2 T \tag{6.6}$$

And in the frequency spectrum, we indeed get:

$$\int_{-\infty}^{\infty} |S(f)|^2 \, df = \int_{-\infty}^{\infty} A^2 T^2 \, \text{sinc}^2(Tf) \, df = A^2 T \tag{6.7}$$

This uses the fact that $\int_{-\infty}^{\infty} \text{sinc}^2 x \, dx = 1$ (cf. Appendix A.1.2).

The signals transmitted in underwater acoustics are most often narrowband signals. They are based on the transmission of a pure frequency (called the *carrier frequency* and noted f_0), modulated in time by an envelope function. If the carrier wave is written as a complex exponential, the signal writes:

$$s(t) = a(t) \exp\{j(2\pi f_0 t) + \varphi(t)\} = y(t) \exp\{j 2\pi f_0 t\} \tag{6.8}$$

The term $y(t)$ contains an amplitude term $a(t)$ (e.g., a rectangular function limiting the duration of the signal) and a complex exponential with variable phase $\varphi(t)$, modifying the frequency or the phase of the carrier. The frequency spectrum is then:

$$S(f) = \int_{-\infty}^{\infty} y(t) \exp(2j\pi f_0 t) \exp(-2j\pi f t) \, dt$$

$$= \int_{-\infty}^{\infty} y(t) \exp(2j\pi (f_0 - f) t) \, dt = Y(f - f_0) \tag{6.9}$$

$$Y(f) = \int_{-\infty}^{\infty} y(t) \exp(-j 2\pi f t) \, dt \quad \text{is the frequency spectrum of } y(t)$$

This is equivalent to saying that the spectrum of $s(t)$ is the spectrum of $y(t)$ translated around the carrier frequency $f = f_0$.

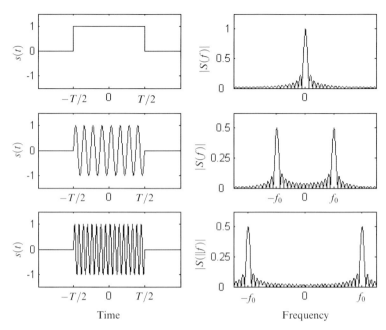

Figure 6.1. Time signal and frequency spectrum modulus of a rectangular signal considered on its own (*top*), and then as the envelope modulating two distinct carrier frequencies f_0 (*middle* and *bottom*). The same spectrum, initially centred on the null frequency, is shifted depending on the carrier frequency.

This result is a very important property of narrowband signals: the useful information present in the signal is not contained in the carrier frequency, but in its modulation. Conversely, this modulation information can be carried at will by any frequency (high enough to be compatible with the spectrum width). The carrier must therefore be considered as a simple physical vehicle for the information, useful until the signal has been physically received. One of the first operations performed by a sonar-processing chain is indeed demodulation: the initial spectrum is brought down around a frequency lower than the carrier frequency, or even around a null frequency, to facilitate the later stages of processing.

In practice, acoustic signals have real values. Writing the time domain signal as $s(t) = y(t) \cos(2\pi f_0 t)$, the frequency spectrum now writes:

$$S(f) = \int_{-\infty}^{\infty} y(t) \cos(2\pi f_0 t) \exp(-2j\pi ft) \, dt$$

$$= \tfrac{1}{2}[Y(f - f_0) + Y(f + f_0)] \tag{6.10}$$

The spectrum of $s(t)$ is therefore the spectrum of $y(t)$ translated around $f = f_0$ and $f = -f_0$ (Figure 6.1).

The complex notation of a signal $s(t)$, or *analytic* notation, is often more bconvenient than its real expression. The envelope of the analytic signal

$s_a(t) = y(t) \exp(2j\pi f_0 t)$ is written as $y(t) = s(t) + \mathrm{HT}(s(t))$, where HT is the *Hilbert transform*, defined by:

$$\mathrm{HT}(s(t)) = \frac{1}{\pi} \int_{-\infty}^{\infty} \frac{x(\tau)}{t - \tau} d\tau \qquad (6.11)$$

6.2.1.2 Time and space resolution of a signal

Let us consider a sonar receiving two signals close to each other in time (e.g., echoes from two targets at almost equal ranges from the insonifying sonar). Each echo has its own finite duration, and they are both assumed identical. If the two signals are too close, they will completely overlap. The receiver "sees" only one waveform, and there is no way to detect that there are in fact two distinct targets. If the time lag between the echoes increases, it becomes possible to detect inside the resulting waveform the presence of two arrivals (two peaks). The minimum time lag enabling the detection of two separate targets is called the *time resolution* of the signal.

The definition of the time resolution is obvious for signals with a rectangular envelope (Figure 6.2); in this case, the resolution is simply the duration of the signal. More explanations are needed for signals with smooth, "bell-shaped" envelopes. Let us consider two closely adjacent signals, with identical envelopes but random phases. Over the same time interval, the two signals are added coherently, with an a priori random phase difference. Therefore, for a given time lag, many different results will be possible, with constructive interference of the signals or without, and depending on the details of the envelope shape. It is admitted that, on average, signals are separable if the time difference between the signals equals the duration of $-3\,\mathrm{dB}$ from the maximum amplitude (i.e., half the maximum energy). Figure 6.3 shows the sum of two Gaussian pulses delayed by their duration at $-3\,\mathrm{dB}$.

When acoustic signals are used to measure the difference in range between sonar and targets, it is often easier to think in terms of *spatial resolution* rather than time resolution. The spatial resolution is then the minimal distance between two targets

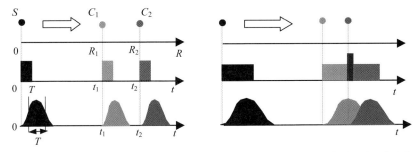

Figure 6.2. Time resolution and the echoes from two targets C_1 and C_2. (*Left*) The time resolution of the transmitted signal (*in black*) is sufficient to detect and separate the two target echoes (*in grey*). (*Right*) The time resolution is too low to separate the signals.

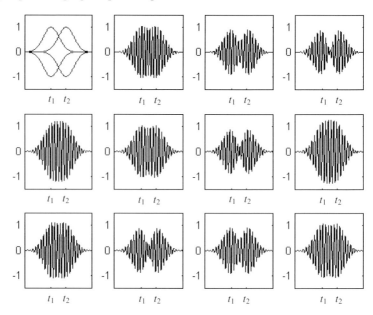

Figure 6.3. Sum of two pulses with Gaussian envelopes (*top left*), separated by their duration at −3 dB. Depending on their random phase difference, the resulting envelope may clearly show the peaks, or merge them completely.

separated by the time resolution T of the signal. The delay of two-way propagation between the sonar and a target i at distance R_i is given by:

$$t_i = \frac{2R_i}{c} \qquad (6.12)$$

The echoes of two targets at distances R_1 and R_2 (as in Figure 6.2) will be separated in time by:

$$\delta t = t_2 - t_1 = \frac{2(R_2 - R_1)}{c} = \frac{2\delta R}{c} \qquad (6.13)$$

They will be distinguished as separate only if $\delta t > T$. The spatial resolution of a signal of duration T is therefore:

$$\delta R = \frac{cT}{2} \qquad (6.14)$$

The time signal must be considered here at the output of a reception processing chain. This is important, as we shall see later (Section 6.4.2) that, for some modulated signals (with a modulation bandwidth B), one should not use the duration of the signal transmitted but its autocorrelation function at the output of the receiver, whose time spread (at −3 dB) is approximately $\delta t \approx 1/B$. The spatial resolution of such signals is then approximated as:

$$\delta R \approx \frac{c}{2B} \qquad (6.15)$$

We shall see that this definition is consistent with Equation (6.14), since for non-modulated signals $B \approx 1/T$ (see Section 6.4.1). It is therefore better to use the formula given in Equation (6.15), which is valid whatever the type of signal.

6.2.1.3   Correlation function

The *cross-correlation* function of two time domain signals measures the degree of similarity between these signals:

$$C_{xy}(t) = \int_{-\infty}^{\infty} x(\tau)y^*(\tau - t)\, d\tau \qquad (6.16)$$

The amplitude of this function increases with the similarity between the two functions x and y. If the two signals are identical, it becomes the *autocorrelation* function:

$$C_{xx}(t) = \int_{-\infty}^{\infty} x(\tau)x^*(\tau - t)\, d\tau \qquad (6.17)$$

Its value is maximal for $t = 0$, and it is then equal to the energy of the signal $\int_{-\infty}^{\infty} |x(\tau)|^2\, d\tau$. It collapses on each side of this maximum, and this evolution depends on the structure of the signal $x(t)$.

For example, for the rectangular function, the autocorrelation function equals:

$$C_{xx}(t) = \begin{cases} \dfrac{T - |t|}{T} & \text{for } t \in [-T; +T] \\ 0 & \text{elsewhere} \end{cases} \qquad (6.18)$$

If the rectangular function modulates a carrier f_0, it is easy to understand intuitively that the autocorrelation function is the triangle function modulating the carrier (Figure 6.4).

The correlation operation allows extraction of a known signal from a noisy acquisition, by applying a filter with characteristics as close as possible to those of the expected signal. This operation is called *adapted filtering* (we shall see an application of this in Section 6.4.2).

6.2.1.4   Frequency resolution: the ambiguity function

Underwater acoustic signals are among other things subject to Doppler effect (see Section 2.5.1). The frequency of the signal received might differ significantly from the frequency of the signal transmitted (by 1%, typically). This effect is prejudicial in most applications, as the receiver is designed for the nominal frequency and thus perturbed by this change. But, in some systems, the Doppler shift is used to measure the relative speed between sonar and target. For a signal of a given type, it is therefore interesting to know how resistant it is to Doppler effects, or, conversely, how high the tolerable Doppler shift can be.

The Doppler effect moves and distorts the initial frequency band. Intuitively, we can therefore imagine that this is more penalising as the signal has a narrower band. A sine signal transmitted for a sufficiently long time, detected with a narrow

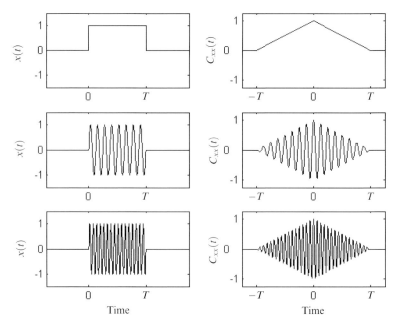

Figure 6.4. Normalised autocorrelation function of a rectangular signal, on its own (*top*) and modulating a carrier frequency (*middle* and *bottom*).

bandpass filter centred on the frequency transmitted, will be heavily penalised by even small frequency changes. Conversely, such a signal, with high frequency sensitivity, has good performances when the Doppler frequency shift (and thus the speed of the target) needs to be measured accurately. On the other hand, signal modulated over a large frequency band will see its power spectrum almost unchanged by Doppler effects, even when they reach a few percentage points.

The time and frequency resolution performance of a signal are jointly expressed using the *ambiguity function*:

$$A(\delta f, \delta t) = \left| \int_{-\infty}^{\infty} s(f_0, \tau) s^*(f_0 + \delta f, \tau - \delta t) \, d\tau \right|^2 \qquad (6.19)$$

where $s(f_0, \tau)$ is the nominal time domain signal at the carrier frequency f_0, and $s(f_0 + \delta f, \tau - \delta t)$ is an alteration of this signal, delayed by the propagation time δt and frequency-shifted by the Doppler shift δf. The result of this cross-correlation operation therefore expresses the similarity between the nominal signal and its delayed (or Doppler shifted) versions (i.e., the ability of the signal to measure time or frequency accurately). It is, of course, maximal if the two terms in the product coincide exactly (i.e., with null delay and null Doppler shift).

Figure 6.5 shows an example of the ambiguity function. The same rectangular envelope modulates two carriers, thus creating two spectra with distinct relative widths. The ambiguity functions have the same triangular time dependence (as in

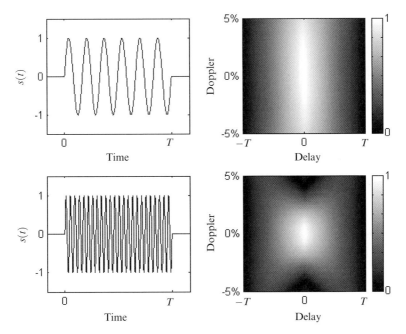

Figure 6.5. Envelopes of the normalised ambiguity function (*right*) for a rectangular function modulating two distinct carrier frequencies (*left*). The lowest frequency, corresponding to the widest spectrum, tolerates the Doppler effect to the limits of the calculation (±5%). The spectrum of the second signal is four times narrower, and tolerates the Doppler effects much less. For a null delay, the ambiguity function collapses for Doppler shifts beyond ±2%.

Figure 6.4), because the durations and envelope shapes are similar. But the frequency resolution changes drastically between the two.

6.2.2   Signal-to-noise ratio

6.2.2.1   *Definitions*

For a given receiver configuration, the *signal-to-noise ratio* (SNR) expresses the relative importance of the (power) contributions of, respectively, the signal expected and the perturbing noise. The SNR is the main parameter affecting the performance of a receiver in all applications of target detection or parameter estimation.

In the following, the (narrowband) acoustic signal $s(t)$ at the input to the receiver is characterised by its amplitude A, assumed constant during the duration T of the signal. The energy of the signal received is then:

$$E = \frac{A^2 T}{2} \qquad (6.20)$$

The average power is:

$$P = \frac{E}{T} = \frac{A^2}{2} \tag{6.21}$$

The noise $n(t)$ added to the signal received is taken to be a *Gaussian noise* with zero mean. For a *white noise*, the power spectral density $n(f)$ (i.e., the power in a 1-Hz frequency band) is constant with frequency in the frequency band analysed:

$$n(f) = \frac{n_0}{2} \tag{6.22}$$

The SNR (in natural values) is represented by r in the following equations. At the input to the receiver, it is classically defined as the ratio of the signal energy to the power spectrum density of the noise:

$$r_0 = \frac{2E}{n_0} \tag{6.23}$$

However, a power SNR will most often be used, as it is closer to experimental reality. It is defined as the ratio of the instantaneous power of the signal to the noise power in a frequency band B, which is, of course, at least as large as the signal bandwidth:

$$r_{0P} = \frac{E}{n_0 BT} \tag{6.24}$$

After processing (see Figure 6.6), the SNR r expresses the ratio between: (1) the power of the output signal $z_{s+n}(t)$ with signal and noise at the input, corrected for the output level upon noise only; and (2) the power of the output signal $z_n(t)$ with input noise only (Figure 6.7) expressed as the signal variance σ_n^2.

To compare the input and output SNR (a common case), one must distinguish two configurations. If the output signal $z(t)$ is physically homogeneous to the input signal $x(t)$ (i.e., the receiver performs linear operations), we will have:

$$r = \frac{(\langle z_{s+n}(t) \rangle - \langle z_n(t) \rangle)^2}{\sigma_n^2} \tag{6.25}$$

Conversely, if the output signal is homogeneous to the energy of the input signal (i.e., in practice, if the receiver performs a signal quadrature), we will use the following definition of the output SNR:

$$r = \frac{\langle z_{s+n}(t) \rangle - \langle z_n(t) \rangle}{\sqrt{\sigma_n^2}} \tag{6.26}$$

Figure 6.6. SNR at the input and output of a receiver.

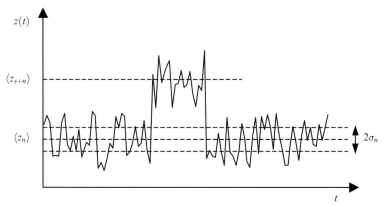

Figure 6.7. SNR for a signal embedded in noise: the difference between the average output level $\langle z_{s+n} \rangle$ and the average output level for noise only $\langle z_n \rangle$ is to be compared with the noise power σ_n^2.

This ensures that the input and output SNR are directly comparable.

With these definitions of the output SNR, the *processing gain* is the ratio r/r_{0P} between the power SNR values at the output and input of the receiver. It reads (in dB):

$$PG = 10 \log \frac{r}{r_{0P}} \tag{6.27}$$

6.2.2.2 Coherent summation

To improve the SNR, a simple and common way is to add several successive signals. The idea is that the variance of output random fluctuations (caused by the noise) will increase at a slower rate, during the summations, than the energy of the output signal, thus improving the SNR.

Let $z_n(t)$ be the output signal of a processing chain with only noise present (i.e., there is no input signal). $z_n(t)$ is assumed to be Gaussian, and is characterised by its mean $\langle z_n(t) \rangle$ and its variance σ_n^2, assumed identical for all realisations. The output of the receiver in the presence of a permanent signal and the Gaussian noise is itself Gaussian, with an average level $\langle z_{s+n}(t) \rangle$. The SNR after reception of a single signal is therefore:

$$r_1 = \frac{(\langle z_{s+n}(t) \rangle - \langle z_n(t) \rangle)^2}{\sigma_n^2} \tag{6.28}$$

It should be remembered now that the sum of N Gaussian processes with identical mean m and identical variance σ^2 is itself a Gaussian process, with mean Nm and variance $N\sigma^2$. Applying this result to the sum of N successive outputs from the processing chain, we get the following results:

- the output from the noise only has a mean amplitude $N\langle z_n(t) \rangle$ and a variance $N\sigma_n^2$;
- the output with a signal present has a mean amplitude of $N\langle z_{s+n}(t) \rangle$.

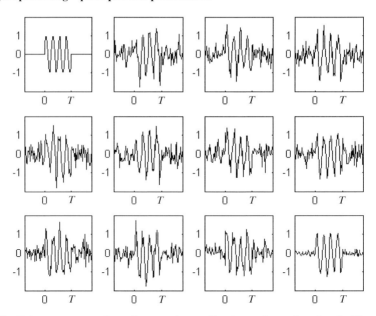

Figure 6.8. Coherent summation of successive realisations of a noisy signal. (*Top left*) The initial signal. The next 10 diagrams show 10 different realisations of the signal with noise added. The last diagram (*bottom right*) shows the average of the noisy signals: its SNR is much better than for any of the individual noisy signals.

The final SNR will therefore be:

$$r_N = \frac{(N\langle z_{s+n}(t)\rangle - N\langle z_n(t)\rangle)^2}{N\sigma_n^2} = Nr_1 \qquad (6.29)$$

The processing gain equals $r_N/r_1 = N$ (i.e., $PG = 10\log N$ when expressed in dB). Figure 6.8 shows an illustration of this processing.

During reception of a continuous signal (e.g., for a passive sonar receiver), the sum of the successive realisations is replaced by a continuous integration during the duration T. The number of summed processes is now the number of independent time series taken into account during integration. The time interval between two independent time series can be considered equal to the -3-dB width of the main lobe of the autocorrelation function. This equals the inverse of the signal's bandwidth B (i.e., $1/B$). There are therefore $N = BT$ independent series, and the processing gain is now:

$$PG = 10\log(BT) \qquad (6.30)$$

For a quadratic receiver (as with passive sonars, see Section 6.3), it becomes:

$$PG = 5\log(BT) \qquad (6.31)$$

6.3 SIGNAL DETECTION WITH PASSIVE SONARS

6.3.1 Non-coherent detection

A passive sonar is used to detect two types of signal, which correspond to noise radiated by the target: (1) spectral lines (i.e., signals with a very narrow band); (2) broadband noise (e.g., cavitation, hydrodynamic flow). The origin and the main characteristics of radiated noise are described in detail in Chapter 4.

In this case, one uses *"non-coherent" reception*: the signal's energy is detected in the frequency band of interest, and cumulated during the period of observation. The receiver (Figure 6.9) processes the data in several steps: bandpass filter, quadrature and integration of energy over a time T_i. The last step aims at improving the SNR (as presented in Section 6.2.2.2). These processing steps are performed after array processing and beamforming (see Chapter 5), which have already improved the SNR.

A few points specific to the operations performed should be noted:

- The characteristics of the bandpass filter depend on the application. This filter aims at selecting the useful portion of the noise spectrum radiated. To detect broadband noise, one commonly works with octave or one-third-octave bands.[1] For narrowband signals, one uses spectral analysis with numerical algorithms such as Fast Fourier Transforms (FFTs). In this case, each FFT channel can be assimilated to a very narrow bandpass filter, and the typical order of magnitude of the filtering width is 1 Hz.
- Time integration aims at reducing the variance of output noise (after processing). Its duration T_i is as long as possible to be compatible with the stability of the integrated signal. It always leads to values of $B \times T_i$ much larger than 1.

6.3.2 Broadband detection performance

For broadband detection, the processing gain of the passive sonar can first be approximated in the following fashion. Let $n_0/2$ and $n_1/2$ be the spectral power densities respectively associated with ambient noise and signal noise. The input SNR is therefore simply written as:

$$r_0 = \frac{n_1}{n_0} \tag{6.32}$$

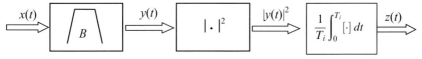

Figure 6.9. Structure of a typical passive sonar receiver. From *left* to *right*: bandpass filter, quadrature and finally integration over a time T_i.

---

[1] A frequency band $[f_1, f_2]$ is one octave if $f_2 = 2f_1$; a third of an octave corresponds to $f_2 = 2^{1/3}f_1$ (i.e., to $f_2 = 1.26f_1$ approximately).

It can be demonstrated that, if the input signal $x(t)$ is a Gaussian noise with a spectral power density of $n/2$, the output after integration can be assimilated to the sum of random Gaussian variables and squared, and thus follows a χ^2 distribution (see Section 4.6.3.5), with the number of degrees of freedom given by the number of independent samples of the signal output after bandpass filtering (bandwidth B) and integration (time T_i). We saw this number was given by the product $B \times T_i$. If this parameter is high enough, the output of the processing chain can be assimilated to a Gaussian variable of mean nB and variance n^2B/T_i. The output is then:

- in the presence of noise only, of power spectral density $n_0/2$, the mean output of the processing chain is $\langle z_n(t) \rangle = n_0 B$, and its variance is $\sigma_n^2 = n_0^2 B/T_i$;
- in the presence of a signal, of power spectral density $n_1/2$ superposed on the noise, the mean output of the processing chain is $\langle z_{s+n}(t) \rangle = (n_0 + n_1)B$.

With the definition for a quadratic receiver, the detection SNR is defined by:

$$r = \frac{\langle z_{s+n}(t) \rangle - \langle z_n(t) \rangle}{\sqrt{\sigma_n^2}} \qquad (6.33)$$

Then:

$$r = \frac{n_1}{n_0} \sqrt{BT_i} = r_0 \sqrt{BT_i} \qquad (6.34)$$

The processing gain of the entire chain is therefore:

$$PG = 10 \log(\sqrt{BT_i}) = 5 \log(BT_i) \qquad (6.35)$$

This is important as the integration time is a priori independent of the bandwidth observed, and the product $B \times T_i$ can be large.

6.3.3   Narrowband detection performance

Similar reasoning is used for narrowband detection. If the power of the spectral line detected is A^2, the input SNR, considered in a 1-Hz wide band of analysis, equals:

$$r_0 = \frac{A^2}{n_0} \qquad (6.36)$$

At the output of the narrowband filter with bandwidth δf, the SNR becomes:

$$r_1 = \frac{A^2}{n_0 \delta f} \qquad (6.37)$$

At the receiver's output, if the signal and noise are mixed, we have:

$$\langle z_{s+n}(t) \rangle = A^2 + n_0 \delta f \qquad (6.38)$$

The output SNR of a quadratic receiver is therefore:

$$r = \frac{A^2}{n_0 \delta f} \sqrt{\delta f T_i} \qquad (6.39)$$

The processing gain, calculated between the output y of the narrowband filter and the output z of the receiver, is:

$$PG = 10 \log \left(\frac{r}{r_1} \right) = 5 \log(T_i \delta f) \qquad (6.40)$$

This is similar to the processing gain of the broadband case (Equation 6.35).

Considering now the processing gain, referenced to the input SNR (in a 1-Hz band), we get the following expression:

$$PG = 10 \log \left(\frac{r}{r_0} \right) = 5 \log \left(\frac{T_i}{\delta f} \right) \qquad (6.41)$$

Note that the *frequency resolution* δf obtained after FFT of a time signal is a function of the duration of signal analysis T, the sampling period T_s (or sampling frequency f_s) and the number of points N in the FFT:

$$\delta f = \frac{1}{T} = \frac{1}{NT_s} = \frac{f_s}{N} \qquad (6.42)$$

6.3.4   Other operations in passive sonar processing

After these processing steps, aimed at detection of a target, several other steps are possible, aimed at localisation and identification of the target.

Doppler analysis is frequently used for target tracking. It needs to be applied after narrowband filtering and the selection of significant spectral components, whose frequency variations are followed as a function of time. Unfortunately, there is no absolute radiated frequency reference at this stage, and so the target speed cannot be estimated directly. But sudden frequency variations, affecting all spectral lines simultaneously, make it possible together to detect either a change in speed or azimuth of the target, or rapid variations of the sonar–target range (e.g., at closest approach, when the Doppler shift changes in sign).

If interference patterns due to guided propagation are observable as modulations on range–frequency plots, they may be exploited to infer the multipath structure. Also the cross-correlation function of the received signal may feature several peaks; the delay between them may be used to estimate the multipath structure, and reconstruct the target position.

Finally the target identification is obtained from narrowband spectral analysis: splitting into spectral components is a strong indicator of the acoustic signature of the target. Beyond the classical Fourier analysis (adapted to stable harmonic signals), *time–frequency methods* are powerful tools for studying quick variations in the various spectral components. Several methods are now widespread for this purpose (short-term Fourier transform, Wigner–Ville transform, wavelet decomposition: see, e.g., Cohen, 1995). Figure 6.10 presents time–frequency plots (or sonograms) for ship-radiated noise.

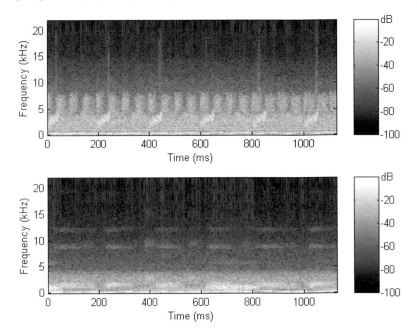

Figure 6.10. Time–frequency plots of radiated noise from two different ships. The low-frequency spectral tones are hardly distinguishable (horizontal lines along the time axis). Regular modulations with time correspond to motor noise; vertical striping on the whole height is due to cavitation noise generated by propeller blades.

6.4 ACTIVE SONAR SIGNALS AND PROCESSING

6.4.1 Narrowband pulse signals

6.4.1.1 Signal characteristics

The narrowband pulse (continuous wave pulse, or CW pulse, also sometimes called a "burst") is the most commonly used in underwater acoustics, in most sounders, sonars and positioning systems. This signal consists in a sine wave of frequency f_0, transmitted during a limited time T (Figure 6.11), usually with constant amplitude A:

$$S(t) = A\sin(2\pi f_0 t) \quad \text{if } 0 < t < T \tag{6.43}$$

The frequency spectrum of the signal is a sinc function ($\text{sinc}(x) = \sin \pi x/\pi x$) centred on the carrier frequency f_0:

$$S(f) = \frac{AT}{2}\left[\frac{\sin(2\pi T(f - f_0))}{2\pi T(f - f_0)} + \frac{\sin(2\pi T(f + f_0))}{2\pi T(f + f_0)}\right] \tag{6.44}$$

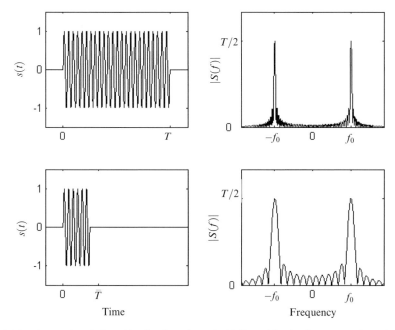

Figure 6.11. Narrowband signal in the time domain (*left*) and in the frequency domain (*right*), for two different times of transmission at the same carrier frequency. The bandwidth occupied is inversely proportional to the duration.

The 3-dB bandwidth is given by:

$$\frac{\sin(2\pi T(f - f_0))}{2\pi T(f - f_0)} = \frac{1}{\sqrt{2}} \tag{6.45}$$

It now becomes:

$$\delta f \approx \frac{0.886}{T} \tag{6.46}$$

This is commonly approximated to $\delta f \approx 1/T$.

   Usual frequencies employed with sounders range from 12 kHz to 200 kHz, for corresponding pulse durations of typically 10 ms to 0.1 ms (i.e., bandwidths of 0.1–10 kHz). The ratios $\delta f/f_0$ are therefore small; these signals are to be considered as narrowband.

   The resolution of these signals is given by their duration. Two CW pulses of duration T with rectangle envelopes can only be separated if they are separated by a time of at least T. After reflection from a target, this corresponds to a spatial resolution of $cT/2$ (0.75 m for a 1-ms pulse).

   The ambiguity function of a CW pulse reads:

$$A(\delta f, \delta \tau) = \left| \left(1 - \frac{|\delta \tau|}{T} \right) \operatorname{sinc} \left[\delta f T \left(1 - \frac{|\delta \tau|}{T} \right) \right] \right|^2 \tag{6.47}$$

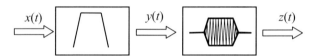

Figure 6.12. Basic receiver structure for a CW pulse: bandpass filter (*left*) and envelope detection (*right*).

Its -3-dB width corresponds in frequency to $0.886/T$, and in time to $0.585T$. The latter is smaller than T; it would thus be the time resolution of a CW pulse just as if it were received through a correlator (which is never the case in practice).

6.4.1.2   Receiver and processing gain

Two operations[2] are necessary to process a CW pulse (Figure 6.12):

- bandpass filtering, as close as possible to the spectrum of the signal transmitted, and possibly corrected for the expected Doppler shift;
- envelope detection of the filtered signal.

The process finally consists in just detecting the echo energy inside the signal's frequency band. No processing gain can then be expected, and the SNR is the same at the output and at the input of the receiver. In performance evaluations for sonars using CW pulses, a zero processing gain is assumed ($PG = 0\,\mathrm{dB}$).

However, the SNR might decrease if using a non-optimal filter. Too narrow a bandpass filter will not account for the entire spectral power of the signal, whereas too wide a filter will integrate additional noise outside the signal's useful bandwidth. The optimal filter will maximise, in the filtered band B_f, the energy of the signal (given by the integral of $|S(f)|^2$ over the band B_f), relative to the energy of noise (proportional to B_f). It can be shown that in the case of a CW rectangular pulse this ratio is maximum for:

$$B_f = B_{opt} \approx \frac{1.37}{T} \tag{6.48}$$

This value optimises the processing gain. A different value of B_f will decrease the SNR by $-10|\log B/B_{opt}|$. The correct matching of the receiving filter is therefore important.

6.4.1.3   Relative merits of narrowband signals

The main interest of CW pulses is their simplicity of transmission and processing, along with performances that are usually good enough for many applications. Their narrow frequency band is well suited to narrowband transducers, which are efficient, cheap and easy to design.

---

[2] Before the generalisation of the digital processing stages shown here, the traditional processing chains used by analogue systems were slightly different (see Appendix A.6.2 for details).

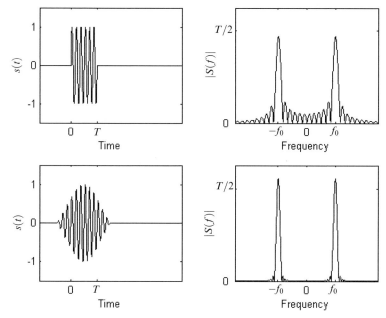

Figure 6.13. Effect of a bell-shaped amplitude modulation (*bottom*) on a CW pulse (*top*): for a given duration (at −3 dB in the second case), the frequency spectrum is improved, with a strong decrease in the level of sidelobes.

The major defect is their very poor spectral content, which decreases interest in their use for advanced processing to characterise targets. Furthermore, they require relatively high input SNRs (i.e., transmission with high instantaneous levels).

The characteristics of a CW pulse can be slightly improved by using a more refined envelope. For example, a bell-shaped modulation amplitude (e.g., the main lobe of a $\sin x/x$ function, or a Gaussian), with a width T at −3 dB, will give a more compact frequency spectrum with fewer sidelobes than a rectangular burst of duration T (as in Figure 6.13).

6.4.2 Frequency-modulated pulse (*chirp*)

6.4.2.1 *Signal characteristics*

A *chirp* in its simplest form consists in a carrier frequency modulated linearly with time, limited by a rectangular envelope of duration T (Figure 6.14):

$$S(t) = A \sin\left(2\pi\left[f_0 + m\frac{t-T}{2}\right]\right) \quad \text{if } 0 < t < T \tag{6.49}$$

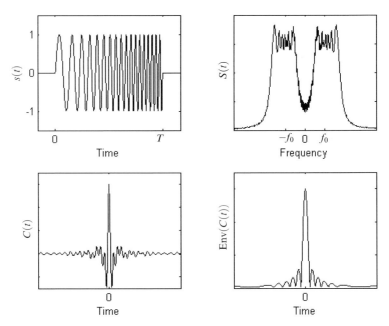

Figure 6.14. (*Top*) Chirp signal, represented in the time domain (*left*) and in the frequency domain (*right*). (*Bottom*) Autocorrelation function of the chirp signal (*left*) and its envelope (*right*).

The instantaneous frequency is given by the time derivative of the phase $\Phi(t) = 2\pi[f_0 + m(t - T)2]t$, in $f(t) = 1/2\pi\, d\Phi(t)/dt = f_0 + m(t - (T/2))$. It thus varies between $f_0 - m(T/2)$ and $f_0 + m(T/2)$ (i.e., with a bandwidth $B = mT$).

The frequency spectrum of the chirp can be roughly assimilated to a rectangle of width B centred on f_0 (see Figure 6.14).

6.4.2.2 Receiver and processing gain

Frequency-modulated signals are processed on reception by correlating the received signal $x(t)$ (somewhat distorted by noise) with a copy $s(t)$ of the expected[3] signal:

$$y(t) = \int_0^T x(t + t')s(t')\, dt' \tag{6.50}$$

---

[3] The quality of the correlation depends, of course, on the similarity between the signal received and the reference copy used in the correlator. It is therefore desirable to use on reception some reference signals that take into account, as much as possible, the deformations of the physical signal during its propagation (Doppler effect, frequency selection by attenuation, etc.).

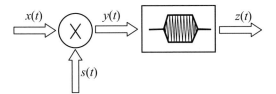

Figure 6.15. Coherent processing of a modulated pulse (time correlation and envelope detection).

An envelope detection is then applied to the output of the correlator (Figure 6.15). This series of operations is often called *coherent processing*.

Assuming for the sake of simplification that $x(t) = Qs(t - \tau)$ (i.e., the signal received $x(t)$ is just a τ-delayed Q-attenuated version of the expected signal $s(t)$), it can be shown that at the time of arrival of the signal the output of the correlator is at its maximum:

$$y(t = 0) = \int_0^T x(t')s(t') \, dt'$$

$$= Q \int_0^T s^2(t') \, dt'$$

$$= Q \frac{A^2}{2} = QE \tag{6.51}$$

If there is only noise with a probability density function n_0, y is a centred random variable with variance $n_0 E/2$. At the output of the receiver, the ratio between the r.m.s. value of the maximum energy (at signal reception) and the average power (for noise) now writes:

$$r = \frac{\dfrac{Q^2 E^2}{2}}{\dfrac{n_0 E}{2}} = \frac{Q^2 E}{n_0} \tag{6.52}$$

At the input to the receiver, using the signal power $Q^2 E/T$ and the noise power $n_0 B$ over the signal bandwidth B, we get the power SNR:

$$r_{OP} = \frac{\left(\dfrac{Q^2 E}{T} \right)}{n_0 B} = \frac{Q^2 E}{T n_0 B} \tag{6.53}$$

The processing gain (for power) is therefore expressed as:

$$PG = 10 \log \left(\frac{r}{r_{OP}} \right) = 10 \log(BT) \tag{6.54}$$

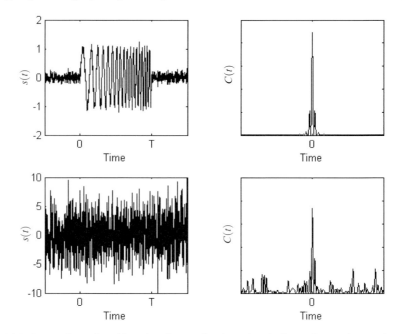

Figure 6.16. Extraction of a chirp signal out of noise: signal plus noise at the receiver input (*left*) and cross-correlation function at the receiver output (*right*), for SNR values of $+20\,\text{dB}$ (*top*) and $-10\,\text{dB}$ (*bottom*).

6.4.2.3   The concept of "pulse compression"

We can also look, after processing, at the envelope $z(t)$ of the autocorrelation function:

$$z(t) = \frac{\sin(\pi Bt)}{\pi Bt} \tag{6.55}$$

The -3-dB width of its main lobe is:

$$\delta t = \frac{0.886}{B} \tag{6.56}$$

One can see that:

- the time resolution δt of the receiver is now independent of the duration of the transmission, and only depends on the frequency modulation width B;
- the signal's energy, spread over the time T at the beginning, is at the output from processing "compressed" over a duration δt. We find once more the processing gain of $T/\delta t = BT/0.886 \approx BT$ seen earlier. This process is commonly called *pulse compression*.

An example of processing results is given in Figure 6.16, illustrating clearly both the pulse compression in time, and the processing gain allowing extraction of the signal from noise. An item of note is that the linear frequency modulation used in the traditional chirp signal is quite sensitive to Doppler effects. The frequency shift induces a degradation of the correlator output level, and an offset in the measurement of time. These come from the fact that the Doppler frequency shift is equivalent to a time delay, which cannot be distinguished from the actual time delay due to propagation. To combat this effect, hyperbolic frequency modulations are sometimes used instead:

$$S(t) = A \sin\left(2\pi f_0 m \log\left(1 + \frac{t}{m}\right)\right) \qquad (6.57)$$

Another item of note is that, as with narrowband pulses, it can be advantageous to complement frequency modulation with amplitude modulation of the envelope ("bell-shaped"). This diminishes the sidelobe level of the autocorrelation function, which can degrade some applications that require a very accurate time signal.

6.4.2.4   Relative merits of frequency-modulated pulses

Chirp signals (processed by correlators) are clearly advantageous compared with narrowband pulses (bandpass filtered), in terms of SNR gain. The processing gain of chirp signals is more important as the product BT is large, and therefore the signals are long and with a large bandwidth. T and B can indeed be increased independently, and PG can reach high values. Conversely, for narrowband pulses, we always nearly have $B \approx 1/T$. Using a chirp enables transmissions with limited powers to be compensated by lengthening the duration of the signals. This maintains a good time resolution on reception, imposed by the modulation bandwidth, and independent of the transmitted signal duration. But it implies heavier processing than for CW pulses. Its use is not justified except when it is imperative to get a high SNR gain. This occurs mainly with:

- military sonars, looking to maximise detection ranges and improve SNR conditions;
- sediment profilers, which must compensate the very strong attenuation of the signal during propagation inside the seafloor, while maintaining a good time resolution.

The BT products used in underwater acoustics are commonly of a few hundreds, with processing gains typically between 20 dB and 30 dB. The limits of this product BT are imposed on the one hand by material constraints (the band is limited to that of the transducers available; duration of the transmitted signal is less than the time of return of the first echo), and on the other hand by the propagation medium. Signal coherence must be sufficient for the duration and bandwidth considered, for the correlation to remain efficient.

6.4.3   Phase-modulated signals

Some applications use signals made of a phase-modulated carrier frequency:

$$S(t) = A \sin(2\pi f_0 t + \varphi(t)) \qquad (6.58)$$

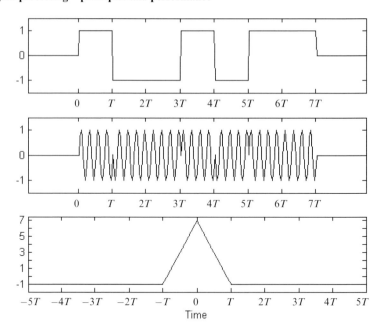

Figure 6.17. MLBS-modulated PSK (Phase Shift Keying) signal: time signal and autocorrelation function. The binary sequence used is [1 0 0 1 0 1 1]. (*Top*) Corresponding amplitude modulation. (*Middle*) Time domain signal transmitted. (*Bottom*) Autocorrelation function of the MLBS.

Most often, $\varphi(t)$ has the discrete values 0 or π, changing every T seconds (this modulation is called BPSK, or Binary Phase Shift Keying, with two states of phase). The carrier can then be easily modulated by multiplying the amplitude by $+1$ or -1 to get phases of 0 or π (see Figure 6.17). Other, richer phase-modulation schemes can be used to get higher communication rates. For example, QPSK, or Quadrature Phase Shift Keying, uses four phase states: $\varphi(t) \in [0, \pi/2, \pi, 3\pi/2]$. The modulations are more efficient as the number of states is higher (they allow transmission of more data in a particular bandwidth). But they are correspondingly sensitive to perturbations of the signal, through propagation and noise.

Such modulation can be used to transmit data (for BPSK, 1 bit every T seconds, worth 0 or 1 depending on whether $\varphi(t)$ is worth 0 or π). Processing then consists in tracking the carrier frequency using a *phase-lock loop*[4] and detecting the phase changes carrying the digital message. This modulation technique is used in many digital data-transmission systems.

Phase-modulated signals are also used for detection and time-measurement

---

[4] In a phase-lock loop, the processed signal is multiplied by a sine wave at the same frequency, generated by a voltage-controlled oscillator. The phase of the resulting signal is low-pass filtered, and the result is used to command the oscillator frequency. This process allows us to follow the phase variations of the received signal precisely.

applications. The principle is the same as for the chirp, as described above, except that the frequency modulation is replaced by phase modulation. The series of $\{\varphi_i\}$ being known a priori, the processing at reception consists in correlating the signal transmitted and the signal received. One generally[5] uses *maximum length binary series* (MLBS), which are interesting in that they suppress the secondary lobes of the signal's autocorrelation function. Using an MLBS with M bits ($M = 2^N - 1$, where N, an integer, is the order of the series), one can obtain:

- an autocorrelation function of duration T (at half-height) equal to one digit, with no secondary lobes if the MLBS was transmitted recursively and with no interruption; and
- an SNR-processing gain of $10 \log M$ compared with the use of only one digit.

The use of this technique is justified for very accurate measurements of propagation time, requiring an excellent time resolution and a high processing gain (e.g., for the pulse response of the environment, in oceanic tomography).

6.5   STRUCTURE OF SONAR RECEIVERS

6.5.1   Typical sonar receiver

A modern receiver typically performs the following series of operations:

- *Preamplification* of the hydrophone's signal. It has several functions: to match as best as possible the electrical signal from the hydrophone to the dynamics and the sensitivity of the processing units; to compensate the sensitivity differences of the hydrophones, if there are several of them; to increase the level and prevent the SNR degradation of hydrophone signals being transmitted from the arrays toward the processing systems due to cable attenuation and electrical noise. This preamplification is complemented with first bandpass filtering, to limit the acoustic perturbations possibly present in the received signal.
- *Time-Varying Gain* (TVG). This function equalises the time variations of the signal. It aims at decreasing the initial dynamic range of the physical time signal, which may be too large for the rest of the processing chain (see Section 6.5.2 for more details).
- Frequency shift through *demodulation* around the carrier frequency. The aim is to bring the signal's useful bandwidth (Section 6.2.1) back around the null frequency, or a very low frequency. This will later be used to limit the digitising frequency, and thus the number of samples of the digital signal. This operation is

---

[5] Other series, obeying constraints other than MLBS, can be used to transmit digital data for communication or positioning. There are, for example, codes that generate families of sequences with low cross-correlation levels, used to transmit distinct signals simultaneously. Other codes are more adapted to beam-spreading techniques, used to ensure discrete transmissions, or transmissions that resist perturbations in the propagation channel (see, e.g., Proakis, 1989).

all the more interesting as the signal has a narrow bandwidth and a high carrier frequency.

- *Analogue/Digital Conversion* (ADC). The dynamic range of the converter should withstand the high instantaneous variations of underwater acoustic signals; 8-bit converter dynamics are usually too small, and one should use 12- or 16-bit converters (or even higher).
- *Filtering* suited to the signal transmitted. This filtering aims at maximising the SNR, by filtering the signal received according to the bandwidth of the signal transmitted (narrowband signals) or its phase (pulse compression, for modulated signals).
- *Array processing.* For systems with several receiving hydrophones placed on an array, the previous operations are applied independently for each receiving hydrophone channel. After signal digitisation and adapted filtering, various array processing schemes can be applied (the most common being beamforming: see Chapter 5).

These processing steps are physically performed by analogue electronic circuits for the first stages, or by Digital Signal Processing (DSP) boards. All these functions, including adapted filtering and beamforming, are generic functions found under similar form in all sonar systems. The digital signals coming from adapted filtering and array processing are then directed towards post-processing machines (usually personal computers), performing operations specific to each type of sonar (e.g., detection; measurement of arrival time, the Doppler shift or the angle of arrival; sonar imagery; spectral analysis).

6.5.2   Time-varying gain

The echo level from a target depends, of course, on the distance between the target and the sonar system (e.g., following a variation in $-40 \log R - 2\alpha R$). At the receiving processor output, the amplitudes of echoes will be functions of the sonar–target distance, rather than actual target strengths. To compensate, a common solution is to correct the signal received using the law expected for propagation loss, transposed into the time domain (from the simple relation $R = ct/2$). This correction is known as TVG for time-varying gain (Figure 6.18). It is used in all systems that employ target echoes (sounders and sonars).

For this compensation to be accurate, its coefficients must be adapted to the physical law of target echo level variation. We saw in Chapter 3 that this geometrical decrease can be in $-20 \log R$ for volume targets, in $-30 \log R$ for surface targets, or in $-40 \log R$ for point targets. Sonar systems must therefore propose the relevant options. This correction must also compensate for the extremely large dynamic range of the signals. This often leads to either limiting the use of TVG, or truncating it into several intervals. Critical for analogue TVGs, this problem of dynamic range is less crucial for digital signal-processing systems.

Besides its role in target level equalisation, TVG processing offers another benefit. Indeed, by reducing the dynamic range of the physical signal, it prevents

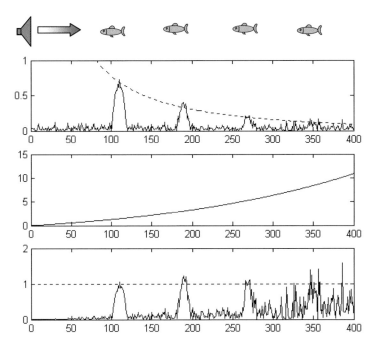

Figure 6.18. Principle of TVG. (*Top*) Physical signal corresponding to identical targets at varying ranges. (*Middle*) TVG compensation law. (*Bottom*) TVG-corrected signal. The identical targets now have quasi-similar echoes; but on the other hand the noise level has increased.

it from exceeding the dynamics of the converter, since it is usually applied at the input to the processing chain, before ADC. Increasingly better processors are now used in processing chains, currently with 12 or 16 bits; this makes this functionality of the TVG correction less crucial. It is, however, still necessary in some systems in which signals undergo very large fluctuations (e.g., multibeam echo sounders). Note that it is not always necessary to adhere strictly to a very accurate law of evolution of the signal with time; other, simpler corrections can be adequate, as long as they effectively limit the dynamic range of signal input to the ADC. Once digitisation has been performed, it is always possible to compensate for the law applied, in order to return to the physical values of the original signal.

An important point to stress is that, if the TVG corrects the amplitudes to similar levels, it does not increase the SNR in any way. The amplification performed affects the noise as well as the signal, and the contrast between the two is not changed by the compensation law (see Figure 6.16).

6.5.3   Signal demodulation

The demodulation of a signal around its carrier frequency aims at keeping only the meaningful frequency content of the signal, and thus at limiting the amount of

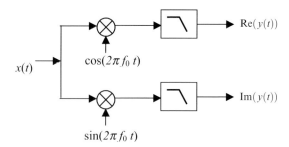

Figure 6.19. Complex signal demodulation.

digital data to process at later stages. If the signal has a bandwidth B centred around its carrier frequency f_0, the Shannon criterion means the time sampling must have a frequency at least twice the higher frequency of the spectrum (i.e., $2f_0 + B$). If the signal is demodulated around f_0, it becomes possible to sample it at frequency B.

Demodulation of analogue signals is performed by multiplying the signal to demodulate with a sine signal at the demodulation frequency. The spectrum of the resulting signal is thus brought back around the null frequency. This is balanced by the apparition of a parasite copy of the spectrum around the double frequency, which is removed by low-pass filtering.

In fact, to keep the phase characteristics of the signal, this operation must be performed twice, in parallel (Figure 6.19). The two sine signals used for demodulation are in phase quadrature: the two outputs then correspond to the real and imaginary parts of the complex demodulated signal.

A very efficient digital implementation of the demodulation operation presented in Figure 6.19 consists in sampling the original signal at $4f_0$, instead of multiplying it by sines and cosines of the demodulation frequency f_0. It should simply be observed that the samples of a cosine $\cos(2\pi f_0 t_i)$, where $t_i = i/4f_0$, give the cyclic sequence $\{1, 0, -1, 0, \ldots\}$. So, to multiply a $4f_0$-digitised signal by $\cos(2\pi f_0 t)$ (i.e., by this cyclic sequence), it is enough to take every other sample and alternate the signs. And since the corresponding sine sequence is $\{0, 1, 0, -1\}$, the samples discarded for the cosine channel will be used similarly for multiplying by $\sin(2\pi f_0 t)$, hence giving the quadrature channel.

6.5.4   Adapted filtering

When receiving a signal of known characteristics, adapted filtering consists in correlating the signal received to a copy of the signal transmitted. The cross-correlation function thus realised goes through a very clear maximum when the signal received coincides with the signal transmitted (provided that the coherence between the transmitted and received signals is sufficiently high). Practically, this operation is only performed for modulated signals. In the case of non-modulated signals, a bandpass filter is often enough to isolate the frequency contribution of the signal expected.

Adapted filtering is expressed as:

$$y(t) = \int_0^T x(\tau)s^*(\tau - t)\, d\tau \tag{6.59}$$

It is nowadays always performed digitally. In practice, *sensu stricto* correlation of the time samples of the signal is not necessary. It can be shown that correlation (in the time domain) is identical to frequency domain filtering with a filter whose transfer function is the conjugate of the spectrum of the reference signal. As it is often preferable to work in the frequency domain, the (digital) processing follows:

$$x(t) \xrightarrow{\text{FFT}} X(f) \xrightarrow{\text{Filter}} Y(f) = X(f)S^*(f) \xrightarrow{\text{FFT}^{-1}} y(t) \tag{6.60}$$

Its advantage is the gain in processing time due to the use of FFT calculations.

The received signal $x(t)$ may not only be delayed, it may also be phase-shifted. The output of the correlator is then degraded by the phase shift, and the correlation maximum is no longer at the time of arrival of the signal (time lag). The solution consists in twinning the correlation of the signal received: (1) with the signal transmitted; (2) with the signal transmitted in quadrature. The outputs from the two correlators are then summed quadratically. This process eliminates the time offset and gives the maximum output level. It is, however, associated with a widening by a factor 2 of the correlation function main lobe (i.e., a loss in time resolution).

6.6   SONAR SYSTEM PERFORMANCE

6.6.1   The sonar equation

All the effects mentioned earlier (propagation, noise, reverberation, antennas, signals, processing, …) can be synthesised in the *sonar equation*. This equation can be written in a very simplified form, to give the probability of detection or measurement by an underwater acoustic system:

$$\text{Signal} - \text{Noise} + \text{Gain} > \text{Threshold} \tag{6.61}$$

The sonar equation is an equation of energy conservation used for evaluation, with some assumptions, of the performance of a system:

- The *signal* considered here is the signal level received by the system. It is therefore equal to: (1) for active sonars, the transmitted signal level, corrected for two-way propagation loss and target backscattering strength; (2) for passive sonars, the noise level radiated by the target, corrected for transmission loss.
- The *noise* affecting reception is the sum of the contributions from ambient noise and the self-noise of the sonar. It is characterised by its power spectral density. In some cases, for active sonars, one must also add the reverberation level.
- The *gain* features the array directivity index, and the receiver processing gain.
- The *reception threshold*, considered *at the output* of the receiver, depends on performance determined by the ultimate detection or measurement operations.

The power budget of an active underwater acoustic system can be written in more detail:

$$SL - 2TL + TS - NL + DI + PG > RT \tag{6.62}$$

where SL is the transmitted source level; $2TL$ is the two-way transmission loss; TS is the target strength, quantifying its capacity to reflect the acoustic energy received; NL is the noise or reverberation level; DI is the directivity index due to the array structure of the system; PG is the processing gain of the signal and the receiver used; and RT is the reception threshold.

For a passive sonar, Equation (6.62) simplifies to:

$$RNL - TL - NL + DI + PG > RT \tag{6.63}$$

where RNL is the level of noise radiated by the target and TL is the transmission loss along one way only. The terms corresponding to the target strength and reverberation disappear.

6.6.2   Note about the reception threshold

In the following presentation, the reception threshold, considered at the receiver's output, is the minimum SNR value making possible a given level of performance. It determines limit conditions for correct functioning of the system, which may be defined, according to the type of system considered, by:

- the *probabilities of detection and false alarm* for target detection;
- the *probability of error* for data transmission;
- the *estimation accuracy* for a measurement operation.

The reception threshold is linked to the output SNR r by:

$$\begin{cases} RT = 10\log r & \text{(linear receiver)} \\ RT = 5\log r & \text{(quadratic receiver)} \end{cases}$$

as defined and explained in Section 6.2.2.1.

This presentation is slightly different from those found in previous classical sonar books (Urick, 1983; Burdic, 1984a, b). In the latter, a *detection threshold* is defined at the input of the receiver; hence it includes both the processing gain and the *detection index*, which is the minimum SNR at the output; also the array directivity index is included inside the SNR at the input of the receiver. Although this approach has justifications of its own, it is more intuitive to separate clearly, in the sonar equation analysis:

- the input SNR, imposed by the physics of the problem;
- the various SNR gains brought by the receiver (array processing, filters, ...);
- the output SNR, to be linked directly with the expected performance.

6.6.3   Performance in detection

The *detection* process is the validation of the presence of a sonar echo in ambient noise or reverberation. The decision about detection is then taken if the received signal level after processing exceeds a predetermined threshold.

In detection mode, which is the basic function of a sonar, reception quality is characterised by two indicators:

- the *probability of detection* p_d: this is the probability that a processed signal exceeds the detection threshold (for a "true" signal present at the receiver input);
- the *probability of false alarm* p_{fa}: this is the probability that a noise peak exceeds the detection threshold (in the absence of a "true" signal at the input).

The entire problem lies in the correct definition of the threshold. Too high a threshold will only be reached with signals of high energies (p_d is then small). Conversely, too low a threshold may raise decisions about detection caused by noise peaks exceeding the threshold (p_{fa} is then high). Generally, the aim is to reconcile high values of p_d and low values of p_{fa}, which are contradictory. A compromise must be chosen (Figure 6.20), depending on the operational imperatives of the system,

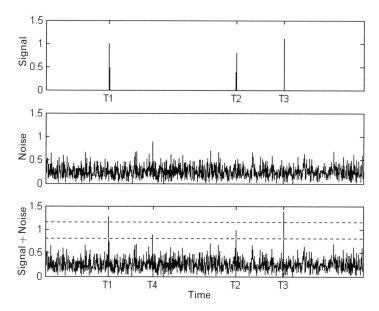

Figure 6.20. Illustration of the concepts of detection and false alarm. The expected signal (*top*) shows three arrivals, respectively, at times T_1, T_2 and T_3. It is perturbed by noise (*middle*). The sum of signal + noise (*bottom*) is analysed by two distinct detection thresholds (horizontal dashed lines). With the upper threshold, which is too high, only T_1 and T_3 are detected. With the lower threshold, the three arrivals are detected, but a noise peak at T_4 is also detected. The optimal detection threshold will therefore need to lie between these two values.

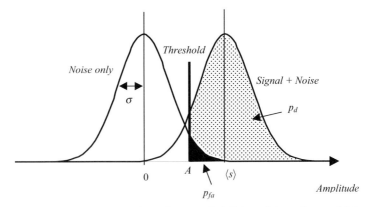

Figure 6.21. Graphical representation of the probabilities of detection and false alarm.

favouring one criterion or the other. For example, for a naval sonar detecting a threatening target (submarine or mine), a *non-detection* may be fatal, whereas a *false detection* will only waste time and effort.

The detection criterion traditionally used in sonar or radar applications is the Neyman–Pearson criterion: it consists in maximising the probability p_d of detection for a given upper limit of the false alarm probability p_{fa}.

Let x be the amplitude of a received signal, to be compared with a detection threshold A. In the case of a Gaussian noise and a Gaussian signal + noise, with identical variance σ and respective averages of 0 and $\langle s \rangle$ (the signal is simply added to the noise), the expression of p_d and p_{fa} can be written as a function of the selected detection threshold A.

The *probability of detection* is the probability that the signal + noise exceeds the selected threshold:

$$p_d = \frac{1}{\sigma\sqrt{2\pi}} \int_A^{+\infty} \exp\left(-\frac{(x - \langle s \rangle)^2}{2\sigma^2} \right) dx = \frac{1}{2} \left(1 - \mathrm{erf}\left(\frac{A + \langle s \rangle}{\sigma\sqrt{2}} \right) \right) \qquad (6.64)$$

and the *probability of false alarm* is the probability that the noise exceeds this threshold:

$$p_{fa} = \frac{1}{\sigma\sqrt{2\pi}} \int_A^{+\infty} \exp\left(-\frac{x^2}{2\sigma^2} \right) dx = \frac{1}{2} \left(1 - \mathrm{erf}\left(\frac{A}{\sigma\sqrt{2}} \right) \right) \qquad (6.65)$$

where erf is the error function.

Figure 6.21 shows a graphic interpretation of Equations (6.64) and (6.65).

Recalling Equation (6.25), the output SNR is defined as the average power after processing $\langle x_{s+n} \rangle$ with signal + noise as input, corrected for the average output level for noise only $\langle x_n \rangle$, and divided by the variance σ_n^2 of just the noise:

$$r = \frac{(\langle x_{s+n} \rangle - \langle x_n \rangle)^2}{\sigma_n^2} \qquad (6.66)$$

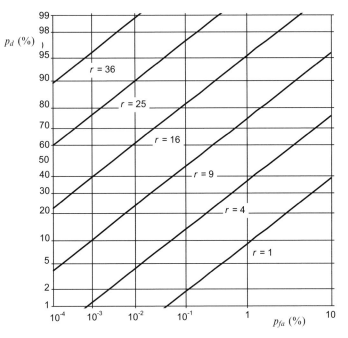

Figure 6.22. ROCs in the Gaussian case.

The relation between the output SNR and the probabilities of detection and false alarm is finally given by:

$$\sqrt{r} = 2(\mathrm{erf}^{-1}(1 - 2p_d) - \mathrm{erf}^{-1}(1 - 2p_{fa})) \tag{6.67}$$

Graphical representations linking values of r, p_d and p_{fa} are known as the Receiver's Operational Characteristics (ROCs). The plots corresponding to the elementary case (Equation 6.67) are shown in Figure 6.22.

The basic result (Equation 6.67) for the Gaussian case can be used in many configurations. But some interesting variations are noteworthy:

1   In the case of a fluctuating signal, the detection performance is, of course, degraded. The probability of detection will be a function of the amplitude density probability of the signal. If it follows a Rayleigh distribution (corresponding to a maximal fluctuation rate), it can be shown that the probability of detection can be approximated with the threshold A and the detection index r:

$$p_d = \frac{\exp\left(-\dfrac{A^2}{2 + r}\right)}{1 + \dfrac{2}{r}} \tag{6.68}$$

Practically, this expression is valid if $p_d > 0.1$ and $p_{fa} < 0.01$.

2   In the case of an optimal receiver processing a signal of unknown phase (reception by correlation of an out-of-phase signal, or non-coherent processing in active or passive sonar), the ROC curves are obtained through formally more complex equations:

$$p_d = Q(r, -2 \ln p_{fa}) \tag{6.69}$$

$Q(a, b)$ is the *Marcum function*, defined as:

$$Q(a, b) = \int_0^{+\infty} z \exp\left(-\frac{z^2 + a^2}{2}\right) I_0(az)\, dz \tag{6.70}$$

where I_0 is the modified Bessel function of the first kind and order 0. In practical configurations with high enough SNR, this expression leads to numerical results close to the ideal Gaussian case.[6]

6.6.4   Error probability in digital communications

The information transmitted in digital communications is coded with distinct signals affected by predefined values. For example, a binary coding can be achieved easily with two frequencies f_0 and f_1 corresponding to the values 0 and 1 of a bit of information (BFSK modulation for binary frequency shift keying). The receiver must now be configured to receive two distinct signals (e.g., by coupling in parallel two processing chains like the one in Figure 6.12), and decide which is the correct one (by comparing the output levels). Generally, systems with M modulation states are used, and the processing complexity increases with M.

The performance of a transmission system can be characterised somewhat differently from a target detection system. The notions of probability of detection and probability of false alarm, as defined earlier, are no longer really adapted to this case. They are usually grouped as a probability of error p_e in the transmission of information. The *Maximum A Posteriori* (MAP) criterion maximises this probability of error. For binary transmission (restricted here to transmission of 1 bit), it is easily perceived that p_e is the probability that the noise output level is larger on one of the receiving channels than the other. More simply, the noise output level is larger than the signal output level, leading to the error probability:

$$p_e = \frac{1}{2}\left(1 - \text{erf}\left(\sqrt{\frac{r}{2}}\right)\right) \tag{6.71}$$

or inversely:

$$\sqrt{r} = \sqrt{2}\, \text{erf}^{-1}(1 - 2p_e) \tag{6.72}$$

---

[6] This is explained by the fact that the Marcum function (Equation 6.70) is related to the Rice distribution (Equation 4.58), in the same way that the error function is related to the normal distribution. Hence, for sufficient values of r, the result tends to the Gaussian case in the same way as the Rice distribution tends to a Gauss law when its non-random component dominates the random component.

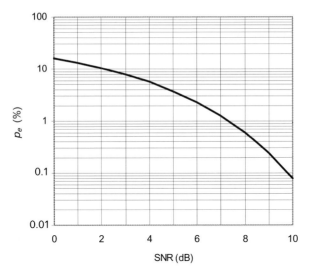

Figure 6.23. Bit error probability p_e for BFSK (Binary Frequency Shift Keying) transmission, as a function of the SNR.

This elementary result for a one-bit transmission (Figure 6.23) can be generalised to M modulation states (coding with M symbols):

$$\begin{cases} p_e = \dfrac{M-1}{2}\left(1 - \operatorname{erf}\left(\sqrt{\dfrac{r}{2}}\right)\right) \\[2mm] \sqrt{r} = \sqrt{r}\,\operatorname{erf}^{-1}\left(1 - 2\dfrac{p_e}{M-1}\right) \end{cases} \tag{6.73}$$

There are many other possibilities, depending on the type of modulation and channel considered. The interested reader can find more details in theoretical textbooks about digital communications (e.g., Proakis, 1989).

6.6.5 Parameter estimation errors

Underwater acoustic systems are usually not limited to the detection of targets. They also estimate some parameters, like:

- The range between the sonar and the target. In active sonars, it is measured from the two-way travel time of the signal. In passive sonars, it is measured from the delays between arrival times on several receivers.
- The speed of the target. In active sonars, it is measured using the Doppler shift of the signal backscattered.
- The angular direction. It is measured using the directivity patterns of the receiving arrays.

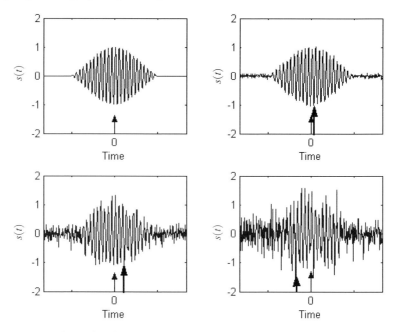

Figure 6.24. Error in estimation of arrival time for a noisy signal. The quality of the estimation (here a simple detection of a maximum) decreases with SNR. The estimated time of arrival (thick arrow) moves increasingly more relative to the "true" time of arrival (thin arrow).

6.6.5.1 The Cramer–Rao bound

The measurement accuracy of these values will evidently depend on the characteristics of the signal used (e.g., its time resolution) and the performance of the receiver (i.e., its directivity pattern). But the signal is usually noisy, and the measurement should in fact be considered as random (Figure 6.24 illustrates the influence of noise on estimation errors). The theory of parameter estimation allows definition, in general, of the accuracy limits as a function of SNR. It can be shown, in particular, that when a parameter θ is estimated by measurement of a random signal ν with a conditional probability distribution defined by $p(\nu, \theta) = p(\theta)p(\nu/\theta)$, the variance of the estimated θ is larger or equal to an absolute lower limit, the Cramer–Rao Bound (CRB):

$$\mathrm{var}(\theta) \geq \frac{1}{E\left(\left[\dfrac{\partial \ln(p(\nu/\theta))}{\partial \theta}\right]^2\right)} \tag{6.74}$$

where $E(\ )$ is the expected value (or statistical average) of a random variable. The CRB gives the furthest limit of estimation accuracy.

6.6.5.2   *Measurement of arrival time*

The time of arrival of an active signal contaminated by noise (assumed Gaussian) is frequently measured in sonar. As in detection, theory shows that the optimal receiver correlates the signal received to the reference signal, the time of arrival is measured without bias and it can be shown that the CRB has been reached. The time variance thus obtained is minimal. The same receiver is therefore optimal for the two functions of detection and estimation. The variance of the arrival time τ is given by the *Woodward formula*:

$$\text{var}(\tau) = \frac{1}{r(2\pi B_e)^2} \tag{6.75}$$

with $r = 2E/B$, and B_e is the "effective" bandwidth of the signal defined by:

$$B_e^2 = \frac{\displaystyle\int_{\Delta f} f^2 |S(f)|^2 \, df}{\displaystyle\int_{\Delta f} |S(f)|^2 \, df} \tag{6.76}$$

where B_e^2 is the variance of the spectrum around the central frequency. For a rectangular spectrum of width Δf, $B_e = \Delta f/3.46$. For a sinc spectrum of width Δf at $-3\,\text{dB}$, then we have $B_e = \Delta f/2.66$.

The Equation (6.75) value of $\text{var}(\tau)$ quantifies the optimal accuracy available for a given SNR.

The constraint on measurement accuracy is less often considered in sonar literature than the detection performance, possibly because of the applications usually envisaged. However, this constraint is actually a priority for all applications dealing with accurate measurement of time (positioning, tomography, bathymetry). The Woodward formula should then be used to define the minimum SNR, rather than the ROC plots for detection.

6.6.5.3   *Measurements of angle and frequency*

Similar relations can be obtained for the accuracies of angle and frequency measurements.[7] They are derived from the general expression of the CRB.

The relation corresponding to the measurement of an angle ϕ writes:

$$\text{var}(\phi) \geq \frac{1}{r\left(\dfrac{2\pi L_e}{\lambda}\right)^2} \tag{6.77}$$

where r is now the SNR at the output of the antenna and therefore includes its

---

[7] A model for phase-difference measurement accuracy is given in Appendix A.8.1.

directivity index. L_e is the "effective" length of the antenna, defined by:

$$L_e^2 = \frac{\displaystyle\int_L y^2 |g(y)|^2 \, dy}{\displaystyle\int_L |g(y)|^2 \, dy} \qquad (6.78)$$

where $g(y)$ is the weighting law of the array receivers.

For a non-weighted $(g(y) = 1)$ linear array of length L:

$$\begin{cases} L_e = \dfrac{L}{2\sqrt{3}} \\[4mm] \operatorname{var}(\phi) \geq \dfrac{12}{r\left(2\pi\dfrac{L}{\lambda}\right)^2} \end{cases} \qquad (6.79)$$

With weighting, the effective array length will decrease, typically by a factor of 1.2 to 1.4.

Similarly, for estimations of frequency f, one can use the "effective" duration T_e of the signal to write:

$$\operatorname{var}(f) \geq \frac{1}{r(2\pi T_e)^2} \qquad (6.80)$$

6.6.5.4   Case of a multiple estimation

If the same parameter is estimated on several realisations, the variance decreases with the inverse of the number of samples used. For example, for an estimate of arrival time using N independent signals, the variance reads:

$$\operatorname{var}(\tau) = \frac{1}{Nr(3\pi B_e)^2} \qquad (6.81)$$

This variance decrease corresponds in fact to the improvement (in N) of the SNR, as a function of the amount of signals used in the estimation (see Section 6.2.2.2).

7

The applications of underwater acoustics

This chapter aims to present albeit briefly the current applications of underwater acoustics. They have been grouped in different categories (navigation, military, fishery, mapping and geology, physical oceanography, underwater intervention), which although seemingly artificially separated, allow us to define very distinct types of acoustic system. Seafloor mapping systems will be more detailed in Chapter 8.

7.1 NAVIGATION

Navigation and safety (e.g., sounding, speed measurements, obstacle avoidance) have contributed in a major way to the initial development of underwater acoustics, and have been the starting points of many systems (transponders, sounders, logs, sonars, ...). Even though some of these applications may seem a bit marginal now, in terms of technological interest, and have often been finally replaced by other techniques, they still remain an important and active part of underwater acoustics systems.

7.1.1 Acoustic beacons

Among the first underwater acoustic systems developed, in the early 20th century, were beacons designed to signal zones hazardous to navigation or to equip light-ships. These systems simultaneously transmitted an acoustic signal (underwater or aerial) and an optical or radio signal. The delay between the two times of arrival, because of the two different propagation velocities, enabled observers equipped with ad hoc receivers to measure their distance from the beacon.

Nowadays, simpler, autonomous acoustic beacons (*pingers*) are commonly used to signal the location of a submerged obstacle, or to help in the positioning of a

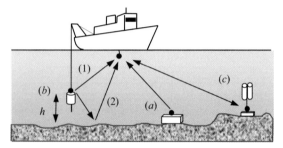

Figure 7.1. Acoustic beaconing configurations: (*a*) simple marking of an object on the seabed; (*b*) use of a pinger to measure the height above the seabed of a submerged system, by measuring the delay between paths (1) and (2) close to the vertical, $h \approx c(t_2 - t_1)/2$; (*c*) use of a transponder to locate and remotely operate a submerged instrument.

submerged system. The general purpose of these beacons is to facilitate detection and location of the platforms they are on. So they are usefully installed, as safety devices, on costly or dangerous loads or systems going to sea, to facilitate future retrieval in case of accidental loss.

They are also extremely helpful for operations at sea. For example, it is possible to measure the depth above the seabed of a system equipped with a pinger and dipped or deep-towed behind a ship, from the time delay between the direct signal and the signal reflected on the bottom (Figure 7.1).

Most often, underwater beacons just emit short Continuous Wave (CW) tones with a stable ping rate; the 8–16-kHz octave band is very frequently used, but higher frequencies are possible as well. Typical emission levels are around 180 dB to 190 dB re 1 µPa at 1 m. The simplest systems are programmed to transmit at a fixed ping rate. More complex devices transmit only upon reception of an interrogation signal. Named *transponders*, they allow a conventional "dialogue" using coded signals, hence increasing the reliability of the link.

These latter elaborate systems are used for underwater operations. Very often, for example, submerged scientific instruments are equipped with an *acoustic release*, to trigger their release back to the surface remotely. This type of application obviously demands high reliability, in order to avoid non-triggering upon command, as well as unwanted triggering due to noise.

7.1.2   Echo-sounding

Single-beam bathymetric sounders used in navigation and mapping are designed to measure the instantaneous water depth under a hull. They are without doubt the most common underwater acoustic systems, whatever the domain of application (Figure 7.2(*a*)). Their functioning is presented in more detail in Chapter 8. The most common frequencies range from around 10 kHz to several hundreds of kHz, depending on the water depth to explore. Normal systems, working in the upper part of the frequency range, deal with small water depths, in areas where a shallow

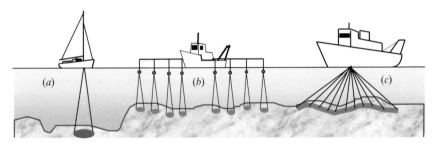

Figure 7.2. Different configurations of bathymetric sounders: (a) single-beam echo sounder; (b) swathe sounding; (c) multibeam echo sounder.

seafloor can indeed be a navigation hazard. Conversely, low-frequency sounders are more often used in mapping surveys and oceanography. The simplest sounders now belong to the domain of consumer electronics, and are very widespread in pleasure boats. Professional navigation sounders provide an accuracy of around 1% of water depth. Their emission levels typically range from 200 dB to 230 dB re 1 µPa at 1 m.

Hydrographic bathymetry sounding is subject to very strict accuracy standards, particularly for coastal areas, harbours and estuaries. Most often, marine charts are nowadays using sounding data from specialised multibeam sounders. Two geometric configurations are employed: either a series of monobeam sounders are installed on horizontal poles (Figure 7.2(b)) (this method is mainly used in river and harbour hydrography), or several acoustic beams are transmitted in a fan shape (Figure 7.2(c)). The latter configuration is much more convenient and will be detailed in Section 7.4.2 and in Chapter 8.

Sounders are generally made to detect the seabed, and are therefore steered downwards. But they can also be oriented upwards (e.g., on nuclear submarines sailing under the polar ice pack).

7.1.3   Speed measurements

Doppler logs are used to estimate the velocity of their carrying platform relative to the surrounding environment. They use the Doppler frequency shift of backscattered echoes (Section 2.5.1), proportional to the speed of the sonar relative to the target (Figure 7.3(a)). The reference medium from which the echoes rebound can be the seabed, if it can be reached by the signals. This case is the most favourable, yielding the most accurate navigation. Conversely, if the seabed cannot be reached, the signal is backscattered by the water column itself, and the navigation, measured relative to the water mass, must then be corrected for currents.

These systems work at very high frequencies (between 100 kHz and 1 MHz). They transmit several beams with distinct orientations in the horizontal plane. The Doppler shift measured in these different directions provides the spatial coordinates of the velocity vector. Acoustic logs are now mostly useful on submersibles,

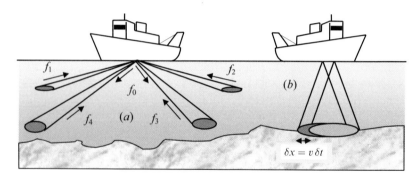

Figure 7.3. (*a*) Doppler log. The frequency shifts of the different beams are used to determine the velocity components in each beam direction, by $v_n = c(f_n - f_0)/2f_0$, $n = 1$ to 4. (*b*) Correlation log. The delay δt between the signals of the two beams is a function of the spacing δx between the transducers and the ship's speed v.

since surface vessels have easier means of accurately measuring their speed using positioning systems.[1]

A relatively recent improvement of acoustic velocity measurement systems, the *acoustic correlation logs*, use seabed echoes received separately by the transducers of a horizontal antenna. The structure of a signal received by a "fore" element of the array will be received, identically, by an "aft" element when the array (and the platform) moves forward. The time delay δt is simply linked to the displacement speed v and the spacing δx of the transducers by $\delta x = v \delta t$. So, by looking for the maximum cross-correlation function of the two signals, it is possible to estimate the delay, which is then used to calculate the platform speed (Figure 7.3(*b*)).

7.1.4   Obstacle avoidance

Submarines and underwater exploration platforms are often equipped with obstacle avoidance sonars. Such systems scan horizontally over a large angular sector, usually at a very high frequency (several hundreds of kHz) achieving narrow beams with a limited array size. These sonars are used to detect, locate and identify obstacles a few tens or hundreds of metres from the platform. The technological solutions for the acoustic sweeping of the horizon have evolved with time, from searchlight-type systems with a single beam rotating mechanically, to the modern multibeam systems processing a fan of preformed beams, whose technology is described in Section 8.4.

---

[1] Most often, the Global Positioning System (GPS) by satellite.

7.2 MILITARY APPLICATIONS

Despite the current trend toward relative disarmament and the widespread decrease in navy budgets, military applications still represent the major part of economic activities linked to underwater acoustics. The modern naval systems of submarine detection are both huge and extremely sophisticated; they require design and development means commensurate only with the major player companies in the defence industry.[2]

7.2.1 Passive sonars

Passive military sonars are designed for detection, tracking and identification of submarines. They work at very low frequencies, between a few tens of hertz and a few kHz. It is indeed in this range that the acoustic energy radiated by ships is the most important (in particular their characteristic spectral peaks, identifying the noise sources). The detection ranges are also longest at low frequencies, due to the smaller absorption losses. Their applications during the two world wars were rather anecdotal, but passive sonars have been the preferred instrument for submarine detection since the 1960s (Tyler, 1992). For today's surface warships specialising in ASW (Anti-Submarine Warfare), they provide detection ranges higher than active sonars, and very attractive possibilities of target tracking and identification. Submarines are then submitted to extremely tight constraints on their radiated noise level to insure their covertness. On the other hand, they can use this unique silent detection tool to counter their predators (while the use of any active sonar by a submarine in operation is a priori excluded for clear reasons of stealth, except at very high frequencies).

Modern passive sonars are characterised by the deployment of towed linear arrays, very long (several tens to hundreds of metres), able to efficiently detect and locate low-frequency noise sources (Figure 7.4(*a*)). This principle is known as TASS (Towed Array Sonar System). The arrays are made of a flexible hose-like fitting, enclosing the receiving hydrophones (up to several hundreds), preamplification electronics and the corresponding cables. They are stored and deployed using specific winches,[3] both onboard surface vessels and submarines. The size of these towed arrays make them very efficient both in terms of SNR (Signal-to-Noise Ratio) gain and angular resolution. Towed far behind, they are also reasonably free from acoustic interference caused by the noise radiated by the carrying vessel. Moreover, being submerged at a sufficiently deep level, they benefit from better propagation conditions than surface sonars.

---

[2] Important names in the field of naval sonar industry are Raytheon, EDO, Lockheed Martin, Westinghouse (USA), Thales Underwater Systems (France), BAe (UK), STN-Atlas (Germany).

[3] Note that the same technology is employed in marine seismics (for hydrocarbon prospecting and geophysical exploration), with variants: the arrays then reach lengths of several kilometres, and several arrays may be towed at the same time.

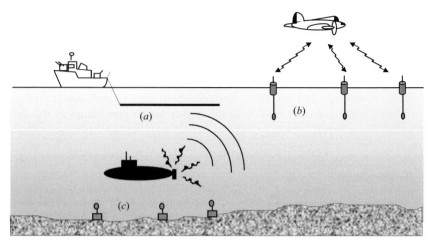

Figure 7.4. Passive detection systems: (*a*) towed linear array; (*b*) airborne buoys; (*c*) permanent surveillance network.

Figure 7.5. Typical array configuration on a modern submarine: (*a*) towed linear array; (*b*) bow cylindrical array; (*c*) flank array; (*d*) interception array.

Other types of array are more specifically used in submarines. They may be spherical or cylindrical, placed in the bow, and able to form many beams in the horizontal and to a lesser extent vertical plane; or large flank arrays, installed alongside the hull or the conning tower; or interception arrays at the top of the conning tower (Figure 7.5). Used as a complement to the towed line array, these are mostly used in tracking applications.

7.2.1.1   Tracking

After detection, a passive sonar must be able to accurately track the noisy target. This is achieved by exploiting the spatial structure of the signals received by the various antennas. Angles are measured very precisely in the horizontal plane by beamforming on the reception arrays, in particular the TASS, which is highly suited to forming very narrow beams. Horizontal bearing angles are plotted on time/angle graphs known as BTRs (Bearing–Time Records), which clarify the position evolutions of the various targets detected simultaneously in the area (Figure 7.6).

However, TASS suffers from two major measurement limitations due to its

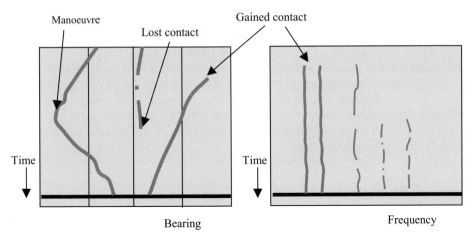

Figure 7.6. Classical data displays for passive sonars. (*Left*) Bearing-Time Record (BTR). The broadband detection of three noisy targets are plotted according to the horizontal angle as a function of time. (*Right*) LOFARgram. The narrowband spectrum, recorded in a given beam, is displayed as a function of time, making clear the changes in the target's signature.

linear shape. First, as the beam patterns are symmetrical around the array axis, they cannot measure angles separately in the vertical and horizontal planes. This is usually not too limiting because, at long ranges, the sonar and the target may be considered to be approximately at the same order of depths. But, even when working in the horizontal plane, the angle measurements are ambiguous, each beam being formed symmetrically on the right and the left of the array: hence each beam detection gives two directions, and the ambiguity must be suppressed by changing the towing heading.

The angles in the vertical plane have to be estimated using the other arrays. Unfortunately, their vertical dimension is much less than the possible horizontal extent of TASS or flank arrays, and their angle measurement performances are not comparable.

Besides "traditional" beamforming, based on a highly intuitive concept, the techniques of angular localisation with linear antennas have been the subject of many theoretical studies in recent years. These resulted in the design of *high-resolution* techniques (see, e.g., Bouvet, 1991), using the cross-correlation properties of the signals received by the different transducers on the array, and yielding remarkable angular resolution performances (although they are quite sensitive to the acoustic field and noise structure characteristics).

Measuring the distance between the sonar and the target is quite straightforward with an active sonar, where the propagation delay is available. This is, of course, more difficult to achieve with a passive sonar. A first technique consists in measuring the phase shift or the time delay (using the cross-correlation function) between signals received at different points of a long array, or on different subarrays spaced apart. The length differences thus obtained between the acoustic paths are used to triangulate and deduce the position of the noise source. Another solution is to measure the delays and possibly the vertical angles between the multiple paths along which the detected signal is coming. A geometric reconstruction of the sound field in the vertical plane can then point to the location of the noise source.

Surface vessels performing passive-sonar tracking have the advantage of getting access to coordinated information provided by specialised ASW aircraft or helicopters, deploying additional transducers (dipped sonars or airborne buoys) (Figure 7.4(*b*)). This greatly improves the coverage of an operational area, and expands the tracking possibilities.

7.2.1.2   Spectrum analysis and identification

Identifying the target is another essential aim of passive sonars, and it is perhaps the most impressive for the public. It is indeed often performed "by ear", by highly trained operators whose skills can distinguish ship classes and even individual ships of a particular class, just by listening to the noise radiated! These specialists, whose training and knowledge are, of course, highly classified, work on "audio" signals pre-processed through frequency filters and demodulators.

"Audio" analysis is also assisted by automated systems that perform very fine spectral analyses. A very common processing in passive sonar is the *LOFARgram* (for LOw-Frequency Analysis and Recording) (Figure 7.6), which represents the evolution of the processed noise spectrum as a function of time. After removing the broadband noise, this accumulation with time reveals characteristic spectral peaks and their evolution in the noise radiated by the target. Another processing, named *DEMON* (for Demodulated Noise), consists in detecting low-frequency modulations caused by machinery and propulsion, by analysing the envelope of the demodulated noise. Automated systems are known to be more efficient at detecting and extracting stable signals with a well-marked harmonic structure from noise, while human operators are more efficient with transitory signals.

7.2.1.3   Interceptors

Sonar interceptors are passive systems specialising in the detection of active signals transmitted by the opponent. They therefore aim at identifying and locating hostile sonar long before it can start detecting. They are able to do this because the signal received by the interceptor has undergone only a one-way propagation loss, two times smaller (in decibels) than the two-way propagation loss suffered by the hostile sonar receiver. The same principle is used with airborne radars (the detection of enemy radar surveillance being vital to military aircraft), and on roads (where drivers want to detect radars to avoid speed checks). The sonar interceptors installed on submarines include specialised arrays, and highly efficient processing systems for locating and identifying active sonar sources as fast as possible. They are particularly useful, even vital, for the detection of homing signals from torpedoes. In this last case, a possible acoustic countermeasure is the release of decoys, in order to confuse the hostile sonar.

7.2.1.4 Surveillance networks

Fixed systems of passive surveillance of marine areas can be considered in the category of passive sonars. Such networks, equivalent to huge arrays, were systematically placed by the USA during the Cold War along its coasts and in passages unavoidable by enemy submarines (*SOSUS* network, for Sound Surveillance System). Partially declassified and decommissioned, parts of this network are now used for oceanographic and geophysical monitoring (studies of whales, small volcanic eruptions or earthquakes, etc).

7.2.1.5 Airborne systems

The buoys used by ASW aircraft and helicopters often work in passive mode (DIFAR buoys, for Direction Finding Acoustic Receiver). They are autonomous listening stations, retransmitting by radio waves the underwater signals they receive. Their deployment enables the creation of a genuine temporary surveillance network. Their interest lies in their ease of deployment, and their low self-noise level. The mediocre directivity performance of their individual antennas because of their small dimensions, compared with the long wavelength probed by passive sonar operations, is compensated by their potential as a network for locating a target.

7.2.1.6 Mines and torpedoes

Underwater mines with acoustic triggering are equipped with a system of automatic detection of the noise from a vessel. It triggers the ignition after processing the acoustic signature of the target, with varying degrees of sophistication in the analysis of the noise received. To decide on whether detection has been achieved or not, these systems are often complemented with detectors of magnetic field and hydrostatic pressure variations. Similarly, torpedoes can be equipped with a passive homing system, to direct them toward the main source of noise in the target (i.e., its propelling system).

7.2.2 Active sonars

The active sonars of surface vessels are used for surveillance, tracking and identification of submarines, depending on their characteristics. The frequencies in use have decreased with the years, from a few tens of kHz down to a few kHz, and they are now around 1 kHz, or less. The transducers are installed in a dome below the hull (Figure 7.7(*a*)), or in a bow bulb or in a "fish" towed at variable depths (Figure 7.7(*b*)).

7.2.2.1 Variable depth sonars

For a sonar transmitting near the surface, performance is highly dependent on the local conditions of propagation. In deep waters, the main limitation is due to the downward refraction of acoustic waves caused by the negative velocity gradient close to the surface. This refraction creates shadow zones, where direct insonification is

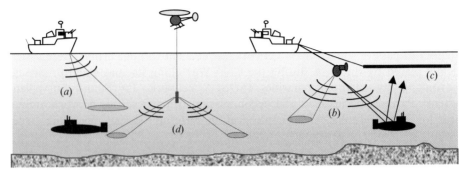

Figure 7.7. Active detection of submarines: (*a*) hull-mounted sonar; (*b*) towfish-mounted sonar, with (*c*) TASS receiver; (*d*) helicopter-dipped sonar.

impossible for tens of kilometres (see Section 2.8.3.2). Mastering the characteristics of these shadow zones is paramount in ASW tactics.

For a long time, the only solution to combat shadow zones, for surface warships using active sonar, was to transmit at an oblique incidence and to use the "bottom bounce" path reflected from the seabed. The performance of this method relies heavily on the topography and reflectivity of the seabed.

Since the horizontal extent of shadow zones decreases when the source depth increases, the use of Variable Depth Sonars (VDS) is a much better solution to fill the detection gaps caused by shadow zones. Towing the sonar under the surface thermocline dramatically increases the range capacities of detection. This led to the development of specific sources that are able to withstand high hydrostatic pressures; these sources are towed at depths that depend on the cable length and the ship's speed.

7.2.2.2   *Low-frequency active sonars*

Active sonars transmit at very high power levels, around 240 dB re 1 μPa at 1 m. Obtaining such levels in the 1-kHz frequency band is a challenge; it required the emergence of original transducer technologies (described in Section 5.1.2.3); and it led to the building and deployment of sources of huge dimensions onboard specially equipped ASW vessels.

Reception may be achieved either on the transmitting array itself, or on a long towed array. The latter solution is increasingly encountered with the use of very low-frequency active sonars. These sonars require receiving arrays that have dimensions that are comparable with passive sonars, and the same tools can then be used for both detection modes. The receiver and the source may also be quite distant from one another (e.g., deployed from distinct vessels). These multi-static configurations clearly improve the potential for detection and tracking.

The evolution of active sonars toward lower frequencies came from the need to increase detection ranges and to compensate for improvements in the stealth of submarine hulls (e.g., by anechoic coatings designed to absorb incident acoustic waves, which are less efficient at long wavelengths). Moreover, pulse-compression techniques (see Chapter 6) are systematically used in detection mode, to increase the useful power further and improve the detection range. The signals used in practice are of several types: modulated signals (usually in frequency) for detection and range measurement; and narrowband signals for estimating the Doppler shift (i.e., the target's speed).

7.2.2.3   Other active sonar types

- Specific active sonars can be dipped from ASW helicopters (Figure 7.7(c)). These can move quickly from one area to the next, and deployments are easier, which makes them highly efficient.
- Some types of buoy deployed by aircraft and helicopters are actually active sonars (*DICASS*, for Directional Command Activated Sonobuoy System).
- Torpedoes are equipped with high-frequency active homing systems, to track their target.
- Finally, submarines themselves carry active sonars, but they usually avoid using them as much as possible in operations, to retain their covertness.

7.2.2.4   Active sonar limitations

The use of active sonars for submarine detection is, of course, affected by ambient noise and ship's self-noise. But, more specifically, these systems are prone to limitations due to the reverberation of signals at the interfaces and on heterogeneities inside the water column, which can often mask weak echoes from targets. Minimising the effect of reverberation can be obtained by improving the spatial selectivity of the signal used (wideband, very narrow directivity lobes on reception), or by using (by means of narrowband frequency filtering) the differences in the Doppler shifts of the target and reverberation echoes. Shallow waters and continental slopes are particularly unfavourable areas, because of the problems raised by interface reverberation.

7.2.2.5   Minehunting sonars

Underwater mines have long proven their remarkable and deadly efficiency, leading to the specific development of a particular type of active sonar for *Mine Counter-measures* (MCM), or minehunting. These systems work at very high frequencies (typically between 100 kHz and 500 kHz). They provide acoustic images of the seabed with extremely high resolutions, and are used on dedicated vessels. As for active ASW sonars, the main constraint lies in the reverberation level from the water–seabed interface, rich in micro-reliefs whose echoes can very easily mask the signals backscattered by mines. This is even more true as modern mines are shaped and coated with anechoic materials to be as unobtrusive as possible.

Minehunting systems aim at detection and identification of targets that are proud on the seabed or floating in mid-water. They are placed under ship hulls or on self-propelled towfish (PVDS, for Propelled Variable Depth Sonar), and scan the seabed and water column in front of the ship.

Surveillance sonars are towed sidescan sonars with very high-resolution specifications, used to map very accurately a restricted seafloor area, and to provide extremely detailed images. Being towed, they cannot ensure the safety of the supporting platform.

Whatever the type of system, identification of the type of target detected is a crucial step, as is confirming whether it is really a mine or not; processing is most often based on analysis of the acoustical shadow projected on the seabed.

Finally, the destruction of acoustically triggered mines can be achieved by using *acoustic minesweeping* systems. These feature low-frequency broadband sources able to mimic, in a realistic way, the noise radiated by specific vessels.

7.3   FISHERIES ACOUSTICS

Marine fisheries make intensive use of underwater acoustic techniques (see McLennan and Simmonds, 1992; Mitson, 1983; Diner and Marchand, 1995). Modern vessels for industrial or small-scale fishing are always heavily equipped with sonar systems, as are oceanographic vessels that specialise in halieutic surveys. The first function of these acoustic systems is to detect and locate fish schools, and to aid in their capture. Scientific applications, instead, aim at identifying species and evaluating the quantitative importance of the biomass.

7.3.1   Fishery sounders and sonars

The fundamental acoustic tool of fishermen is the single-beam echo sounder, which transmits vertically below the support vessel. It is used to detect and locate fish schools, and even isolated individual targets. The main acoustic characteristics of a standard fisheries sounder are the same as a bathymetric sounding system. But high energy levels are required at transmission, because of the low values of target strengths, and echo processing and display are necessary throughout the entire water column. Different frequencies are used, depending on water depth, and they are sometimes packaged into one multi-frequency system. Frequencies commonly used range between 20 kHz and 200 kHz. The angular and time resolution characteristics are typically of the order of $10°$ and 1 ms, respectively.

The echoes obtained by fisheries sonar (Figure 7.8 and Plates 1 and 2 in the colour section) are made either of very intense spots, corresponding to very dense schools, or of smaller, less reflective dots. The dimensions of dense schools are typically between a few metres and a few tens of metres. At night, the schools disperse, and the echogram shows clouds of small, less reflective patches, corresponding to individual echoes. The basic understanding of such echograms is evident, but their detailed analysis requires more experience, as there are many sources of interpretation error.

Panoramic-scanning fisheries sonars (Figure 7.8(*b*)) are much more complex and expensive systems. They detect and locate fish schools in the horizontal plane, which

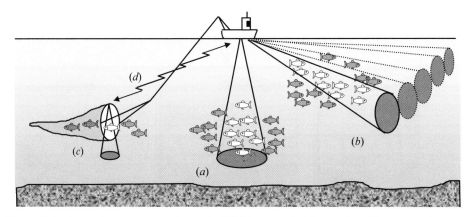

Figure 7.8. Acoustic equipment of a typical fishing vessel: (*a*) sounder; (*b*) panoramic sonar; (*c*) netsonde; (*d*) trawl positioning and remote controls.

allows for more efficient exploration. This function may be fulfilled in a simple way by a searchlight-type single-beam sonar, mechanically rotating so that the beam scans the horizontal plane. However, modern elaborate systems use a cylindrical array for panoramic insonification, forming beams in the horizontal plane with apertures of a few degrees. These beams can also be inclined in the vertical plane (creating an "umbrella-shaped" directivity pattern), increasing the potential for target detection. They usually work at slightly lower frequencies than the sounders (typically 20–80 kHz), and their range can reach several kilometres. They provide much more spatial information than vertical sounders, but are also subject to more complex perturbations (e.g., seabed or sea surface reverberation). Their results are therefore all the more difficult to interpret and exploit.
\cr

7.3.2 Other equipment onboard fishing vessels

Trawls are also equipped with a particular type of echo sounder, called a *netsonde* (Figure 7.8(*c*)). This system is installed at the entrance to the trawl's pocket, on the top rope, and transmits downwards. It allows instant visualisation of the position and the opening of the trawl (Figure 7.9(*a*)), and at the same time can detect and evaluate the catch entering the net. These data are transmitted to the ship and displayed for an instantaneous monitoring of the fishery operations. Note also that trawls can be equipped (Figure 7.8(*d*)) with acoustic positioning systems, in order to locate them relative to the towing vessel, and also with acoustic remote control systems, for example, to close part of the net.

7.3.3 Scientific uses

Beyond their use by fishermen, acoustic detection systems are valuable tools for scientists studying the biomass. The main application consists in quantifying fish

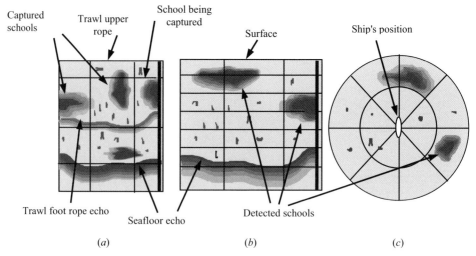

Figure 7.9. Schematic visualisation of the data from different fisheries acoustic systems: (*a*) netsonde; (*b*) sounder; (*c*) panoramic sonar.

populations. In the presence of isolated fish, this is done by counting the individual echoes detected. This requires the sounder having a sufficiently high spatial resolution, enabling narrow angular filtering. Besides counting, the detection of individual echoes also allows individual target strengths to be accessed. Since the measured value of one point-like target inside a single beam is strongly dependent upon the target's angular position (due to the beam's directivity pattern), this determines the use of specific techniques, such as *split-beam*[4] or *dual-beam*,[5] that are encountered in echo sounders specially developed for halieutic research.

In the presence of schools or dense layers of fish, whose individual echoes cannot be separated, population measurement is done through *echo-integration*. It consists in quantifying the biomass of an area from the cumulative acoustic energy reverberated by all the targets in the zone sampled, and knowledge about average, individual target strengths (obtained, e.g., from measurements of isolated individuals of the same population). However, this technique, seemingly very simple, requires much caution, concerning the instruments used (calibration of the sounder, in directivity and sensitivity) and the methodology used. Furthermore, the total energy is not always proportional to the number of targets (due to shadowing by the upper targets, or multiple scattering inside the school). Echo-integration is usually

---

[4] A split-beam sounder uses different parts of the transducer, as in interferometers, in order to determine the angular position of the target relative to the beam's axis; it is then possible to compensate accurately the measured echo level from the directivity pattern value for this angle.
[5] A dual-beam sounder uses two coaxial receiving beams with different apertures. The echo level difference between the signals from the two beams gives an estimation of the angular position of the target.

employed for fish stock monitoring, for scientific or economic purposes, or for measurement surveys on dedicated oceanographic vessels. It is a very demanding technique in terms of measurement precision. For instance, the transmission and reception sensitivities of the sonar, or its beamwidth pattern, must be fine-tuned to obtain a very precise accuracy (typically 1 dB to 2 dB; a 3-dB bias in level estimations would lead to an error by a factor of 2 for biomass quantitative estimation).

Apart from measuring the number of individuals, it is, of course, appealing to try to identify the species, for monitoring and managing populations as well as for selective fishing strategies. Presumably, acoustic signals can in some way help in the identification of species, either from the characteristics of individual echoes or from parameters associated with the structure and behaviour of schools. While it is difficult to imagine that much information can be extracted from individual fish echoes using today's sounder signals, the global approach based on the geometrical characteristics of the fish school (dimensions, shape, depth, altitude) does seem to be much more promising, provided that these parameters are accurately measurable. Experimental research in this area is made extremely difficult by the near impossibility of validating targets in real situations with sufficient accuracy. Although interesting results have been obtained under controlled conditions, it seems we are still today far from identifying species automatically, even using the best performing sounders. Nevertheless, more can be expected from the recent or forthcoming technical improvements in sounders.

The scientific applications mentioned above have justified several innovative technological developments. The traditional, single-beam, fisheries echo sounders are subject to a series of structural limits, linked to the geometry of the beam, that yield a mediocre horizontal resolution (see Section 8.2.2.4). This makes detection of targets close to the seabed, and in particular, estimation of school sizes alongside the ship more difficult. Such limitations could be overcome by using multibeam sonars like the ones used in bathymetry, but with specific constraints: they should use signals from the entire water column, which means they should process amounts of data much larger than those, already large, of bathymetry systems. And the need to detect targets close to the bottom means directivity patterns must be highly constrained (sidelobes with very low levels).

Also, the relatively poor spectral content of normal narrowband signals means they are not appropriate for identification of targets. It is quite likely that fisheries research will look in the years to come at the use of multibeam sounders with a broad spectrum of frequencies. These will be used to extract much more information from target echoes, using techniques that combine spatial, temporal and frequential analysis.

The ongoing modernisation of industrial fisheries techniques, regarding capture tools and acoustic detection systems, has led in just a few decades to a worrying decrease in natural resources. This situation means we should focus technological development toward selective fishing tools and techniques, for which acoustics is, of course, a cornerstone. This is a major challenge, in the years to come, for scientists and engineers in the field, as a preliminary to their larger use by fishermen.

7.4   MARINE GEOLOGY AND SEAFLOOR MAPPING

Advances in marine geosciences have been largely secondary to the development of tools dedicated to acoustic investigation. Depending on their structure and processing principles (see, e.g., de Moustier 1988, 1993; Tyce, 1988; Somers, 1993), these tools can either provide "sonar images" of seafloor features (see Blondel and Murton, 1997), or quantitatively access specific parameters about the seabed: impedance contrast between the water and the bottom, topography at various scales ranging from large geophysical structures to microscale textures, presence and structure of subbottom layers. While not replacing the direct methods of seabed study made by geologists, which is based on the analysis of samples, acoustic systems provide a *quasi-instantaneous wide observation* of the differences in morphology between the water–seabed interface and sedimentary layers. These systems are thus complementary to point sampling and *in situ* geotechnical measurements, as they allow us to generalise parameters measured locally. Three types of acoustic system are extensively used in marine geosciences: sidescan sonars, multibeam sounders and sediment profilers. In addition to these systems, which are all based on sonar technology, seismic measurements help in visualising structures within the seabed.

7.4.1   Sidescan sonars

The working principles of sidescan sonars will be detailed in Section 8.3. They are designed to provide "acoustic images" of the seabed, with high definition. In marine geoscience, they are therefore used to give a near-visual representation of the geological facies. Sonar images also give general indications about the nature of the water–bottom interface, directly linked to the "reflectivity" of the signal. A soft sedimentary bottom (silt, mud) will send back little energy because of its low impedance contrast with water and its interface smoothness. A rocky or gravelly bottom will have the opposite effect, with strong impedance contrast and high roughness. Sidescan sonar images faithfully reproduce even the small details on the interface (depending, of course, on the resolution achievable). The presence of some interface relief, even of low amplitude, will produce echoes from the faces "visible"to the sonar, with intensities modulated by the local angle of incidence; acoustic shadows will appear when the relief is important enough to mask some areas. The resolution of sidescan sonars is typically of a few tens of centimetres,[6] thus adapted to the sedimentary small-scale relief. Acoustic images can differentiate between very fine geological structures (e.g., ripples, dunes, rocky outcrops). The quality obtained by the most recent sidescan sonars is often impressive, and comparable with photographs taken under a grazing artificial lighting (Figure 7.10).

Sidescan sonars are usually lightweight systems, easily transportable and designed for shallow waters, with typical ranges of a few hundreds of metres.

---

[6] Although it can reach several tens of metres for certain low-frequency sonars (see Blondel and Murton, 1997).

Figure 7.10. Acoustic images of rocky outcrops on a sandy, rippled seafloor, obtained using a Klein-5500 sidescan sonar (450 kHz). The image width is 100 m. Low acoustic reflectivity is coded with dark grey levels, strong reflectivity with light grey levels (data provided courtesy DCE-GESMA).

Their small size, their ease of use and the quality of their data make them highly attractive – and widespread – for many coastal applications.

Deepwater imaging adds specific constraints. The use of a high-frequency sidescan sonar close to the bottom yields excellent results, because of the small values of the absorption coefficient and the low ambient noise level. But it needs to be made of material suited to great depths (i.e., high pressures) and requires electric cables several kilometres long. This calls out for the use of a large fish towed at low speed. Several companies now propose such systems, usually also equipped with a sediment profiler and a positioning system. At the other end of the range, it is possible to use lower frequency sonars close to the surface, with a very large swathe width. The British sonar *GLORIA*,[7] working at 6.5 kHz, is an extreme case: it can cover areas 30–60 km wide, and resolves large structures with a resolution of around 60 metres. Interest in this second type of sidescan sonar has been lessened by the emergence of imaging multibeam sounders, which are not only easier to use but also provide bathymetric data.

Mainly designed for imagery, sidescan sonars are increasingly frequently complemented with a bathymetric function, based on interferometric measurements

---

[7] For Geological LOng-Range Inclined Asdic. Other examples of this concept are the SeaMarc II (USA), a 12-kHz sidescan sonar towed at the surface; or the intermediate TOBI (UK), a 30-kHz deep-towed sonar, which is closer to classical, high-frequency, sonar technology.

between two superposed receiving antennas (following the principle explained in Section 5.3.7). Although their quality is less than that acquired with specially designed multibeam sounders, these measurements are still a very useful complement to sonar imagery.

7.4.2   Multibeam sounders

Multibeam bathymetric echo sounders appeared[8] at the end of the 1970s; their potential has evolved considerably since then, and they are now in general use. The first models were limited to a small number of beams with modest angular performances (16 beams of 2.7° for the first *SeaBeam*). They now typically feature between 100 to 200 beams, over an angle sector of 75° on each side of the vertical.[9] Large arrays installed below ship hulls allow reaching directivity lobes of around 1°. These sounders can map wide areas rapidly and accurately; a low-frequency multibeam sounder used in deep water can, for example, cover around 10,000 km^2 per day. The success of this new concept has been huge: around 700 systems have been built today, of all sorts.

The bathymetry measurement is made inside each beam, formed at reception, by looking for a (time, angle) couple which reconstitutes the wave trajectory in the water column and the position of the impact point on the seabed. Different measurement techniques can be used (see details in Section 8.4). Close to the vertical, where the echo is very short, the time of maximum amplitude is detected; this corresponds to the axis of the beam considered. At oblique incidence, the same method can be used, but an interferometric measurement is usually preferred: each beam is divided into two sub-beams, and the phase difference between the two signals is measured over time, to determine the moment where it is null, corresponding to the aiming direction of the main beam. In all cases, beams must be very accurately formed so that they are as spatially selective as possible.

The accuracy of oblique-beam bathymetry is highly sensitive to perturbations of the angular measurements, and movements of the support vessel must be very accurately compensated. This requires the use of a motion reference unit, whose data are recorded along with data from the acoustic transducers. Finally, refraction of acoustic paths by the sound velocity profile can be quite troublesome. Accurate knowledge of hydrologic conditions is therefore imperative to perform quality bathymetric surveys with a multibeam sounder, especially as the angular aperture is larger. Finally the sounding data must be georeferenced, using the navigation data provided by an operational positioning system.

Multibeam sounders also use their large angular aperture and grazing-angle signals, comparable with those of a sidescan sonar, to record acoustic images,

---

[8] The first scientific system was the SeaBeam instrument installed on N.O. *Jean Charcot*, operated by IFREMER, in 1977.

[9] This figure corresponds to a swathe width 7.5 times the water depth. Practically, however, because the quality of the most oblique data are difficult to maintain, this ratio is limited to 4 or 5.

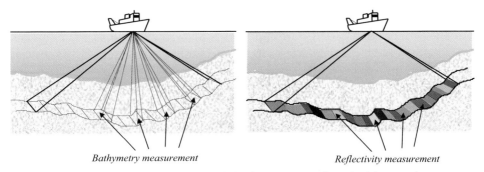

Bathymetry measurement   *Reflectivity measurement*

Figure 7.11. Multibeam sounders: (*left*) bathymetry sounding; (*right*) sonar imagery.

using the same principle (Figure 7.11). The performance is, however, not as good as with a deep-towed system, because of the movements of the platform, the incidence of the signals (not grazing enough) and the imperfect suitability of the antennas for this particular function. Despite this restriction, with these systems, marine geologists have at their disposal integrated tools that can acquire at the same time the bathymetry and reflectivity (an example is given in Figure 7.12; see also Plates 3–6 in the colour section). Moreover, simultaneous use of seismic and sediment profiling data frequently gives a very complete picture of sedimentary structures.

Along with the traditional exploitation of bathymetry and imagery measurements for pure mapping purposes, the acoustic data from sounders and sonars (provided they are correctly calibrated) can be used to assess objectively some physical properties of the seabed (its reflectivity being linked to its impedance and microscale roughness), and hence to get indications about its nature. This is evidently similar to the techniques used in satellite remote sensing, using spaceborne radar to map the Earth's surface. As with radar, several techniques can be used with sonar data, sometimes complementing each other: variations of the backscattering strength with the angle of incidence, quantification of the textures of high-resolution images, analysis of raw backscatter, etc. The main difficulty with these approaches lies in their subsequent interpretation: the processes observed are very complex physically, because of the very nature of the sediments; they are imperfectly investigated by acoustic systems, and the final data is often ambiguous. Furthermore, the ground-truth operations to "calibrate" the acoustic measurements are difficult to achieve, especially in deep water. Despite these difficulties, these methods of investigation (a very active research area) are expected to provide interesting features for geological interpretation. But one should not forget that, ultimately, acoustics cannot directly access parameters such as granulometry or mineralogical composition, which are the basis of geological analyses of sediments.

7.4.3   Sediment profilers

Sediment profilers (or SBP for SubBottom Profilers) aim at exploring the first layers of sediment below the seafloor, over a thickness commonly reaching several tens of

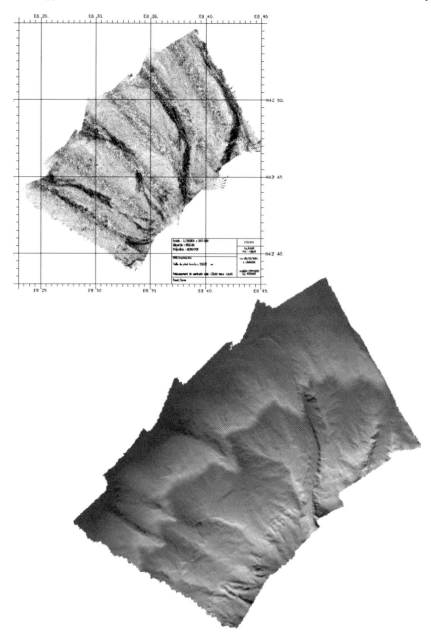

Figure 7.12. Bathymetric map and sonar imagery obtained using the same multibeam sounder (Simrad EM300, 30 kHz). Area off the west coast of Corsica, spreading over 15 nautical miles from north to south; water depths range from 1,100 m (SE) to 2,800 m (NW). (*Top*) Sonar image with its geographic coordinates; dark tones correspond here to high reflectivities. (*Bottom*) Corresponding bathymetric data, shown after artificial shading of slopes. Note the high reflectivity levels associated with the bottom of canyons.

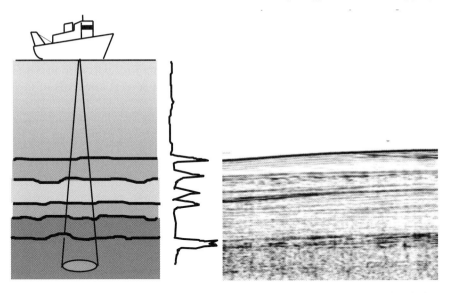

Figure 7.13. Profile obtained with a sediment profiler. (*Left*) Sediment layering below the seabed. (*Middle*) Time domain signal for this echo. (*Right*) Final cross-section of the seafloor, obtained from the time domain signals plotted consecutively along the ship's track.

metres. Technologically speaking, they are single-beam sounders working at a very high level and low frequency (in the 1–10-kHz range, 3.5 kHz is often used as the nominal signal frequency). Sediment profilers mostly use frequency-modulated signals and pulse-compression techniques to improve their penetration range. Some models use non-linear parametric sources, which provide a very narrow directivity lobe despite the low frequency thus generated.

The echo signal comes from the reflection (not the backscattering, see Section 8.5.3) on the interfaces between layers, which correspond to acoustic discontinuities (Figure 7.13). The echoes gathered while the ship is moving are juxtaposed graphically, reconstituting a vertical cross-section of the sediment layer discontinuities. For years, these records have been limited to a pictorial representation of the sediment layering. Today, echo amplitude processing with calibrated SBPs makes it possible to retrieve the reflection coefficients and absorption coefficients associated with the sediment layers crossed by the signal.

7.4.4 Marine seismics

The exploration of the deep structure of the seafloor makes use of several specific techniques, using seismic waves, as in land-based geophysics (see, e.g., Claerbout, 1976). Although these *marine seismics* techniques are traditionally considered as outside underwater acoustics *sensu stricto*, the similarities of the physical processes, the instruments and the processing techniques are evident and warrant a brief mention.

The fundamental assumption of seismics is that the subsurface is made of successive interfaces between layers, discontinuous enough to reflect or refract acoustic waves significantly. The structure of the layering is of the utmost interest for geologists, geophysicists and oil companies searching for hydrocarbon deposits. These investigations must be conducted over depths of several hundreds of metres, or even several kilometres, below the seabed. Because of the strong absorption of acoustic waves in the sediments, this is only possible using very low frequencies, with very high energies at transmission. The acoustic sources are therefore not the traditional transducers, transmitting a controlled signal, but devices generating pulse signals with high powers. Apart from simple explosive charges, the sources most often used are imploding gas bubbles, mechanically (*airguns*) or electrically (*sparkers*). Some sources generate shock waves through mechanical percussion (*boomers*).

The reception of the seismic signals thus generated is performed by one or several long linear arrays towed behind the vessel. These arrays can be as long as several kilometres, and include several hundred hydrophones grouped as subarrays, or *traces*. As with the military applications of detection, interest in this technique lies in the improvement of the SNR and in the potential for specific spatial processing.

There are two types of seismic measurement technique, called *reflection* and *refraction* seismics (Figure 7.14). Reflection seismics uses the echoes at different interfaces at near-vertical incidence. These echoes are recorded as a function of their time of arrival, and superposed shot after shot to form an image similar to the image obtained with a sounder. The thickness of each layer is calculated from the times of arrival of the echoes, and assumed values of the sound velocity. Post-processing takes advantage of the high redundancy of data, due to the recording geometry: each point on the seafloor is insonified and its echo is recorded a number of times, due to the array length. The various echoes issued from one seafloor point, geometrically determined and corrected (by *migration* algorithms), are therefore stacked in order to improve the SNR.

Refraction seismics use a peculiar property of acoustic wave propagation in layered media. An incident wave with a sufficiently large grazing angle will give rise to several interface waves. The main interface wave will propagate with the velocity of the layer immediately below. The times of arrival of the different paths, corresponding to the different discontinuities below the seabed, are recorded by the many transducers carried on the reception array. They are used to reconstruct the structure of underlying discontinuities. Increasing the number of points of measurement can resolve ambiguities about the respective thickness and sound velocity of each layer.

Both refraction and reflection seismic methods can be extended to non-plane or 3-D structures, but the reconstruction algorithms are ponderous and complex, and use very large amounts of data. One has to keep in mind that seismic records relate to phenomena generated inside an along-track vertical plane, contrary to swathe sonars which exploit backscattering. The gain this configuration provides in terms of redundancy and data density is paid for by poor coverage efficiency. This short-coming may be countered by towing several parallel arrays together. However,

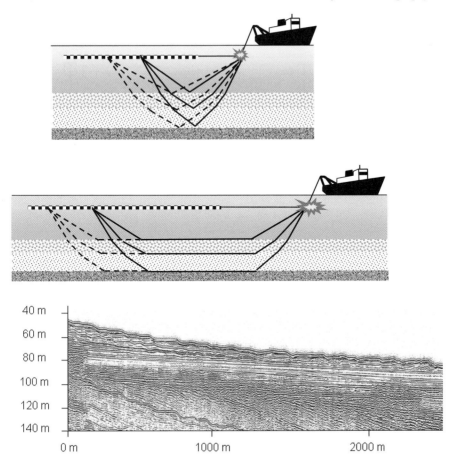

Figure 7.14. (*Top*) Reflection seismics principle. (*Middle*) Refraction seismics principle. (*Bottom*) Typical example of a reflection seismic profile, recorded in shallow water.

coverage strategies and efficiencies are very different for seismics and swathe sonar surveys.

Because they are complementary to surface investigation tools (sounders and sonars), seismic instruments are indispensable tools for the marine geosciences. They are also essential for offshore hydrocarbon exploration, and this particular industry has been and still is the major player in the development of relevant technologies.

7.5 PHYSICAL OCEANOGRAPHY

7.5.1 Ocean acoustic tomography

Modern physical oceanography uses low-frequency acoustic techniques to measure the physical characteristics of water masses at medium scale. This method, known as

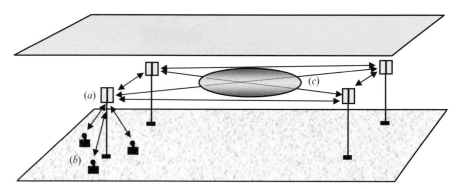

Figure 7.15. Configuration of an acoustic ocean tomography network: (*a*) low-frequency transmitter–receiver; (*b*) high-frequency acoustic positioning system; (*c*) region of sound velocity perturbation.

Ocean Acoustic Tomography (OAT), was first proposed in the 1970s (see Munk *et al.*, 1995). The idea is to measure the propagation times of multiple paths between a transmitter and a receiver on each side of the oceanic area to study. The presence in this area of sufficiently large perturbations (gyres, thermal fronts, current, etc.) will induce local variations in the acoustic velocity field, and thus in the propagation times associated with the different paths. Measuring the fluctuations of propagation times, throughout the experiment, gives details about the characteristics of the perturbation. This inversion is performed by comparing the experimental propagation times with those predicted by a model of the local environment. In practice, this method allows us to monitor mesoscale oceanographic processes (i.e., with characteristic dimensions of a few tens of kilometres), over durations of several weeks or months. It is then possible to detect and assess local velocity anomalies (i.e., temperature or salinity anomalies) relative to an average value, or to investigate currents (by comparison of acoustic propagation in each direction). The major interest of acoustic ocean tomography is that it allows quasi-instantaneous monitoring of a large area, without the problems of spatial and temporal sampling of the traditional techniques (involving depth-profiling probes deployed from a ship).

Technologically speaking, these experiments use transmitter–receiver networks working at very low frequencies (250 Hz and 400 Hz), thus reaching ranges of several hundreds of kilometres. The design and building of these sources is very intricate, as they must reconcile a low nominal frequency with a bandwidth that is broad enough to get a reasonable resolution in time (a few milliseconds), a transmission level that is high enough to be detectable at far ranges, a high electro-acoustic ratio because they must function autonomously with a limited energy supply and hydrostatic constraints because they must function at varying depths. Another imperative is the use of extremely reliable clocks, to measure propagation times accurately enough, over quite long times. The SNR in reception is improved by using long-duration modulated signals, yielding a high processing gain (see Section 6.4.3). The receivers record the signals, which are processed later. The accurate positioning of the instruments, deployed on submerged mooring lines, must be ensured by using local positioning systems (long-baseline type).

7.5.2   Global acoustics

The discovery in the 1940s of the SOFAR (SOund Fixing And Ranging) channel and its opportunities for long-distance sound propagation (see Section 2.8.3.1) was the starting point of new techniques to investigate the ocean using acoustic waves of very low frequencies.

The historic Australia–Bermuda Experiment (1960) demonstrated the possibility of transmitting acoustic signals over thousands of kilometres. A series of explosions on the Australian coast were recorded by a receiver in Bermuda, 20,000 km away (Figure 7.16), with a good SNR upon reception. The difficulty of interpreting this type of propagation lies in determining the "horizontal" path of the acoustic wave, taking account of the curvature of the Earth and diffraction by the continents (Munk et al., 1988).

The international scientific programme called *Acoustical Thermometry of Ocean Climate* (ATOC) proposes monitoring the average temperatures of water masses at the scale of the large oceanic basins (Munk and Forbes, 1989). This uses a principle that is very similar to tomography together with properties of acoustic propagation over very large distances demonstrated during the Australia–Bermuda Experiment. Sources at the centre of the areas studied transmit very low-frequency signals (around 75 Hz). These signals are recorded on various receivers, after propagation over thousands of kilometres in the SOFAR channel. The monitoring of propagation times on a scale of months enables monitoring of possible temperature increase in the oceans due to global warming. This technique is particularly interesting, as it can detect extremely small variations in temperature, translated into time delays that are large enough to be detectable over such long distances. Furthermore, the geographic scales in play here smooth out any local fluctuation in the sound velocity.

As a prelude to the ATOC experiments, the Heard Island Experiment (Munk *et al.*, 1994) was conducted at the beginning of the 1990s (Figure 7.16). It consisted in transmitting signals, modulated around 57 Hz, from a source in the South Pacific

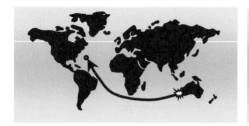 

Figure 7.16. (*Left*) The Australia–Bermuda Experiment involving very long-distance acoustic propagation. (*Right*) The Heard Island Experiment concept.

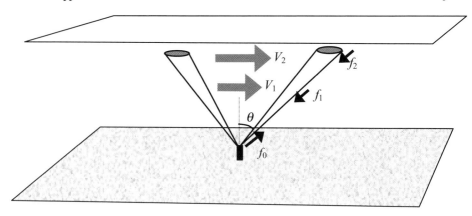

Figure 7.17. Acoustic Doppler Current Profiler (ADCP). The local current velocities V_i are used to build a vertical velocity profile, and are deduced from the frequency f_i of the echo at a particular altitude (determined by the time of reception): $f_i = f_0(1 + 2V_i/c \sin \theta)$.

Ocean, and recording propagation times over thousands of kilometres. Despite the polemics raised by concerns about the influence of high-power acoustics emissions on marine life, this experiment still went ahead and confirmed the potential of large-scale acoustic monitoring techniques.

7.5.3 Other physical oceanography applications

The technology of acoustic signal transmission over very large distances has also been applied to the positioning of deep drifting floats, used to follow deep underwater currents. The area to study (on the scale of an oceanic basin) is equipped with 260-Hz sources transmitting at known intervals in the SOFAR channel. While drifting, the floats record the times of arrival of signals from each source, used a posteriori to reconstruct their movements. The range of these localisation devices commonly reaches 1,000 km.

At a much smaller spatial scale, physical oceanographers use Acoustic Doppler Current Profilers (ADCPs) to measure local underwater currents (Figure 7.17). These are based upon the same principle as the Doppler logs presented in Section 7.1. The current speed at altitude z is linked to the Doppler shift by $V(z) = \delta f(z)c/(2f_0 \sin \theta)$. The shift $\delta f(z)$ is measured as a function of the vertical position in the water slice analysed. This allows us to reconstruct the vertical structure $V(z)$ of the current field (Figure 7.17). To get sufficient backscatter levels in the water column, ADCPs use very high frequencies (often several hundreds of kHz), thus limiting their useful range.

7.6 UNDERWATER INTERVENTION

The means available for direct intervention in increasingly deeper waters have steadily progressed during the last 50 years. The scientific community interested in

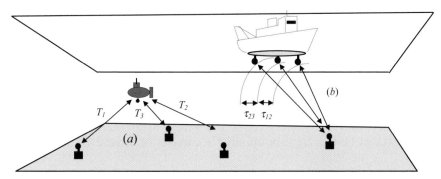

Figure 7.18. Acoustic positioning systems. (*a*) Long-baseline: the travel times between the three beacons and the mobile are used to determine its position at the intersection of the three spheres of radii $R_i = cT$; (*b*) SBL: the differences τ_{ij} of the travel times are used to determine the position of the bottom of the ship relative to the reference beacon.

the deep ocean wanted dedicated instrumentation that could be used at depths of several kilometres, and the development of appropriate deployment tools. At the same time, the offshore industry increasingly looked at the possibility of deepwater hydrocarbon exploitation and shipwreck investigation. All these applications, some of which have important economic implications, led to the development of the original underwater acoustic techniques, for the local positioning of ships and submersibles on the one part, and for the transmission of data on the other.

7.6.1 Acoustic positioning

Realisations in this domain have been numerous and varied. Three types of system, corresponding to different types of measurement, answer all these needs today:

- *Long-baseline* systems use a network of acoustic beacons (at least three), widely spaced over the area to be covered (Figure 7.18(*a*)). Their position must be accurately determined prior to using the system. The position of the moving object[10] (e.g., submersible or torpedo) that needs to be located is deduced from the travel times of the signals received from each beacon. Measurement of the absolute durations requires the use of clocks that are synchronous with the moving object and the beacons, or a system of interrogation of the beacons (transponders) by the moving object. After calibration, long-baseline systems can yield localisation accuracies of around a metre. A recent interesting variant consists in installing the acoustic beacons below drifting buoys whose positions can be tracked by GPS (see Thomas, 1994).
- *Short-baseline* systems (SBLs) use a single transmitter and a series of receivers placed close to each other (Figure 7.18(*b*)). The relative position is determined

---

[10] This technology was also used some years ago for positioning drilling vessels vertically above the drill hole. It has now been replaced by satellite positioning systems.

by time differences between the paths received at different points on the antenna. They are easier to use than long-baseline systems, but accuracies are not as good.

• Conceptually similar to SBLs, *ultra-short-baseline* systems (USBLs) use a single receiver featuring a small array. Measuring the phase differences between the different points on the array determines the direction of arrival of the acoustic waves from the transmitter placed on the moving object. Depth can be measured using a pressure sensor, and transmitted acoustically; alternatively, it can be assessed acoustically if the receiver has access to absolute travel times. The most modern USBLs have positioning accuracies of around 10 metres in deep water.

7.6.2   Acoustic data transmission

Underwater acoustic data transmission applications are very varied. In the scientific, naval and industrial domains, one uses them mainly:

• between a submersible and the support vessel, or between two submarines, for standard communications and the transmission of measurement data;
• between a vessel and an autonomous measurement station or an automated system, to get data available at the bottom (recordings, or status information) without having to recover a submerged instrument physically;
• for transmission of data to an automated system;
• for voice communication between divers and a submersible (underwater telephone).

Similarly to other modern transmission systems, underwater acoustic data transmission is generally performed using digital signals (see, e.g., Proakis, 1989). The data are coded as binary symbols, each type of symbol being transmitted with different acoustic signals. For example, the symbols "0" and "1" can correspond to the transmission of two pulses with differing frequencies (Frequency Shift Keying, or FSK, modulation), or to phase changes of a sinusoid (Phase Shift Keying, or PSK, modulation). The design of acoustic digital data transmission systems can therefore benefit from the powerful techniques already developed in telecommunications. However, today's international standard for underwater telephones uses an analogue modulation around a 8-kHz carrier frequency (with a very mediocre quality of the reconstructed sound).

Underwater acoustic data transmission presents some problems that are difficult to avoid. The first is the achievable data rate: the frequencies usable are a few tens of kHz at most, to get acceptable ranges. The available bandwidths are thus reduced, and therefore the amount of information that can be transmitted. And the vagaries of propagation strongly degrade the quality of the signals transmitted, in particular through multiple paths and reverberation, as well as rapid amplitude fluctuations due to interference and scattering ("fading" effect). The performance of a given system will therefore depend a lot on its conditions of use.

To counter propagation effects, one uses directive antennas, decreasing the effects of multiple paths and reverberation. The signals transmitted must be

Figure 7.19. Digitised picture of the RMS *Titanic* wreck, transmitted acoustically from the deep-sea submersible *Nautile* (IFREMER) to its support vessel, using the TIVA system (© IFREMER).

optimised to counter certain processes; the same signal can be transmitted at different frequencies to decrease the risks of fading; successive signals can be transmitted at time intervals in excess of the expected spread of multiple paths, etc. It is also possible to code digital signals, in order to detect and correct a posteriori some transmission errors. Finally, there are many techniques of signal processing that can be applied at reception. The best known is *equalisation*: an adaptive filter is applied at reception to compensate for the response of the transmission channel, estimated from previous "learning" signals.

The performance and degree of sophistication of each system is highly variable, depending on the application and the techniques used. Data transmission towards automated systems (e.g., in offshore oil exploitation) require total reliability, despite an often complex acoustic environment, but it does not require high transmission rates. Conversely, the transmission of measurements or digitised images (e.g., in scientific applications) can occur in well-controlled acoustic conditions, but requires very high transmission rates. In favourable conditions of propagation, such as vertical transmission at large depths, it is possible to reach transmission rates of 10 kbit/s at depths of 5,000 m. For example, the TIVA (for Transmission d'Images par Voie Acoustique; see Leduc and Ayela, 1990) system installed on the IFREMER deep-sea submersible *Nautile* to transmit digitised video images at a rate of one frame per second, was designed to monitor deepwater interventions from the surface. Figure 7.19 illustrates the unique potential of such transmission systems for quasi-instantaneous control from the surface of deep-sea operations conducted by untethered vehicles.

8

Underwater acoustic mapping systems

8.1 INTRODUCTION

Seafloor mapping is one of the most active domains of modern underwater acoustics, meeting the demands of many types of activity. The oldest need is without doubt coastal navigation, for which the production of good-quality bathymetric maps is paramount to safety. Such charts were, of course, made before the advent of acoustic techniques, but they have greatly gained from them, as much in accuracy as in completeness – especially since the emergence of swathe-sounding techniques. In each maritime country, national hydrographic offices and related operational services[1] aim at effecting very accurate surveys of their territorial waters; therefore, these organisations extensively use specialised sonar systems. The order of accuracy sought in shallow water is better than 1% of the water depth. Deeper water may be mapped with lesser accuracy, for objectives that are scientific (geology and geophysics) or industrial (underwater cables, oil industry exploration and exploitation) in nature. In all cases, underwater charts are expected to provide accurate bathymetry, to mention the presence of obstacles on the seabed, and to indicate the nature of the seabed. Modern seafloor mapping therefore results from the merging of different types of information, a great part of which can be gathered acoustically.

Today's techniques of seafloor mapping are dominated by multibeam echo sounders. These sonar systems can, after transmission of a single signal, perform a large number of point measurements on a wide strip of terrain perpendicular to the ship's route. Their arrays, most often hull-mounted, are characterised by very good angular selectivity, receiving echoes from the seabed with excellent spatial resolution. Today's multibeam echo sounder systems can measure simultaneously the water

[1] For example, NOAA, NAVOCEANO, USCG, NIMA (USA), SHOM (France), SHC (Canada), UKHO (UK), RANHS (Australia), IHM (Spain), IHPT (Portugal), ...

depth and seabed reflectivity. They are efficiently used to produce maps of the seafloor, showing its relief (bathymetric function) or characterising its nature (imaging function). Multibeam sounders appeared in the late 1970s and rapidly achieved wide success with their various user categories (hydrographers, geologists, offshore industry, ...). They underwent significant development, in technology and in performance. They are now the most sophisticated systems available for surface mapping of the seafloor.

The dominance of these multibeam sounders, which are rather complex and expensive, is, of course, not complete, and they coexist today with two other types of system, based on older concepts (of which multibeam sounders are an extension and a synthesis):

- single-beam sounders, measuring depths at the vertical of a ship, used since the 1920s;
- sidescan sonars, gathering acoustic images of the seafloor from echoes at grazing angles of incidence, and used since the 1960s.

In addition to these tools, marine geosciences and offshore industries use sediment profilers, which produce vertical cross-sections of shallow sedimentary layering.

These systems were briefly presented in Chapter 7, but the present chapter will investigate them in much greater detail. We will first describe the working of single-beam sounders and sidescan sonars, as this allows us to introduce fundamental notions. We will then introduce the general description of multibeam echo sounders and their use. The different methods for measuring bathymetry will be presented, along with analysis of the sources of errors. We will then look at the use of multibeam sounders to form sonar images similar to the images obtained by sidescan sonars, presenting examples acquired in various settings. Then, a few pages will be devoted to sediment profilers, whose working principle is different from other sonars used for mapping. Finally, this chapter will end with a rapid overview of the techniques used for seafloor characterisation, which aim at identifying the seabed type and some of its parameters.

8.2   SINGLE-BEAM SOUNDERS

8.2.1   Overview

8.2.1.1   *Signal transmission*

A single-beam sounder (Figure 8.1) transmits, vertically below the ship, a short signal (typically 10^{-4}–10^{-3} s) in a beam of average angular aperture (typically 5–15°). The sounder measures the two-way travel time of the signal, which gives the local water depth. Analysis of the echo can also provide information on the type of seabed. Moreover, the same system can detect echoes from targets in the water column (and therefore be used for fishing). Examples of actual data are presented in Plates 1 and 2 of the Colour Section.

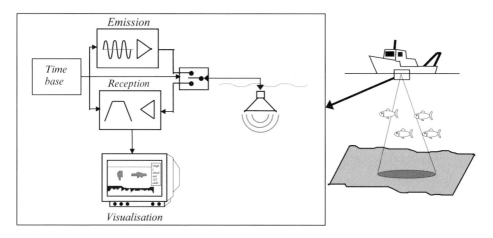

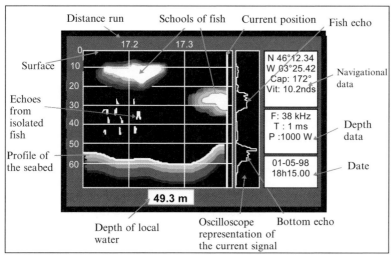

Figure 8.1. General structure of a single-beam sounder (*upper*). Typical data display for visualisation (*lower*).

The frequencies of single-beam sounders depend on the application. They range between 12 kHz for deepwater models up to 200, 400 and even 700 kHz for shallow-water models. Some manufacturers propose dual-frequency systems enabling several types of application with the same sounder. Single-beam sounder signals are generally not modulated; their duration is sufficiently short that the resolution is acceptable. A duration of around a millisecond is usually enough; it is less in shallow water to increase resolution, and more in deep water to increase the instantaneous acoustic power available at reception.

The signal thus conditioned goes through a power amplifier, delivering up to several hundred watts. The levels at transmission are usually larger than 200 dB re 1 μPa at 1 m, and can reach 230 dB re 1 μPa at 1 m.

8.2.1.2   *Signal transmission sequencing*

Signal sequencing is performed by the electronics of the transmitter, and is determined by measurement conditions. The normal way is to transmit a new signal after
receiving the echo from the previous signal (Figure 8.2*a*), in order to avoid ambiguities. The repetition period T_R is therefore:

$$T_R > \frac{2H}{c} + T + \delta T \tag{8.1}$$

where T is the signal duration, and δT its lengthening due to the seafloor reflection.

But echoes of higher order, having travelled several times over the water column,
might still exhibit noticeable amplitudes, especially in shallow water (Figure 8.2*b*).
To avoid their effects, the repetition rate can be reduced by the order n of the
multiples to avoid, and the repetition period is now:

$$T_R > \frac{2nH}{c} + T + \delta T \tag{8.2}$$

Order 2 is usually sufficient (and necessary). However, in certain cases, it is necessary
to go on until $n = 3$ or 4.

Conversely, some deepwater sounders, for which the basic ping rate (Equation
8.1) would be too slow, and sediment profilers (see Section 8.5) use interlaced
transmissions. A new ping is emitted before the previous one arrives, and several
signals are then simultaneously present in the water column. The only constraint is
that the time lags of emission and reception do not overlap; the minimum ping rate is
thus given by $T_R > T + \delta T$ (Figure 8.2*c*). This method implies that there is a way to

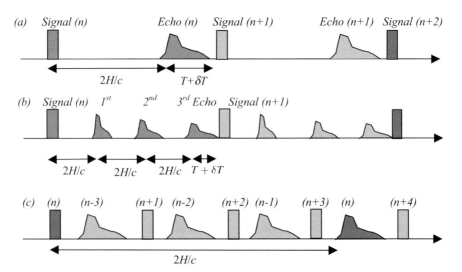

Figure 8.2. Echo sounder signal sequencing: (*a*) adapted to the first echo reception; (*b*)
adapted to the reception of two multiple paths; (*c*) with 4th order interlacing.

remove the ambiguities at reception, which is quite easy to ensure if the water depth does not change too quickly from one ping to another.

8.2.1.3 *Transducers*

In most cases, the same transducer is used for both transmission and reception. It consists in a ceramic disk or rectangle, which may be a single ceramic or an assembly of small, elementary transducers (see Chapter 5). This transducer is installed below the ship's hull, its active face following it (and suitably covered by a tough and acoustically transparent coating), in a place chosen so as to avoid propeller noise and the hydrodynamic perturbations and bubble clouds from the bow. The directivity patterns of single-beam sounder transducers usually show an aperture of a few degrees. No array processing is normally performed, except for particular cases of fisheries *split-beam* or *dual-beam* systems (Section 8.2.2.5).

8.2.1.4 *Receiver electronics*

The basic receiver electronics is quite simple. It is generally founded on non-coherent detection of energy, complemented with a Time-Varying Gain (TVG) device designed for reducing echo level dynamics as much as possible (see Section 6.5.2). Recall that non-coherent reception consists in bandpass filtering around the signal's useful spectrum, and detecting the envelope amplitude or intensity. Reception must be synchronised with transmission, so as to begin only after transmission, since the same transducer is used for both (but even a listening transducer distinct from the transmitter would be saturated by the high signal level during transmission).

The bottom echo is detected at the output from the receiver when the received signal level crosses a threshold, inside a time window imposed by the operator or automatically reset by the system. The arrival time is then measured, using an algorithm based on the detection of the leading edge of the time signal envelope crossing the threshold. The two-way travel time Δt is converted into range, usually with the simple linear relation:

$$H = \frac{c\,\Delta t}{2} \tag{8.3}$$

It may be easily checked that neglecting the actual Sound Velocity Profile (SVP) and using a constant c brings a negligible sounding error.

The value thus measured can be simply displayed digitally (Figure 8.1, *bottom*). More elaborately, the entire signal backscattered can be displayed over time, and the signal of each "ping" added to previous ones; the resulting graph shows a vertical profile of the bottom and the water column. When fishing, it adds interest by showing the echoes from targets in the water column. Such a representation is distorted horizontally by the irregular speed of the ship and the sounder's non-constant horizontal resolution, and vertically by its movements (heave, roll, pitch).

The data acquired by sounders are time-stamped and referenced with the location at which they were acquired. This means that the sounder must be connected to other navigation systems aboard the support vessel (e.g., the position-ing system). The use of an attitude control unit to correct movements of the ship is mostly justified to correct for heave; the compensation of angular motion by beam-forming is usually not possible. The data acquired by the sounder and referenced are then recorded digitally, for later post-processing and use in databases and Geo-graphic Information Systems (GISs).

8.2.2   Performance and limits of single-beam sounders

8.2.2.1   Echo constitution

The acoustic signal incident on the seafloor intercepts an active area that changes with time. We describe here the evolution of this area and the resulting echo, supposing that the seafloor is flat and horizontal, that the sounder beam has a conical directivity pattern, and that only the water–sediment interface generates an echo. The following description is illustrated by Figure 8.3:

(a) At initial instant $t_0 = 2H/c$, the active area is just the impact point.
(b) Then it becomes a disc whose radius increases with time; the instantaneous radius is $r(t) \approx \sqrt{Hc(t - t_0)}$ or $r(\tau) \approx \sqrt{Hc\tau}$ by introducing the time variable $\tau = t - t_0$; the active area increases linearly as $S(\tau) \approx \pi Hc\tau$.
(c) When $\tau = T$ (signal duration), the disk regime stops; the insonified area is then at its maximum πHcT.
(d, e) Beyond $(\tau = T)$ the signal footprint becomes a crown of internal radius $r_{int}(\tau) \approx \sqrt{Hc\tau}$ and external $r_{ext}(\tau) \approx \sqrt{Hc(\tau + T)}$; the active area is thus $\pi(r_{ext}^2 - r_{int}^2) \approx \pi HcT$ and stays constant with time.
(f) The insonified area finally disappears when the crown grows out of the beam

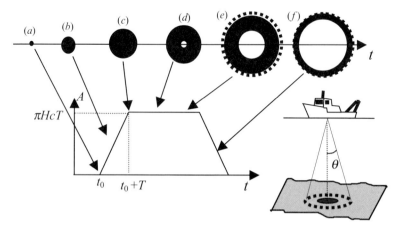

Figure 8.3. Generation of the signal footprint (dark area) for a single-beam echo sounder, and evolution of the active area, in the short-pulse regime. The dotted line outlines the outer limit of the beam's footprint.

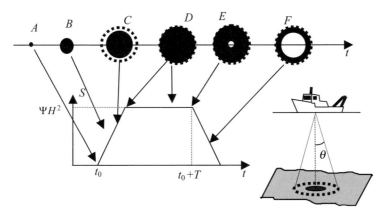

Figure 8.4. Generation of the signal footprint (dark area) for a single-beam echo sounder, and evolution of the active area, in long-pulse regime. The dotted line outlines the outer limit of the beam.

footprint limits, imposing the maximum possible radius $r_{max} = \theta H$; in this regime, the area decreases in $\pi(r^2_{max} - r^2_{int}) = \pi(\theta^2 H^2 - HcT)$.

This happens when the beam aperture is sufficiently wide for the footprint of the time signal to reach its full extent: this is the *short-pulse* (or *pulse-limited*) *regime*.

If the pulse is long enough, the interception by the beam footprint (Figure 8.4) happens before $\tau = T$; the whole beam footprint may then be simultaneously insonified; the maximum backscattering area is equal to ΨH^2 (Ψ being the equivalent solid angle of the directivity pattern); that is, $\pi\theta^2 H^2$ for a conical beam.

This *long-pulse* (or *beam-limited*) *regime* happens at short ranges. The transition range between long- and short-pulse regimes is roughly given by $\Psi H^2 \approx \pi HcT$ (i.e., $H \approx \pi cT/\Psi$).

The backscattered signal level depends not only on the area instantaneously insonified but also on the backscattering strength and the beam directivity pattern. These factors weigh the contributions from the footprint scattering points. Close to the vertical, the backscattering strength varies very abruptly, especially for soft sediments. Qualitatively, the following effects are observable:

- At the echo's leading edge, quickly increasing incident angles are swept, corresponding to a sharp decrease of the backscattering strength values (especially for soft, sedimentary seafloors). The theoretically linear increase is thus damped.
- At the constant footprint plateau, since incident angles increase with time, the backscattered intensity actually decreases.
- The cut-off in directivity pattern is not as sharp as was supposed with a conical pattern hypothesis; a time trail lengthened by the sidelobes is actually observed.

Figure 8.5 gives a qualitative representation of the time variation of resulting echoes and their various components. It makes clear that the echo's trailing edge, due to the directivity sidelobes, will depend on the seafloor backscattering strength in oblique

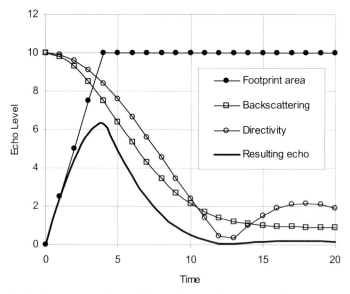

Figure 8.5. Qualitative presentation of the seafloor echo construction as a function of time, detailing components bound to the geometrical signal footprint, the backscattering strength, the directivity pattern (with its first sidelobe) and the resulting echo envelope. The amplitude and time scales are arbitrary.

incidence. Echoes with long trails are thus observed on very rough seafloors with a wide angular response.[2]

8.2.2.2   *Maximum operating range of a single-beam echo sounder*

This performance evaluation is based on the sonar equation analysis presented in Section 6.6.1. The target is here the seafloor, the efficient part of which is limited either by the beam aperture Ψ (long-pulse regime), or by the pulse duration T (short-pulse regime). The seafloor backscattering strength BS_B is approximated by its vertical value, and the signal footprint area by its maximum value:

- in the long-pulse regime, the received echo level is:

$$EL = SL - 20 \log R - 2\alpha R + 10 \log(\Psi) + BS_B \tag{8.4}$$

and the sonar equation:

$$SL - 20 \log R - 2\alpha R + 10 \log(\Psi) + BS_B - NL + PG + DI > RT \tag{8.5}$$

---

[2] We restrained this analysis to backscattering by the sediment interface. However, long trails may also be observed on very soft seafloors with high backscattering by buried heterogeneities (mud with gas bubbles or colonies of living organisms).

- in short-pulse regime, the echo level is:

$$EL = SL - 30 \log R - 2\alpha R + 10 \log(\pi c/B) + BS_B \qquad (8.6)$$

and the sonar equation:

$$EL - 30 \log R - 2\alpha R + 10 \log(\pi c/B) + BS_B - NL + PG + DI > RT \qquad (8.7)$$

In order to determine the maximum range of a single-beam echo sounder, one retains a priori the short-pulse regime (associated with longest ranges), therefore:

$$30 \log R + 2\alpha R = EL + 10 \log(\pi c/B) + BS_B - NL + PG + DI - RT \qquad (8.8)$$

8.2.2.3 *Sounding-measurement accuracy*

The measurement accuracy of a single-beam echo sounder depends on the leading edge detection of the echo time envelope, crossing a given threshold. As the signal envelope is unstable due to fluctuations and additive noise, the detection time fluctuates around its average value. The measurement quality thus depends on the Signal-to-Noise Ratio (SNR) value, on signal processing performed in the receiver (varying between brands and models) and, of course, on the fixed threshold value. A conservative value may be obtained by considering the time measurement as equiprobable over the duration T of the signal's front edge; hence the error standard deviation is $\delta t = T/\sqrt{12} \approx 0.3T$. The corresponding sounding error is $\delta H = c\,\delta t/2 \approx 225T$.

The overall measured sounding ($H = ct/2$) may also be biased by an inaccurate value of average sound speed c. Practically, this should usually lead to negligible errors:[3] the relative errors on H and c being equal, a 1% depth error should imply a 15 m/s error on the average velocity over the water column – which is very unlikely to occur.

One should finally mention the errors due to the supporting platform's motion (heave). Everything considered, the measurement *accuracy* of single-beam echo sounders is usually not a big problem. But much more can be said about the *resolution* of sounding operations.

8.2.2.4 *Sounding-measurement resolution*

The measurement resolution of a sounder is its ability to distinguish two close, distinct targets. In the vertical direction, it is directly given by the emitted pulse duration. It is given by $\delta z = cT/2$, and typically ranges between 0.075 m (for $T = 0.1$ ms) and 0.75 m (for $T = 1$ ms). The use of modulated signals is very rare; in this case the resolution is $\delta z = c/2B$, with B the modulated bandwidth.

The horizontal resolution is related to the angular width of the beam. To be distinguishable, two targets will need to be spaced apart by at least the beamwidth (considered at -3 dB) at the depth considered (Figure 8.6). If $\delta\theta$ is the two-way

---

[3] The SVP problems met with *multibeam* echo sounders are due to *wave refraction*: they have completely different effects and should in no case be disregarded.

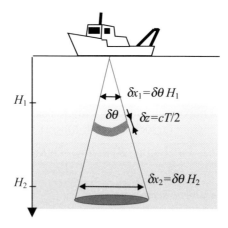

Figure 8.6. Vertical and horizontal resolution of a single-beam sounder.

angular aperture (see Table 5.1 in Section 5.3.2.1), the horizontal resolution is thus:

$$\delta x = 2 \tan\left(\frac{\delta\theta}{2}\right) H \approx \delta\theta\, H \qquad (8.9)$$

For instance, a beam of aperture $10°$ will yield a horizontal resolution of 1.6 m at 10-m depth, 16 m at 100-m depth and 160 m at 1,000-m depth.

The angular width of the beam has another consequence: a target can be observed by the sounder for as long as it stays inside the beam (i.e., for a time linked to the beam aperture, and not to its own size). For example, using the same $10°$ beamwidth, a fish 100 m deep will be observed by the sounder for 16 m, but it should not be deduced that its size is 16 m ... And as its oblique distance to the sounder will change as the ship moves, the echo recorded with time will have a parabolic shape, given by:

$$z = z_0 + \frac{(x - x_0)^2}{2z_0} = z_0 + \frac{V^2(t - t_0)^2}{2z_0} \qquad (8.10)$$

where V is the ship's speed, and t_0 is the time that the sounder is directly above the target located at (x_0, z_0).

This produces characteristic crescent shapes on the echogram (Figure 8.7). Because of this mediocre horizontal resolution, the mapping of seabed relief with a single-beam sounder causes many artefacts. A local depression will not be detected if its width along the track of the vessel is smaller than the beam footprint, since its edges will reflect a masking echo (Figure 8.8). At most, one will witness an increase in the echo's time trace, very difficult to interpret. Conversely, a topographic high will be easier to detect, but its precise shape and extent will be strongly biased by the hyperbolic process mentioned earlier (Figure 8.9).

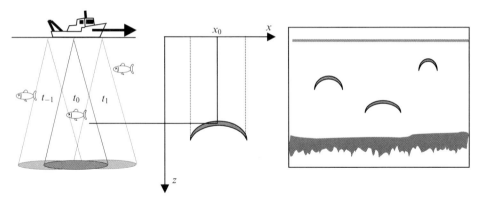

Figure 8.7. Formation of a target echo by a single-beam sounder. At t_{-1}, the sounder starts detecting the target, at an oblique distance larger than the depth. At t_0, the target is detected with the correct distance and depth. At t_1, the sounder ends detecting the target. The resulting echoes in the (x, z) space have characteristic crescent shapes.

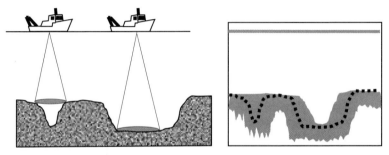

Figure 8.8. Geometric configuration (*left*) and echogram (*right*) for detection of seafloor troughs. The leftmost is narrower than the beam, and is not recorded. The rightmost is detected by the sounder, but its horizontal dimensions are underestimated.

8.2.2.5 *Improvement of single-beam echo sounder performance*

These various types of artefact are unavoidable, because they stem from the very nature of the single-beam sounder and its inability to discriminate transversally inside the beam. However, some aspects can be improved.

Split-beam echo sounders

By dividing the transmitting face of the transducer into several sectors in some elaborate fisheries sounders, it is possible to create several secondary beams artificially. They all detect the target present in the main beam, since their individual lobes are wider. The phase differences between the signals from the various secondary beams enables computing the target position angles and determining its location in the horizontal plane. This technique, called "split-beam" was designed for fisheries sonars, to locate isolated fish and correctly measure their target strength;

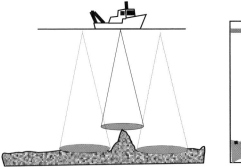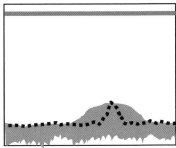

Figure 8.9. Geometric configuration (*left*) and echogram (*right*) for detection of a local relief on the seabed with a single-beam sounder. The relief is detected at the correct depth, but its dimension is overestimated because of the poor horizontal resolution of the sounder.

it may also be used in seafloor mapping, to determine the local slope from the echo phase differences between the various parts of the receiving array.

Sweep sounders

Some seafloor mapping systems make use of several single-beam echo sounders installed on a transversal support (swinging booms deployed on both sides, see Figure 8.10). This method increases the number of sounding points, and has some advantages over multibeam echo sounders (Section 8.4): the problems associated with SVP correction are avoided, since vertical soundings are not very sensitive to sound velocity variations; the sounding points are equidistant whatever the water depth; the intrinsic quality of vertical measurements is simpler and better than when sounding obliquely. But, on the other hand, their practical deployment raises specific constraints. Such systems are thus reserved to local applications (hydrography in harbours and navigation channels) for which maximum accuracy is demanded in restricted areas.

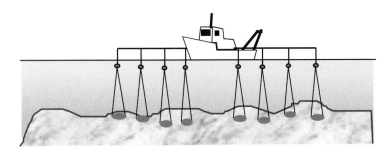

Figure 8.10. Simultaneous deployment of several single-beam echo sounders for hydrography (*sweep echo sounder*).

8.3 SIDESCAN SONARS

8.3.1 Overview

8.3.1.1 Signal transmission

Sidescan sonars are first and foremost considered as visualisation tools, providing *acoustic images* of the seabed. They are usually installed on a fish towed near the bottom (Figure 8.11). This configuration makes them work in good stability and noise conditions, and at grazing incidence. A sidescan sonar insonifies the bottom with two side antennas with a very narrow horizontal directivity (around a degree or less). The signal backscattered by the seabed, recorded along time, reproduces the structure of small irregularities on the seabed, which are better imaged at grazing angles of incidence (see Section 3.6.2.6).

The very simplicity of the working principle makes it an invention of genius: a narrow sound beam is transmitted at grazing incidence and will intercept the bottom along a thin strip spreading out with distance (Figure 8.12). Inside this strip, the very short signal transmitted will delimit the area insonified, of very small dimensions, sweeping over the entire area covered. The echo received along time will represent the bottom reflectivity along the swathe, and particularly the presence or irregularities or small obstacles. The signal is recorded laterally, hence the name "sidescan sonar". It is added to the signals recorded at previous positions of the towfish and, line by line, it creates a genuine image of the seabed.

The basic system structure is the same as for single-beam echo sounders. Frequencies are usually in the range of hundreds of kHz, with pulse durations as short as possible (typically 0.1 ms or less).

8.3.1.2 Transmission sequencing

The inter-ping delay for sidescan sonars obeys a simple principle: a new ping is emitted after the echo from the previous one has been fully received (i.e., its level has sufficiently decreased). The delay between two transmissions is hence

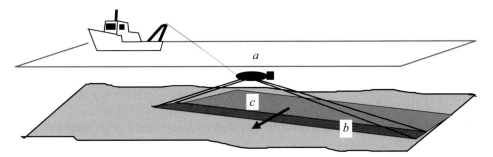

Figure 8.11. Sidescan sonar deployment: (*a*) towfish; (*b*) instantaneously insonified area; (*c*) area covered by previous transmissions.

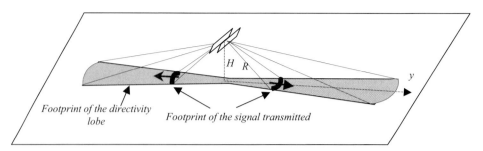

Figure 8.12. Geometry of sidescan sonar insonification.

$T_R > 2R_{\mathrm{max}}/c$, where R_{max} is the maximum reachable oblique range; when the sonar is working at grazing angles this oblique range equals the horizontal distance.

Contrary to single-beam echosounders, there is no easy alternative solution. However, it should be pointed out that some sophisticated sonars can emit to several along-track sectors at different frequencies simultaneously (Section 8.3.2.4).

8.3.1.3   Transducers

The system is based on very long rectangular antennas, with a resulting directivity pattern that is:

● largely open in the vertical plane (several tens of degrees), slightly inclined downward, in order to insonify at far ranges (angles up to ±80–85° from the vertical are commonly reached), to avoid the sea surface, and to limit downward vertical insonification;
● very narrow in the horizontal plane (around 1° or less, to get the best possible spatial resolution) (Figure 8.12).

These antennas are installed on each side of a towfish. Its hydrodynamic characteristics must ensure good stability when moving and a small grazing angle of the signals transmitted. Since the frequencies in use are generally high (typically a few hundred kHz), this ensures correct directivity patterns are achieved with antennas of reasonable lengths. The bandwidths of a few tens of kHz provide a good range resolution (a few centimetres). The ranges are limited to hundreds of metres. Array signal processing is quite simple, as usually no antenna correction is needed (their geometry *per se* being sufficient to ensure expected directivity performances); however, in some cases of high-frequency and narrow-beam sonars, array focusing must be applied.

8.3.1.4   Reception processing

The receiver's basic structure is similar to that of single-beam echo sounders. The processing chain is twinned in order to process both sides. A typical structure for first level receiving operations is as follows:

● TVG equalisation of the received echo level;

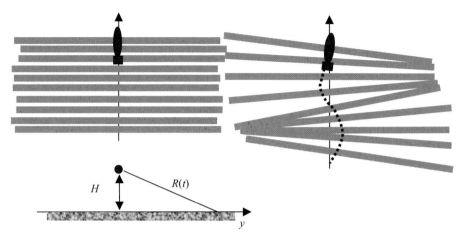

Figure 8.13. Sonar image construction. (*Left*) The pixel lines are simply piled up without trajectory correction. (*Right*) Fish navigation and movements are compensated in order to create a mosaic.

- demodulation and filtering;
- signal A/D conversion;
- non-coherent reception processing: bandpass filtering and envelope detection (a few systems use modulated signals and pulse compression);
- processing gain compensation;
- data storage.

For the second level, the echoes are recorded and displayed as a function of time: for every ping, the backscattered intensity is drawn as a line of pixels perpendicular to the towfish track. Several modes are possible:

- As a first approach, one simply displays a pile of successive echo lines (Figure 8.13), recorded along time. The raw image built this way permits us to control the data (and in many cases it is of surprisingly good pictorial quality, see Figure 8.16 on p. 263). However, it is a poor physical representation, insufficient to get an accurate image of the seafloor.
- Geometrical problems arise when trying to reconstruct an undistorted image of the seabed, using these time records. As time and distance on the seabed are not proportional, equidistant time samples do not correspond to regular seabed samples. To replace the time samples in a spatially correct fashion, a geometric correction must be applied. If the seabed is flat, this is a simple trigonometric relation, using the oblique distance $R = ct/2$, related to the altitude H and the transverse distance y as $R^2 = y^2 + H^2$. The altitude H is known (from the first bottom echo), and one gets:

$$y = \sqrt{\frac{c^2 t^2}{4} - H^2} \qquad (8.11)$$

- When the seabed is not flat, geometrical correction requires simple a priori assumptions about the topography (e.g., regular slope) or additional recording of bathymetry (possibly obtained by the sidescan sonar itself, if it is equipped with an interferometer).
- If the towfish is equipped with a navigation and motion data unit, its movements may be compensated and the pixels relocated in a precise geographic frame, where several tracks of the sonar may be added; the result is named a *sonar image mosaic*. Finally the image is interpolated between the various pixels.

8.3.1.5   Echo construction

The spatio-temporal structure of the signals received by a sidescan sonar is shown in Figure 8.14. The signal transmitted will first propagate in water (Figure 8.14, Point *A*); the sonar will only receive background noise, and possibly echoes from mid-water targets (e.g., fish, bubbles). The actual seafloor echo will only start taking shape when the pulse strikes the nadir of the sonar (Figure 8.14, Point *B*). This reflection creates a first, very intense echo (as the backscatter strength is maximal and the transmission loss minimal). This cannot be used for the imaging itself, but is very useful to estimate the altitude of the sonar above the bottom. As it propagates with time, the signal will then explore the angular zone close to the vertical. This first part, with high reflectivity and poor horizontal spatial resolution, is generally of bad quality. Finally, the signal reaches oblique and grazing incidence, and then becomes really useful for imaging. Its average level will then depend on the local type of seafloor, with possible modulations by the array directivity patterns.

An interesting effect is the formation of shadows on the seabed. A large enough obstacle will intercept part of the angular sector transmitted, and prevent back-scattering from the bottom at times normally associated with these angles. The

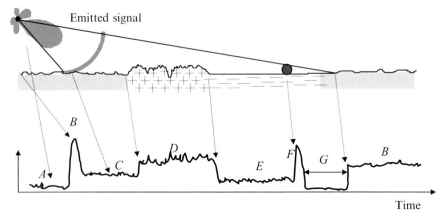

Figure 8.14. Echo generation for a sidescan sonar: (*A*) noise and reverberation in the water column; (*B*) first bottom echo; (*C*) sand area; (*D*) rock; (*E*) silt; (*F*) target echo; (*G*) shadow from the target. See text for details.

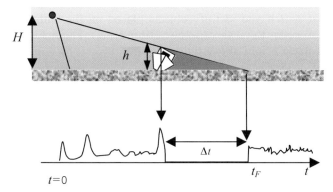

Figure 8.15. Geometrical description of a sonar shadow projection.

Figure 8.16. Images from a Klein 5400 sidescan sonar, illustrating the effects of shadowing by an obstacle. Lighter grey levels correspond to higher reflectivity. No geometrical correction is applied; time-recorded pings are just piled up along-track. The wreck is the *Swansea Vale* (a 77-m long cargo ship, sunk in 1918) at the entrance to the Bay of Brest, on a rippled, sandy seafloor (data provided courtesy of DCE-GESMA).

echo received will thus be very low for a duration depending on the grazing angle and the height of the masking object. This will create on the sonar image the shadow commensurate with the object's shape (Figure 8.15 and 8.16). Its analysis will provide estimates of the size and shape of the object. This is very interesting in all applications that search and identify objects placed on the seabed (e.g., mines or shipwrecks), or to assess the scale of some bottom reliefs. The shadow duration gives access to the obstacle height above the seafloor:

$$\frac{h}{\Delta t} = \frac{H}{t_F} \quad \Rightarrow \quad h = H\frac{\Delta t}{t_F} \tag{8.12}$$

where t_F is measured from the transmission instant.

8.3.1.6   Sidescan sonar systems

Sidescan sonars are usually lightweight systems, easily movable and primarily destined for shallowwater use. The most commonly used frequencies lie in the range 100–500 kHz, making it possible to restrict array length below one metre and to reach ranges up to hundreds of metres. Angle apertures are typically 0.2–1°. Transmitted signals have a typical duration of 0.1 ms, thus a spatial resolution around 0.1 m. Modern systems are often bi-frequency, permitting us to work simultaneously with different resolutions. Towfishes are of small dimensions and weight, easily deployed by small teams from coastal boats.

Deepwater seafloor mapping demands specially designed sonars: either high-frequency systems towed close to the seafloor (e.g., the IFREMER *Sar*, working at 170–190 kHz down to 6,000 m), or low-frequency systems towed closer to the surface and covering several kilometres (e.g., the British sonar *GLORIA*, working at 6.5 kHz). *GLORIA* is an extreme case, as it can map swathe widths of up to 30–60 km, and can observe large structures with a resolution of around a hundred metres. A small number of low-frequency sonars designed for regional surveys in deep water (*TOBI*, *SeaMarc*, *Okean*) have been put into service (see Blondel and Murton, 1997).

8.3.2   Performance and limits of sidescan sonars

8.3.2.1   Resolution

Across-track resolution is given by projecting the equivalent length of the signal used:

$$\delta_y = \frac{cT}{2\sin\theta} \tag{8.13}$$

One can remark that as $\theta \to \pi/2$, $\delta y \to cT/2$, and the resolution is then exactly the resolution of the signal transmitted. When $\theta \to 0$, the approximate Equation (8.13) is not valid (it would yield $\delta y \to \infty$), and the effective resolution is then the same as a single-beam sounder in long-pulse regime:

$$\delta y = \sqrt{HcT} \tag{8.14}$$

Alongtrack resolution is defined by the projected beamwidth $\delta x = R\varphi$, where R is the oblique distance from the sonar to the bottom, and φ is the aperture of the

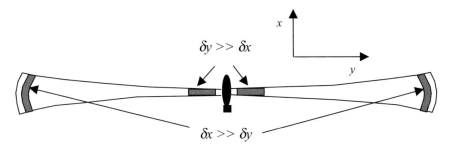

Figure 8.17. Evolution of the horizontal resolution of a sidescan sonar across-swathe.

directivity lobe in the horizontal plane. As in the case of the single-beam echo sounder, resolution worsens proportionally to the distance travelled by the signal.

The resolution of a sidescan sonar is therefore on the one hand inhomogeneous, both across-track and along-track, and on the other hand varying along the enso-nified swathe (Figure 8.17):

- At small distances $\delta y \gg \delta x$, across-track resolution is the worst, whereas along-track resolution is best, because the distance travelled is minimal.
- At large distances $\delta x \gg \delta y$, across-track resolution is best, whereas along-track resolution is bad, because the distance travelled is maximal. The areas insonified by successive pings overlap.

8.3.2.2 Maximum range

Let us consider a sidescan sonar that is working on extended targets and is limited by noise. The echo level to compare with the noise level is thus:

$$EL - NL + DI > RT \qquad (8.15)$$

The echo level is due to surface backscattering:

$$EL = SL - 30\log R - 2\alpha R + 10\log(\Phi c/2B) + BS_B \qquad (8.16)$$

hence the sonar equation is:

$$SL - 30\log R - 2\alpha R + 10\log(\Phi c/2B) + BS_B - NL + PG + DI > RT \qquad (8.17)$$

and the maximum range is given by:

$$30\log R + 2\alpha R = SL + 10\log(\Phi c/2B) + BS_B - NL + PG + DI - RT \qquad (8.18)$$

(it is supposed that the incidence angle is sufficiently grazing to neglect the projection of the signal footprint on the seafloor).

8.3.2.3 Complete coverage

A performance constraint of sidescan sonar is the need to ensure 100% coverage of the insonified area. The condition that needs to be fulfilled is the removal of gaps between insonified areas from two successive pings.

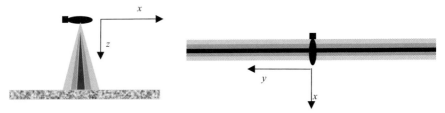

Figure 8.18. Formation of several transmission planes by a sidescan sonar. This method allows us either to increase the survey speed proportionally to the number of sectors, or to obtain an interesting data redundancy.

It is easily seen that this condition is critical close to the vertical, since the insonified strip is narrowest there. With a sonar directivity lobe aperture ϕ, the insonified along-track width is then $\delta x = H\phi$. The sonar advances $\delta x = VT_R$ between two pings, with V its speed and T_R the inter-transmission delay obtained from the sonar maximum reachable range R_{max} by $T_R = 2R_{max}/c$. Finally, one may write the condition as $\delta x = H\phi = 2V(R_{max}/c)$ and, for example, impose a maximum towing speed at:

$$V = \frac{cH\phi}{2R_{max}} = \frac{c\phi}{2}\cos\theta_{max} \qquad (8.19)$$

where θ_{max} is the maximal insonification angle in the across-track plane. The 100% coverage condition is then independent of H. As an order of magnitude, a lobe aperture $\phi = 1°$ with a maximum tilt $\theta_{max} = 70°$ limits the ship's speed to 4.5 m/s (8.7 knots).

One solution to performing high-speed surveys while still fulfilling the 100% coverage condition is to transmit N along-track sectors simultaneously (at slightly different frequencies), in order to insonify, for every ping, N swathes. This makes it possible to increase the survey speed by a factor N.

8.3.3   Measuring bathymetry with a sidescan sonar

The traditional sidescan sonar cannot measure bottom relief, except a rough estimate of its altitude from the echo at nadir. It is, however, possible to compensate for this by adding interferometric capabilities. This is done by using a second antenna, which receives parallel to the main antenna, and has characteristics as close as possible to the main antenna. The time signals from the two receivers are acquired simultaneously. At each moment, the phase shift between the two signals is used to deduce the angle between the direction of the acoustic signal and the axis of the interferometer (Figure 8.19).

The phase difference between the two receivers is:

$$\Delta\varphi_{AB} = k\,\delta R = k(\overline{MA} - \overline{MB}) = ka\sin\gamma \qquad (8.20)$$

It is used to access the angle γ:

$$\gamma = \arcsin\left(\frac{\Delta\varphi_{AB}}{ka}\right) \qquad (8.21)$$

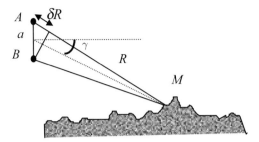

Figure 8.19. Interferometric measurement of bathymetry.

Knowing the oblique distance R (from the propagation time), we can deduce the coordinates of the impact point $M(x,z)$.

This yields a series of angular measurements as a function of time. It only involves a couple of basic receivers, with no beamforming in the vertical plane. The only constraint is that the phase difference between the two receivers, as a function of angles, be correctly calibrated. This method seems very advantageous, but there are several shortcomings:

- The measurement of the phase difference is ambiguous (modulo 2π); the solution for the angle γ is therefore made of a series of values $\gamma_n = \arcsin((\Delta\varphi_{AB} + 2n\pi)/ka)$ with $n = \ldots, -2, -1, 0, +1, +2, \ldots$ This leads to unacceptable errors when this ambiguity cannot be resolved.
- The measurement of the phase difference is very sensitive to noise: additive noise (which induces fluctuations of the signals received), amplitude variations and loss of coherence between the two receivers. A relation between the phase fluctuation and the SNR is given in Appendix A.8.1.
- The possible absence of a directivity diagram for the receivers degrades the output signal SNR, hence the measurement of phase shifts and arrival angles. This problem can be alleviated by averaging the measurements over several successive points, losing the advantage of the high density of measurement points.
- The measurement quality seriously degrades for echoes close to nadir.

Phase ambiguities can be countered in different ways. A first solution is to use an interferometer with the appropriate structure (i.e., of dimension smaller than $\lambda/2$); the problem is that the angular accuracy is then mediocre (since ka is small). A second solution is to use two ambiguous interferometers with different dimensions, and intersect the angular solutions to remove the ambiguities (*vernier* method). A third solution is to measure the phase difference, no longer with a 2π modulo, but by "unwrapping" it; this requires knowing a reference phase value, following its evolution and imposing continuity at phase jumps. Finally, the phase ambiguity may also be suppressed by getting a first-guess estimation of the delay (obtained from the cross-correlation function) between the two receiver time signals.

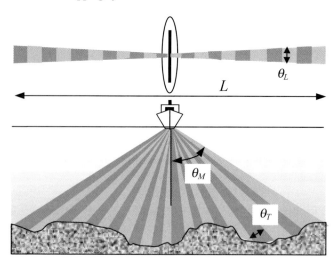

Figure 8.20. Multibeam echo sounder geometry. (*Upper*) View from above, with the overall swathe width L and the along-track (longitudinal) aperture θ_L. (*Lower*) Vertical view, with the across-track transversal beam aperture θ_T, and the maximal beam tilt angle θ_M.

8.4   MULTIBEAM ECHO SOUNDERS

8.4.1   Overview

The multibeam echo sounder at first consists in an extension of the single-beam echo sounder. Instead of transmitting and receiving a single vertical beam, the multibeam sounder transmits and receives a fan of beams with small individual widths (1–3°), across the axis of the ship (Figure 8.20). The immediate interest is, of course, the possibility of multiplying the number of simultaneous depth measurements (typically 100–200), sweeping a large corridor around the ship's path (a total width of 150° covers up to 7.5 times the water depth). The most recent multibeam sounders use their large angular width to record acoustic images using the same principle as the sidescan sonar. But the resulting performance is worse than that of a deep-towed system (towfish), because of movements of the support platform and because incidence angles are not sufficiently grazing. With such systems, geologists have at their disposal integrated tools giving, at the same time, bathymetry and reflectivity measurements. The simultaneous gathering of seismic and sediment profiler data can help in providing a very complete and thorough investigation of sedimentary structures.

Since their appearance in 1977 (with the historic *SeaBeam* on R/V *Jean Charcot*, see Renard and Allenou, 1979), multibeam echo sounders have greatly evolved, and have become very varied, widespread[4] and essential for seafloor mapping operations. These systems can survey large areas rapidly and accurately, and they are essential

---

[4] As evidenced by the list of multibeam sounders regularly compiled (Cherkis, 2001).

for the study of geological morphology and facies. A deepwater multibeam sounder with a 20-km swathe, used on a ship sailing at 10 knots, can cover around 10,000 km^2 per day.[5] Current improvement efforts are targeting the resolution and the quality of bathymetry and imagery measurements, rather than the extension of the coverage.

Multibeam sounders have very varied uses, and several types of system are offered by manufacturers. These can be grouped into three main categories:

- Deepwater systems (typically 12 kHz for the deep ocean, 30 kHz for continental shelves), designed for regional mapping. The large dimensions of their arrays limits their installation to deep-sea vessels.
- Shallow-water systems (typically 100–200 kHz), designed for mapping continental shelves. These systems are best suited to hydrography.
- High-resolution systems (typically 300–500 kHz), used for local studies: hydrography, shipwreck location, inspection of underwater structures. Their small size makes then suitable for installation on small ships, towfishes or autonomous underwater vehicles (AUV).

Table 8.1 shows, as an example, a few characteristics of the multibeam sounders sold by two major suppliers, covering the whole range of the above categories.

By 2001, around 700 multibeam echo sounders had been put into service (Cherkis, 2001), including about 100 deepwater low-frequency systems. More than two-thirds of them were built by one of the two Scandinavian manufacturers mentioned above.[6]

8.4.2   Multibeam echo sounder structure

A typical multibeam echo sounder features the following elements: transmission and reception arrays, electronics of the transmission stage, reception unit, user interface and ancillary systems. These are described in the next sections.

8.4.2.1.   Transmission and reception arrays

They are designed to ensure (Figure 8.21):

- Narrow beamwidth in the horizontal plane. As with sidescan sonars, one wants to insonify a thin strip of terrain across-track of the support platform. This

---

[5] This order of magnitude is interesting. If the surface of the Earth's deep ocean is around 350,000,000 km^2, the global mapping of the deep oceans would represent around 100 years of survey. This sounds high, but would in fact be achievable reasonably rapidly, as more than 50 deepwater multibeam sounders are currently in service. For medium and shallowwater areas, the efficiency should, of course, be far less evident due to the poor swathewidths achievable. A long-term programme named GOMAP (Global Ocean Mapping Project, see Vogt et al., 2000) has been proposed in this respect.

[6] The other main manufacturers in the field are Atlas (Germany), SeaBeam (USA) and Elac (Germany).

Table 8.1. Main characteristics of some current multibeam echo sounders.

Sounder type	Frequency (kHz)	Water depth (m)	Maximum width (m)	Total aperture (°)	Number of beams	Beam widths	Signal resolution (m)
Simrad EM120	12	50 → 11,000	20,000	144	191	1° * 2°	1.5–7.5
Simrad EM300	32	5 → 5,000	10,000	140	135	1° * 2°	0.5–3.75
Simrad EM1002	95	2 → 1,000	1,500	150	111	2.3° * 2.3°	0.15–1.5
Reson Seabat 8111	100	3 → 1,000	1,000	150	101	1.5° * 1.5°	0.05–0.25
Reson Seabat 8101	240	0.5 → 500	400	150	101	1.5° * 1.5°	0.015–0.15
Reson Seabat 8125	455	0.2 → 120	/	120	240	0.5° * 1°	0.01–0.2

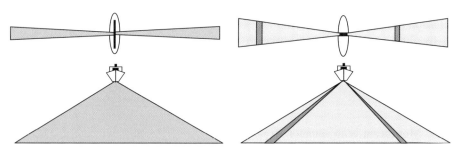

Figure 8.21. Geometry of array directivities for a multibeam sounder. (*Left*) Transmission. (*Right*) Reception (with beamforming in the vertical plane).

implies a long transmission array, which is also useful for transmitting loud signals. This configuration is the easiest to install on the ship's hull, because the array is then along the axis of the supporting platform.

- Large transmission width, to cover as large a swathe as possible. This corresponds to arrays that are narrow across-track. Like the previous bullet point, it is geometrically favourable for setting up the sounder below the hull.
- Good angular discrimination for reception beams, so that the beam footprints are as small as possible. Along-track discrimination is determined by the directivity of the transmission array. Across-track, it is determined by the reception array; this requires a wide array, and may raise installation problems, especially as the frequency is lower. The shapes most commonly used for reception arrays are (Figure 8.22):
  - horizontal linear – this is the simplest configuration, but limits the useful width to 120–140° because of the large beamsteering angle values, and causes installation problems below the hull;

Figure 8.22. Multibeam echo sounder array geometries (from *left* to *right*): horizontal lines, ∨-shaped, ∪-shaped.

- ∘ ∨-shaped – two linear arrays inclined symmetrically, working indepen-
     dently, enable high tilt angles for beamsteering, hence large swathe
     widths, and keep the simplicity of linear array beamforming;
- ∘ ∪-shaped – the array has a circular section, which enables beamforming by
     simple summation of parts of the array.

The reception and transmission arrays of low-frequency multibeam sounders are physically separated; this is to limit the overall size and weight of the array system. In this case the transmission array alone imposes the along-track resolution, and the reception array the across-track resolution, the resulting product imposing the final resolution (Mill's cross principle). For high-frequency sounders, the same array can be used in both transmission and reception.

8.4.2.2 *Transmission electronics*

Transmission electronics are used for signal generation, amplification and impedance matching of the transducers. Typical electric powers range from hundreds to thousands of watts. Transmission electronics must also control the transmission beam, its level, its width and its inclination, according to the configuration parameters and data from the attitude sensors (to compensate for possible roll and pitch).

8.4.2.3 *Reception unit*

This unit performs the following operations:

- level correction as a function of time (TVG), to keep hydrophone signal amplitudes that are in a suitable range as constant as possible;
- digitisation of hydrophone signals, currently between 12 to 16 bits;
- demodulation of the signals down to low frequency, and low-pass filtering;
- beamforming to get the reception beams;
- elementary bathymetry measurement (with one of the techniques described
  below);
- correction of the platform's movements, to control transmission and reception
  beams;
- correction of acoustic paths as a function of the SVP;
- conditioning of imagery signals.

Besides analogue processing and digitisation modules, the use of specific hardware is increasingly rare, and most of these functions are now integrated into dedicated personal computers or workstations, which are much more adaptable.

8.4.2.4   *User interface*

This is the part of the system that presents the user with:

- System control options (operation, settings for the sounder and the ancillary systems, calibration tools, data archival).
- Real-time processing results, which are most often:
  - the control display of the raw signals, by pings;
  - the instantaneous bathymetric profile for all beams of the current ping;
  - backscatter levels over the entire swathe;
  - the raw sonar image;
  - geo-referenced bathymetry and imagery maps (in the most recent systems).

8.4.2.5   *Ancillary systems*

The multibeam sounder has to receive and process data from several ancillary systems in order to locate depth measurements accurately (Section 8.4.5.2). These are:

- The positioning system, giving the accurate geographical position of the ship. Today, this is most often GPS (Global Positioning System), preferably in differential mode.
- The attitude sensor unit, giving the heading, roll, pitch and heave values. Its measurements are used to compensate for:
  - orientation of the line of depth measurements relative to the axis of the ship (heading);
  - effective orientation of the beams, their measurement relative to the ship showing errors associated with its movements (roll and pitch);
  - vertical movements of the ship (heave).
- Sound velocity profiles, to correct acoustic paths between the sonar and the bottom.
- Sound velocity measurements close to the array, to improve beamforming.

8.4.3   Bathymetry measurements

8.4.3.1   *Generalities*

The fundamental principle of bathymetry measurement by a multibeam sounder is joint estimation of times and angles. Each couple (t, θ) is used to access the position of one depth measurement (Figure 8.23).

In the particular case where the sound velocity profile is constant in the entire water column and the acoustic paths are rectilinear, $R = ct/2$ and the coordinates (y, z) of the measurement point, taking the sonar position as the origin, are simply:

$$\begin{cases} y = R\sin\theta = \dfrac{ct}{2}\sin\theta \\[2mm] z = R\cos\theta = \dfrac{ct}{2}\cos\theta \end{cases} \tag{8.22}$$

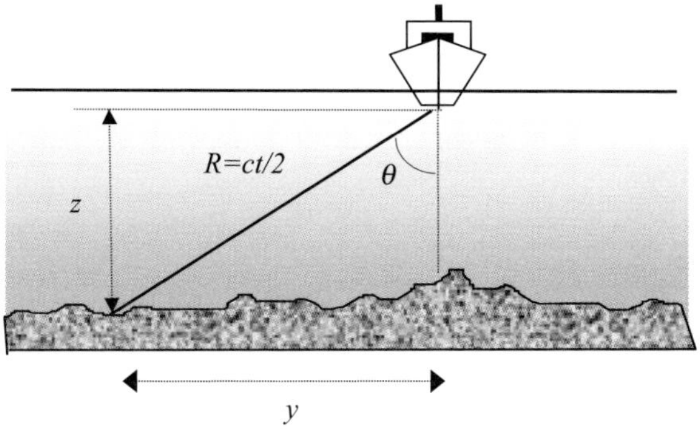

Figure 8.23. Time–angle measurement of bathymetry by a multibeam sounder.

In the more realistic and more complex case where sound velocity varies with depth, the acoustic path must be reconstituted using geometric ray-tracing software (see the main formulae in Section 2.7.2). For a given beam, a ray is launched according to the beam angle, and followed as a function of time along the water column. The computation is stopped when the computed time equals the measured value; the ray extremity defines then the sounding point position.

The measurement is referenced to the position of the sonar. It is therefore imperative simultaneously to know the position and attitude of the support platform, to perform angle corrections (roll, in particular) and associate geographical coordinates with depth measurements. This requires the recording, concurrently with the sonar signals, of measurements from a navigation system (to locate measurement points geographically) and an attitude system (to correct for the platform's movements).

Several techniques may be used to measure the couples (t, θ). Two approaches are possible, each subdivided in two parts:

- For a given angular direction, one looks for the instant of arrival of the signal from this direction. This can be achieved by estimating the maximal amplitude instant of the signal time envelope, or the instant of null phase difference for the signals received on two subarrays.
- At a given instant, one estimates the angle of arrival of the signal on the array. This direction can be obtained by measuring the phase difference of the signals received on two receivers, or by searching the maximal amplitude direction, selected from many beams formed at slightly different angles.

Three measurement methods are presented in the next sections. The *maximal amplitude instant* technique is most commonly used around the vertical, where the signals backscattered are short and the envelope relatively well defined in time. All systems measuring bathymetry in the central area actually use it. The *null phase difference instant* technique is mostly employed outside the central beams around the specular direction; it is very widespread today. The *maximal amplitude direction* technique is used in only a minority of sounders.

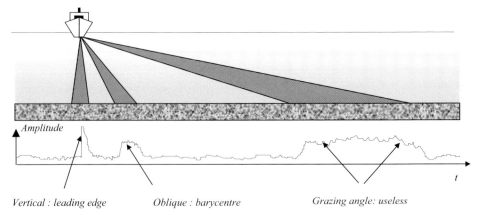

Figure 8.24. Bathymetry measurement considering the instant of maximal amplitude and the influence of the beam incidence angle.

8.4.3.2   *Maximal amplitude instant*

This technique is closest to single-beam measurements. It uses the time envelope of the signal received in a given beam. Assuming this beam to be perfectly vertical and the seafloor horizontal, the depth would correspond to the leading edge of the signal envelope (Figure 8.24); this is the method employed in single-beam echo sounders. For steep oblique beams, the echo envelope is in fact modulated by the directivity diagram of the array, projected onto the bottom. The estimation of the (time, angle) couple then corresponds roughly to the maximum of the signal received. This can be detected by looking for the barycentre of the time envelope. At grazing angles, the echo time spread becomes too wide and its fluctuations too important for determining an accurate arrival time with this method.

This technique provides good results close to the vertical, where the echo envelopes are not much spread in time and show a good SNR. The main difficulty of the measurement is that the signal is then highly perturbed by the reflection perpendicular to the bottom, with very high intensity, which can strongly bias the instant of detection. This specular process[7] is a major source of perturbations for narrow beams with small inclinations. One can try to reduce them by defining the narrowest directivity diagram for the reception beams, and lowering as much as possible the level of the secondary lobes susceptible to receiving this parasite specular echo.

8.4.3.3   *Null phase difference instant*

This widespread technique looks for the time when the phase difference is null between the two signals, at two nearby arrays pointing in the same angular

---

[7] It is in fact advantageous for single-beam measurements, as it guarantees local detection of the highest point.

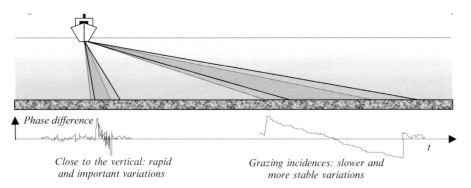

Figure 8.25. Bathymetry measurement considering the instant of null phase difference and the influence of the incidence angle of the beam.

direction. This determines the (time, angle) solution couple. For this purpose, the reception array is artificially divided into two smaller subarrays, which must be long enough to maintain a sufficient angular discrimination, and spaced far enough to optimise the measurement of phase difference. Each of these two arrays forms a beam in the nominal direction of observation (Figure 8.25). The phase differences between the two received time series are computed at each instant, and the detection algorithm must determine accurately the time of null phase difference, corresponding to the arrival of a signal from a target exactly on the beam axis. The problem is that the phase difference is noisy, and the null time must be averaged over an entire time series to get enough accuracy. This degrades the measurement resolution, because the solution is no longer "local" enough and series of neighbouring points are used.

This technique is efficient while the time signals last long enough, and thus corresponds to signals that are sufficiently inclined away from the vertical (Figure 8.25). Compared with the interferometric technique presented in Section 8.3.3, the use of signals after beamforming allows the SNR to be improved because of the directivity gain of the array at reception, and avoids the ambiguities in phase difference measurements by limiting processing to a non-ambiguous sector around the axis of the interferometer.

8.4.3.4 Maximal amplitude direction

This method aims at determining, for a given time of measurement, the angle of arrival of the signal that corresponds to the maximum acoustic intensity. This requires formation of a large number of beams, with very small angular separations. From the time series at the output of these beams, one extracts the amplitudes of the signals that correspond to the different directions, and one keeps the direction that corresponds to the maximum intensity (Figure 8.26). Because the received signals are highly fluctuating, instantaneous determination of the angle of the signal of maximal amplitude is usually inaccurate. The result must be averaged over enough points to reduce this error to an acceptable variance.

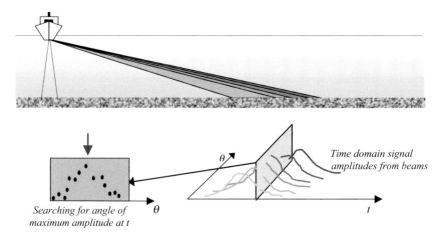

Searching for angle of
maximum amplitude at t

θ

*Time domain signal
amplitudes from beams*

Figure 8.26. Bathymetry measurement with the maximal amplitude direction method.

8.4.4   Sonar imaging with multibeam sounders

The principle of sonar imaging is the same for multibeam sounders and sidescan
sonars. The signal backscattered by the seabed is recorded as a function of time, and
its instantaneous intensity represents irregularities (microscale roughness or seabed
nature) of the ground strip swept by the signal. There are, however, a few differences,
which make matters slightly more complicated.

8.4.4.1   Formation of the sonar image from a multibeam sounder

When constructing a sonar image from multibeam sounder data, time signals are
indeed available after beamforming. They must therefore be recombined so as to
provide a continuous image all along the swathe. Formation of the sonar image is
therefore ideally performed after the digital terrain model has been created by
bathymetry measurements. The central point of a beam is positioned on the
swathe, and the image pixels are distributed around it, until the boundary of the
next beam is reached (Figure 8.27).

8.4.4.2   Map quality

Topographically speaking, the image thus obtained is of much better quality than the
image from a sidescan sonar without bathymetric capability. The pixels are indeed
reported on the (x, y) horizontal plane of the image with minimal geometric distor-
tions. The points corresponding to the centres of the beams are optimally positioned,
and the intermediate points are positioned with a second-order error, linked to the
interpolation between depth measurements. Unlike sidescan sonar, measurement of
the oblique distance alone is now replaced by complete estimation of the position of
the impact points on the seabed (Figure 8.27). This includes the relief of the seabed,

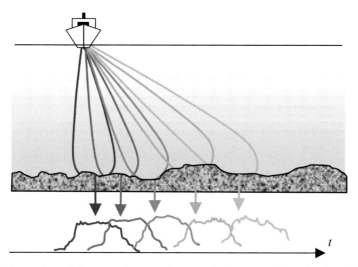

Figure 8.27. Formation of the sonar image using a multibeam sounder. The time signals from each beam are placed along each footprint; the exact geometry is respected by using the depth measurements.

movements of the support platform and refraction of acoustic waves in the water column.

8.4.4.3 Imagery quality

Despite the gain obtained in geometrical accuracy, the quality of multibeam imagery is generally less satisfactory than that of sidescan sonar of equivalent frequency. This stems from several issues:

- The along-track resolution is usually less fine for a multibeam sounder (1–3°) than for a sidescan sonar (less than 1°).
- The beams formed at reception improve the SNR but, on the other hand, their directivity diagrams are prone to modulate the amplitude of the data collected; the resulting image is therefore striated at constant angles, parallel to the ship's track, unless appropriate corrections are applied.
- Finally, the range of incidence angles of a multibeam sounder, used on a surface vessel, is less grazing than for a sidescan sonar, towed close to the seafloor. This is more noticeable as the useful angular sector of a multibeam sounder is most often restricted to ±75°, beyond which bathymetric accuracy is very limiting. Detecting intensity contrasts on microrelief facets is therefore less easy (see Section 3.6.2.6).

One should therefore not expect the same quality of image from a multibeam sounder that can be obtained with a sidescan sonar. Its imaging function should instead be assimilated to *measurement of average reflectivity*, used to map variations in seabed types, rather than a tool for high-resolution mapping.

8.4.5   Performance of multibeam sounders

8.4.5.1   Maximum range

The range of a sonar system is the maximal distance at which one can check that it is functioning well (in general, giving a minimal SNR value[8]). It is obtained from the sonar equation, written for a multibeam echo sounder:

$$SL - 30 \log R - 2\alpha R + 10 \log(\phi c T / 2 \sin \theta) + BS_B(\theta) - NL + PG + DI > RT$$

(8.23)

Considering that $R = H/\cos \theta$ and that the oblique backscattering strength follows a Lambert-type law $BS_B(\theta) = BS_0 + 20 \log \cos \theta$, the maximum range is most often computed as a maximum tilt angle θ for a given water depth H:

$$50 \log \cos \theta - 10 \log \sin \theta - 2\alpha H / \cos \theta$$

$$= 30 \log H - SL - 10 \log(\phi c T / 2) - BS_0 + NL - PG - DI + RT \quad (8.24)$$

For a multibeam sounder, this maximum angle reachable is an essential performance criterion, as it determines the useful swathe width. However, this latter notion is in fact more complex, since it may actually depend on several factors (Figure 8.28):

● the maximum propagation range;
● the maximum angle aperture of the sounder's arrays;
● the effects of refraction on the sound velocity profile.

These three cases are illustrated in Figure 8.28. In the first case (limitation by the propagation range), the transmitted angular sector θ_{max} is restricted to a useful angular sector θ_{eff} effectively covering the resulting swathe. In all cases, the resulting swathe width depends strongly on the water depth (or more accurately the sounder altitude above the bottom).

8.4.5.2   Bathymetry measurement errors

Estimation of the coordinates (y, z) of the depth measurement is affected by inaccuracies in both time and angle. The measurement error can be of two types:

● either a *bias* – the value measured is affected by a systematic error, possibly stable and predictable, which can be corrected during post-processing (this can, e.g., be an angle bias due to misalignment of the arrays during installation on the hull or the effect of SVP refraction);
● or a *random fluctuation* – the value measured fluctuates around an average value, and the quality of the measurement is characterised by its standard deviation. This fluctuation may be related to the outside environment (e.g., noise or

---

[8] Nothing prevents the use of other criteria (measurement resolution, or accuracy); however, without a minimum SNR, there cannot be any detection or measurement, and fulfilling all other criteria would be pointless.

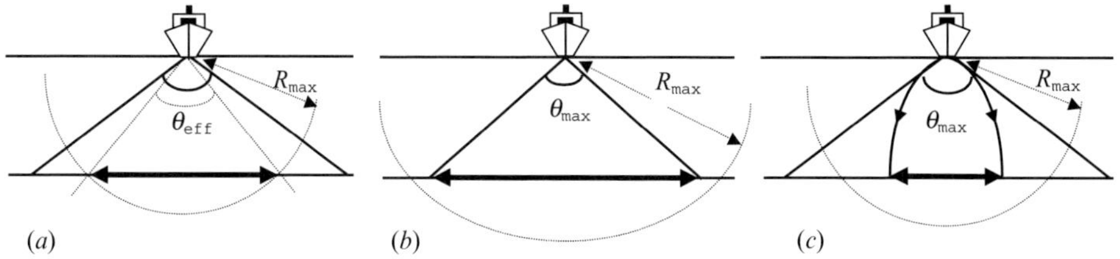

Figure 8.28. Limits of the swathewidth: (*a*) maximum range at oblique distance R_{max}; (*b*) maximum angular aperture θ_{max} of the sounder; (*c*) refraction by the sound velocity profile.

movements of the support platform), or to the target itself (of a fluctuating acoustic character, see Chapter 4).

The sounding estimation error in isovelocity water (from Equation 8.11), corresponding to an inaccuracy δR of the range measurement, is given by:

$$\begin{cases} \delta z_t = \delta R \cos \theta \\ \delta y_t = \delta R \sin \theta \end{cases} \tag{8.25}$$

For an angle measurement error $\delta\theta$, one gets:

$$\begin{cases} \delta z_\theta = R \sin \theta \, \delta\theta = H \tan \theta \, \delta\theta \\ \delta y_\theta = R \cos \theta \, \delta\theta = H \, \delta\theta \end{cases} \tag{8.26}$$

The main error term to consider is often the depth angular error (δz_θ) at shallow grazing angles (i.e., at large values of $\tan \theta$).

Prediction of bathymetry measurement accuracy is an important topic for many multibeam echo-sounder users (see Hare *et al.*, 1995). Three main types of error can affect bathymetry measurements:

- errors in the acoustic measurement itself;
- movements of the support platform;
- inaccuracies in sound velocity corrections.

The global estimation error of the sounder may be modelled as the quadratic sum of these three components, each assumed independent.

Acoustic measurement inaccuracies are due to the fact that, whatever the measurement technique, estimation of the parameter studied (time or angle) is not stable but depends on the fluctuations of the signal used, and therefore the SNR. It can be demonstrated that the parameter estimation variance is at best equal to a limit (Cramer–Rao limit, see Section 6.6.5), inversely proportional to the SNR. On top of the fluctuations caused by noise, one should also account for the fact that the signal backscattered by the seabed is not stable, as it comes from the superposition of individual contributions from scatterers associated with the microscale topography.

Movements of the support platform cause inaccurate positioning of depth measurements. Even if these were determined correctly relative to the ship, badly

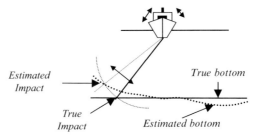

Figure 8.29. Measurement errors associated with movements of the support platform.

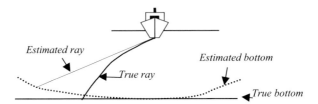

Figure 8.30. Measurement errors associated with the sound velocity profile.

controlled movements (especially roll and, to a lesser extent, pitch, yaw and heave) will give wrong estimates of the final positions of measurements (Figure 8.29).

This problem can be solved by using an attitude sensor that is sufficiently accurate and by continuously measuring the angular position of the sonar. The data from this system are used in two ways: to steer the sounder's beams toward their correct direction instantaneously (mostly in roll for reception, more rarely in pitch for transmission, and even more rarely in yaw), and to correct the position of the estimated bathymetry points geometrically. An accuracy of around 0.05° or better is needed to control the effects of the outer beams.

The sound velocity profile in the water column modifies the acoustic paths (refraction effect), and thus affects the positions of depth measurements (Figure 8.30). All multibeam sounders account for this effect, and calculate the paths of acoustic rays using local sound velocity profiles, measured directly or extracted from geographical/seasonal databases (e.g., the *Levitus Climatological Atlas* database: Levitus, 1982). This process can thus introduce errors if it is not compensated for accurately (i.e., if local velocities are not correct).

Two types of velocity effect should be distinguished:

- Sound velocity on the sounder's array (often measured with a dedicated veloci-meter on the hull close to the array) can influence the direction of the beams formed in the case of a plane array (it can be neglected for a curvilinear array). An incorrect estimate of its value will yield beamforming angle errors. But if the same error affects the sound velocity profile used for calculating refraction of acoustic rays, the angle error can be partially or totally compensated (except for curvilinear arrays).

• The sound velocity profile in the water column is subject to several types of error. Unavoidable fluctuations around its stable average profile have little effect on the refraction of acoustic rays, as the changes in direction around the mean path cancel each other out. But errors in estimation of the average profile will yield appreciable variations, which are uncentred and lead to measurement biases (stable changes of the estimated seabed shape) that are difficult to assess in a few clear cases (∪-shaped or ∩-shaped bottoms). Such cases are met in areas with unstable hydrological conditions.

8.4.5.3   Resolution

The resolution performances of multibeam echo sounders follow the same formulae as sidescan sonars. Along-track, the resolution of bathymetry or imagery is determined by the horizontal beamwidth of the arrays and the oblique distance $\delta x \approx \theta_L R$. Across-track, the resolution of the imagery is given by:

$$\delta y = \frac{cT}{2 \sin \theta} \qquad (8.27)$$

and the limit of this expression at the vertical is $\delta y = \sqrt{HcT}$.

The across-track resolution of bathymetry depends on the number of points averaged in the processing that give each depth measurement. It is at best equal (if no smoothing was performed) to the resolution of the time signal, as in imagery (Equation 8.16). Conversely, if all the beam footprint points are used, the resolution is at worst equal to the beamwidth θ_T projected on the seabed:

$$\delta y = \frac{H\theta_T}{\cos^2 \theta} \qquad (8.28)$$

Practically, δy will be in-between the values of Equations (8.27) and (8.28), depending on the processing details.

8.4.5.4   Signal repetition frequency and platform velocity

The frequency at which signals are transmitted must be as high as possible to maximise the density of data collected over the area, but must avoid ambiguities in the reception of successive signals. The optimal repetition frequency usually consists in transmitting the signal $n + 1$ as soon as the signal n has been received, that is, after a time corresponding to the propagation of the outer beam:

$$T_R = \frac{2H}{c \cos \theta_{max}} \qquad (8.29)$$

where θ_{max} is the maximum inclination of the beams, and the duration of transmission of the signal is neglected relative to the propagation time.

Similarly to the sidescan sonar case – since the geometrical configurations are basically the same – the along-track movement of the sounder can cause gaps in coverage if the width of the beam footprint at the vertical is smaller than the distance travelled during the time between two signals. The maximum velocity V_{max} of the support platform is therefore:

$$V_{max} \approx \frac{c}{2} \theta_L \cos \theta_{max} \tag{8.30}$$

Fulfilling this condition is compulsory in hydrographic surveys, for which 100% coverage is required.

8.4.5.5   *Extension of the notion of maximum range*

This brief overview of the performance of a multibeam sounder, for bathymetry and imagery measurements, leads to redefinition of the notion of maximum range. This was presented in Section 8.4.5.1 as the detection limit of the backscattered signal, defined by either the SNR or the geometry of the outer beams. In so far as the multibeam sounder must obey specific data quality criteria (e.g., sounding depth accuracy, or longitudinal resolution), its optimal use domain must be defined as the domain where all these criteria are met.

   Let us suppose, for example, that the main criterion is hydrographic accuracy standards, and the main measurement error of the sounder is due to roll ($\delta z_{roll} = H \tan \theta \, \delta \theta_{roll}$, Equation 8.26). If the standard requires a depth accuracy of 0.5% of the water depth, and the roll measurement accuracy is 0.1°, it is clear that the effective range of the sounder, given by $\delta z_{roll} \leq 0.005H$, imposes a maximal angle of $\tan \theta \leq 0.005/\delta \theta_{roll}$. This corresponds to a limit of 70.7° for θ (independently of achievability of signal detection and processing beyond this angle).

8.4.6   **Examples of data collected by multibeam sounders**

Several examples of multibeam echo sounder bathymetry and imagery data are presented in the Colour Section of this book (Plates 3–6).

8.5   **SEDIMENT PROFILERS**

8.5.1   **Overview**

Sediment profilers, also known as SubBottom Profilers (SBPs), are structurally similar to single-beam sounders, but working at lower frequencies and gathering vertical cross-sections of the inner sedimentary seabed. As explained in Chapter 7, they are based on an interesting hybrid concept between sonar and seismic systems. The general structure of their hardware is similar to bathymetry or fishery single-beam sounders, with a few noticeable differences in working principles, requiring a separate presentation in this book. Their functionalities are very similar to those of reflection seismics.

   The function of a sediment profiler is to record echoes from the interfaces between sedimentary layers; these layers correspond to breaks in acoustic

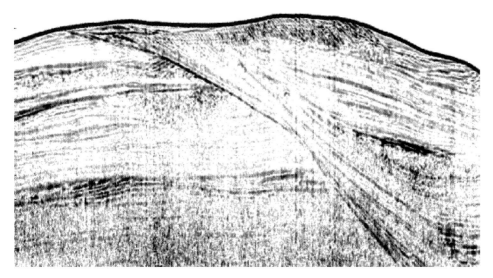

Figure 8.31. Sediment cross-section recorded by a sediment profiler (chirp sounder installed onboard R/V *Suroît*). The vertical scale is ca. 100 ms, corresponding to about 75 m inside the sediment (data provided courtesy of Bruno Marsset, IFREMER).

impedance, generating *reflections* of the acoustic signal (no backscatter is involved here). As with a sounder, the actual movement of the support platform will allow reconstruction of a vertical cross-section of the sedimentary environment being investigated (Figure 8.31).

The cross-section so obtained is at first approximation an image of the boundaries between layers (i.e., morphology of the sedimentary bodies). This image can be quantified by accounting for: (1) along the horizontal direction, the velocity and position of the support vessel; (2) along the vertical direction, penetration of the sediments (often expressed in seconds, following the usage of seismics, but translatable into a distance, which involves making assumptions about acoustic propagation velocities within sediments). This descriptive approach is sufficient for many applications. It was in any case the only one possible with the oldest models of sediment profiler, for which acoustic levels could not be measured.

The latest sediment profilers use a definite improvement: "chirp" frequency-modulated signals and pulse-compression processing (see Section 6.4.2). This technological evolution enabled two crucial improvements:

- The SNR gain from coherent processing is essential to get penetration depths much larger than with traditional signals. To get an order of magnitude, the modulated signals commonly used have *BT* factors of around 100, corresponding to a gain of around 20 dB. This gain would correspond to the absorption loss over 100 m in a (clay-like) sediment characterised by an attenuation coefficient of 0.1 dB/m.

- If the reception processing chain is correctly designed and implemented, one can access the absolute level of the echoes from the bottom, and therefore the characteristics of the seabed that acted on the reception. The imaging of sedimentary structures is then complemented with an estimate of their physical parameters, which are accessible to acoustic investigation (density, velocity, attenuation). This completely changes scientific interest in the use of sediment profilers.

The frequency range of sediment profilers is a few kHz; that is, acoustic wavelengths (in water) typically between 0.2 m and 1 m. The most commonly used nominal frequency is 3.5 kHz; this is a good compromise between the penetration depth and the dimensions of the antenna. Lower (down to 2 kHz) and higher (up to 7 kHz) nominal frequencies are also encountered. The choice of the main working frequency is very important, as it constrains greatly the penetration range of the sounder. Assuming, as a first approximation, that the attenuation in the sediments is proportional to frequency, the range follows approximately the same dependence.

The bandwidth B of the signal is very important, in so far as it determines, in one fell swoop, the processing gain $10\log(BT)$, hence the SNR and the range resolution of the echoes, $\delta z = c/2B$. The usual bandwidths of 1–3 kHz correspond to resolutions between 0.75 m and 0.25 m.

The ping repetition frequency is adjusted to get a high density of data points in the profile, increasing the quality of the representation. In particular, at large depths, it can be much higher than the frequency defined by the time of propagation in the water column. The interesting portion of the signal received (the one that propagated inside the seabed) is often much shorter than the total cycle. It rarely exceeds 0.5 s, whereas propagation in the water column, which does not generate any information, can last much longer (6 s for a depth of 4,500 m). The pings are then interlaced (see Section 8.2.1.2): after a ping has been transmitted, the next pings are transmitted without waiting for its echo. One should only watch that the transmissions do not interfere with the reception windows for the echoes from the seabed. Ambiguities are solved either by using the true water depth after the first non-ambiguous detections, or the data from another sounder.

8.5.2   Sediment profiler performance

8.5.2.1   Maximum penetration depth

The maximum penetration depth is a very important characteristic of a sediment profiler. It is obtained from a particular form of the sonar equation, where:

- the target acts by means of a reflection process, rather than backscatter. Hence, one has to consider a single spherical wave reflected by the seafloor, with a transmission loss of $20\log(2R)$ instead of $40\log(R)$;
- the transmission loss has to be split into two parts (water and sediment), with different absorption coefficient values.

The sonar equation may be written:

$$SL - 20 \log(2H + 2H_s) - 2\alpha H - 2\alpha_s H_s + 20 \log(W_{ws} W_{sw})$$
$$+20 \log V - NL + PG + DI = DT \quad (8.31)$$

where H is the water depth and H_s is the penetration depth inside the sediment; α and α_s are the absorption coefficient, respectively, in water and sediment; $20 \log(W_{ws} W_{sw})$ is the two-way transmission coefficient (in dB) at the water–sediment interface (see Section 3.1.1); $20 \log V$ is the reflection coefficient (in dB) at the buried layer interface considered as the target. Note the particular importance of the PG term (20 dB and more), associated with the use of chirp signals.

The maximum penetration depth for an SBP mainly depends on the sediment absorption coefficient. For a clay-like sediment (α_s around 0.1 dB/λ), typical penetration may reach 50–200 m according to the system characteristics. Since the absorption effect depends strongly on frequency, in order to increase this performance it is far more efficient to decrease the signal frequency rather than to increase the source level.

8.5.2.2 Resolution

The vertical resolution of an SBP has already been introduced as directly depending upon its bandwidth: $\delta z = c/2B$.

Since the echo is caused by reflection rather than backscatter, the horizontal resolution is given by the dimension of the first Fresnel zone $\delta x \approx \sqrt{\lambda H_t}/2$, with $H_t = H + H_s$. Note that this value does not depend on the beamwidth value, and that it is usually much smaller than the footprint size delimited by the beam projection.

8.5.3 Echo formation and consequences

Unlike the three previous types of sonar, the sediment profiler uses reflected echoes, instead of backscatter. This stems from the geometric measurement configuration (the system aims at the perpendicular to the seabed, and can therefore receive the reflected echo that returns in the same direction), and mostly from the low frequency band used: the backscatter is negligible compared with the coherent echo, since the microscale topography amplitude is small compared with the wavelength (Figure 8.32).

This distinct behaviour is absolutely essential when interpreting data from a sediment profiler. Because the echo is generated by specular reflection at the interfaces, it corresponds to a local portion of the seabed limited by the reflection process itself, and not by the footprint of the beam; hence the beamwidth does not influence the echo duration. The consequences are many:

- The resulting image quality does not depend on the beamwidth, which contradicts everything that has been said previously about other sonar mapping systems. Sediment profilers with poor array characteristics can therefore

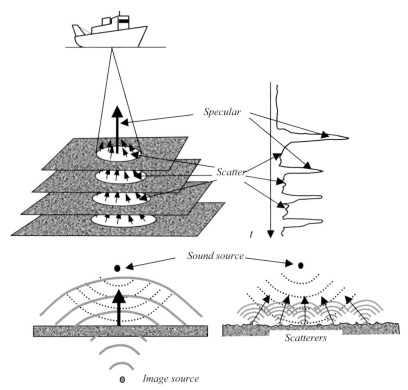

Figure 8.32. Formation of the echo from a sediment profiler: the reflected contribution (*lower left*) dominates the echoes backscattered by the microscale topography of the interface (*lower right*).

produce very good sedimentary cross-sections. The image quality is rather related to the time resolution of the signal used.

- The accuracy of the beam angular steering is of secondary importance. It is enough that the specular direction is included in the main directivity lobe. A sediment profiler can therefore bypass roll and pitch corrections. It can even tolerate bottom slopes. Paradoxically, this tolerance will be better[9] the wider the directivity pattern, thus "bad" when assessed with the usual evaluation criteria.

- The sediment profiler can only work properly at specular incidence. At oblique incidence and with a backscattered signal, it only collects very weak echoes, buried in noise, and reflected specular contributions. Furthermore, the spatial resolution of the backscattered echo is no longer the signal resolution but the footprint dimension, as in a high-frequency sounder; this prevents the separation of close sedimentary interfaces. Inclining the measuring axis increases the footprint, and further degrades the resolution (Figure 8.33).

---

[9] Up to certain limits: too wide a beam decreases the SNR due to a poor directivity index; it also increases the risk of jamming the data by echoes from undesirable lateral targets.

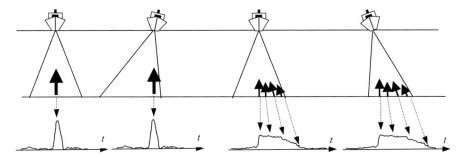

Figure 8.33. Time domain envelope of a specular reflected echo (*left*) and a backscattered echo (*right*). The specular echo does not depend on the inclination of the beam. The duration of the backscattered signal, hence its resolution, depends on the width and inclination of the beam.

8.6 SEABED IDENTIFICATION AND CHARACTERISATION WITH SONARS

8.6.1 Applications and potential

Most application domains of underwater sonar mapping techniques are ultimately concerned with accessing particular characteristics of the seabed. How and to what degree can acoustics fulfil these needs? The answer is not straightforward, and the different levels of requirement need to be distinguished.

In some types of application (navigation, fishery, even hydrography or sediment dredging), the objective can simply be basic identification of seabed type, which can be more descriptive than a real geological analysis. For example, a restricted classification of the seabed type, using a maximum of around ten classes (e.g., rock, boulders and pebbles, gravel, coarse sand, fine sand, silty sand, silt and vegetal cover) may be sufficient for many common applications. Analysis of backscattered signals can in theory provide elements for identification. However, designing such a simplified classification system as a robust and automated tool is not straightforward. The potential customers of this type of application do not always have the finance necessary to get the sonar systems best suited to these tasks (multibeam sonars). Often, they use single-beam sounders, of lesser potential.

A second type of use is concerned with the geological characteristics of the sediments (e.g., granulometry, mineral constituents) and their geotechnical characteristics (e.g., mechanical parameters, such as density, shear and compression modulus). None of these can be identified directly by signals from a mapping sonar – whatever its type. The use of acoustic systems is, however, justified for several reasons:

- Identification of a particular sedimentary facies (in the restrictive way presented in the previous paragraph) is a good first indication of the parameters sought. Mapping the zone under investigation by sonar bathymetry and imagery is often a necessary preliminary to finer local investigations.

- Building a map based on acoustic characteristics allows generalisation, over a complete area, of geological analyses results, which can in practice be completed only for very small areas. Highly used, this approach assumes that any noticeable change in the parameters studied will be visible in the acoustic signals received.

- Finally, acoustic returns can be used to try and extract the parameters sought, by comparison with a model. There are many problems associated with this approach: the validity of acoustic models is limited, and only proven under simplifying assumptions. Even then, the number of input parameters to these models is often too high to identify them all from a small number of parameters extracted from the backscattered acoustic signals. A more realistic methodology consists in local fine-tuning of an acoustic model with local proven data, and in subsequent study, around this reference area, of the evolution of a restricted number of model parameters.

Seabed characterization for military applications is a particular case. Its objective is the building of maps of relevant acoustic parameters (local conditions of reflection and reverberation), making possible performance prediction and analysis of detection systems. The methodology for constituting these databases often uses the same tools as the other domains of acoustic mapping: construction of bathymetry and acoustic reflectivity maps, sampling and analysis, etc. But, even in this case, sedimentary or geophysical study is not a goal in itself, but a starting point for models of acoustic propagation. An important issue today is the gathering of acoustic seabed characteristics that are relevant to military sonars (hence at low frequencies of several hundreds to several thousands of Hz), extrapolated from experimental data collected with mapping tools working in other frequency or incidence ranges.

Two main approaches coexist to characterise the seabed with sonars. The first is definitely pragmatic, and aims at identifying classifying parameters among the data provided by mapping sonars. These parameters may as well be explicitly deprived of any physical significance, provided that they offer a sufficient discriminatory power. The sonar is calibrated as a function of these parameters, over a certain number of validated configurations; new cases are identified by comparing them with this leaving database. Despite the restrictive character of this approach, it must be acknowledged that the only practical applications available nowadays are based on it.

The second approach is more ambitious, and indistinguishable from some applications. It aims at solving the "inverse problem", by extracting objective characteristics of the sediments from the acoustic data available. This implies important modelling efforts, in particular for the scattering of acoustic waves by very complex media (porous media, rough interfaces, layered media, presence of mineral, organic or gaseous inclusions, etc), and for the definition of bridging relations between acoustic characteristics (observed experimentally) and geological characteristics (of interest to the user). This second approach is, of course, theoretical: the inverse problem is highly conditional, with a large number of input parameters and very few observations. Furthermore, ground-truthing is always a difficult matter, especially in deepwater.

8.6.2 Single-beam sounders

Seabed identification using the echoes from single-beam sounders is highly sought by users of these systems for navigation or fishing, as well as hydrography. The problem is not easy to solve. The signals in a restricted angular sector close to the vertical are indeed poorly suited to extraction of classifying parameters typical of the seabed. The backscattered level measured at the vertical is not always a good indicator of the bottom characteristics; it varies only slightly from one seabed type to the next, and the signals received are very unstable (and statistical analysis of these fluctuations does not provide significant advantages). Most information seems to be contained in the average time envelope of the echoes (Figure 8.34), whose shape reproduces the angular sweep of the seabed by the signal, together with some backscattering characteristics of the interface (Pouliquen and Lurton, 1994). But because of the angular structure of the beam (too narrow to sweep a really significant range of incidence, but too wide to perform a fine angular analysis), its potential is much less than for

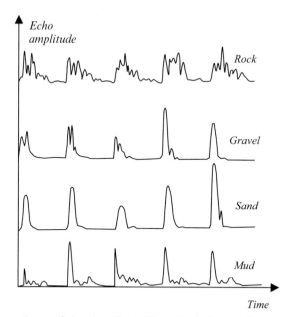

Figure 8.34. Time envelopes of signals collected by a single-beam echo sounder over different types of seabed (Pouliquen and Lurton, 1994). While the gravel and sand echoes reproduce more or less the envelope of the signal transmitted, the rock echoes are widely spread in time and strongly fluctuating, due to the high roughness. Mud echoes are typical, with a strong specular echo followed by contributions from the sediment volume.

systems with large angular coverage and good resolution (sidescan sonars and multibeam sounders).

A classification technique much used in practice is based on comparison of some characteristics of the main echo and the first multiple echo (Burns *et al.*, 1989). Comparing the trails of these two signals defines a couple of parameters character- istic of interface roughness and its impedance contrast, which also depends on the individual characteristics of the sounder used (signal, directivity). A sounder properly calibrated on proven test areas can, by comparison, identify the nature of the seabed encountered in conditions of use similar to its learning period. This method does not actually identify the type of seabed, but segments it into areas based on the characteristics of reference echoes. It is very versatile, simple and robust, and meets with much commercial success.

8.6.3   Sidescan sonars

Seabed characterisation using sidescan sonars is a special matter, because of the mainly imaging role of these systems. The absence of geometrical data that are really exploitable (except for sonars with interferometric capabilities) precludes, for example, studying variation of backscatter with the angle of incidence. With these systems, one tries instead to use directly the information present in the image itself (Figure 8.35). Very often, especially with the most recent systems, the excellent quality of the images obtained at high frequencies gives a near photo- graphic representation of the seabed. No specific processing is then necessary, as soon as some artefacts related to the measurement itself (average level variations due to the angular response of the seabed, influence of the directivity patterns of the antennas, propagation losses) have been corrected. The user can then immediately and visually interpret and segment the image.

Some applications, however, require automated processing of seabed recogni- tion, and several techniques have been proposed in this respect. Some analyse the raw signals from the sonar: the idea is that echo characteristics are closely related to water–bottom interface features, and, for example, that the roughness spectrum modulates the backscattered signal. Another technique analyses the backscattered spectrum (e.g., Pace and Gao, 1988), enabling identification of the seabed among a few types. The interest of this approach lies in the fact it can be applied in real time during signal acquisition. Finally, it is also possible to work on processed sonar images, using texture classification techniques (e.g., Blondel, 1996), which are classical in image processing.

8.6.4   Multibeam sounders

The main advantage of multibeam sounders in seabed characterisation is their potential for selecting the exact signals returning at particular angles. Most fre- quently, this potential is used to assess the backscattering strength as a function of the angle of incidence (de Moustier, 1986). The signal level inside a narrow beam, possibly averaged over several successive pings, is first corrected for different terms in

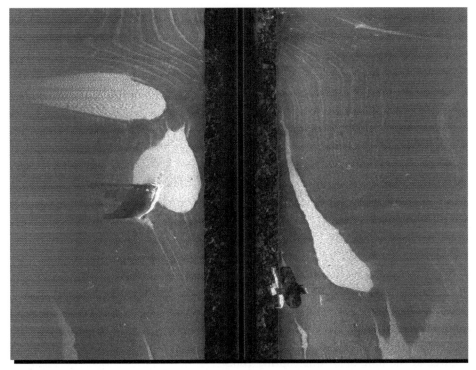

Figure 8.35. Example of a sidescan sonar (Klein 5400) image, showing changes in sedimentary facies. Light colours are for strong reflectivity. Note the differences of sedimentary facies between mud areas (*dark*) and sand ripple areas (*light*). The wrecks are two trawlers, sunk as artificial reefs in the Bay of Douarnenez (data provided courtesy of DCE-GESMA, France).

the sonar equation (transmission level, propagation loss, antenna directivity at transmission and reception) to yield the target strength. This strength is then corrected for the beam's footprint (depending on the geometry, the beam width and the signal duration). This provides the unit backscattering strength as a function of the effective angle of incidence on the bottom (accounting for refraction by the sound velocity profile and bathymetry extracted from the local digital terrain model). These operations give the angular dependence of backscattering, but a short-coming is the assumption that the seabed type is stable over the entire angular sector analysed. This method shows seabed responses that vary quite clearly from the sedimentary facies (Figure 8.36). Accordingly, the information provided is quite coarse. Many ambiguities can occur because fundamentally different seafloors may have similar acoustic energy responses at the same angles. For further analysis, it is possible to use other parameters obtained from the raw signals recorded, or from sonar images. The techniques used for analysis are then the same as those with sidescan sonars. Their advantage is that geometric variation in the data is explicit and can be accounted for. The disadvantage is that the intrinsic quality of the images

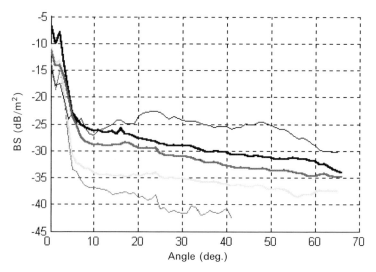

Figure 8.36. Angular backscattering strengths of several seabed types, measured with a multibeam sounder. Data provided courtesy of Michel Voisset, IFREMER.

is often worse, because of the less favourable conditions of acquisition (resolution, angles of incidence). These processing techniques are still very much a domain of research at the time of writing (Spring 2002).

8.6.5   Subbottom profilers

Sediment profilers facilitate investigation of the characteristics of buried sediment layers. Contrarily to sounders and sidescan sonars, signal amplitudes are a function of the impedance contrast and in-sediment absorption, rather than interface micro-roughness. Interestingly, this reduces the number of inflowing acoustical parameters to a few (density, velocity, absorption) which ought to be accessible from the echo amplitude analysis. However, since the echo amplitude from a buried layer is a function both of the local impedance contrast and the absorption and transmission losses in the overlying sediment stack, the problem remains quite complicated. The main difficulty is estimation of the absorption influence; a current approach is to use a frequency analysis to estimate its effect, under a given hypothesis of frequency dependence (linearity). An adapted methodology is to process sequentially, starting at the water–sediment interface:

- the echo amplitude at a sediment layer boundary gives a value for the layer impedance and a first guess at its absorption coefficient;
- the frequency analysis of the next buried interface echo provides more detailed information about the layer absorption coefficient, and allows us to summarise the total transmission loss from the upper interface;

- correction of total transmission loss through the overlying sediments gives access to the next reflection coefficient, etc. Subbottom profilers do represent an invaluable intermediate tool linking sonar mapping with seismic exploration; much is expected of new techniques for quantitative processing of sediment layer echoes.

9

Conclusion

9.1 SYNTHESIS

9.1.1 Propagation

Underwater acoustics uses mechanical waves, the only efficient physical vector for transmission of information through the marine environment. It mostly depends therefore on the intricate physical processes associated with wave propagation in a complex medium such as the ocean. The most noticeable phenomenon is propagation loss, where the amplitude of a wave transmitted in water decreases with distance from the source. This stems not only from geometrical spread, but also, for underwater acoustic waves, from absorption by the propagation medium, dissipating a part of the transmitted wave energy by viscosity or molecular relaxation. Absorption effects increase dramatically with frequency, and they are one of the main factors limiting propagation ranges. This is often compounded by the presence of gas bubbles in sea water close to the surface.

Another characteristic of underwater acoustic signals is that they undergo multiple reflections at the interfaces of the propagation medium (sea surface and bottom). These reflections make the time structure of the echoes received more complex. Very often, these multiple echoes are similar in magnitude to direct arrivals, and perturb their reception. Furthermore, the underwater environment is physically far from homogeneous, and sound velocity frequently varies with depth. Even limited to a few percentage points, these variations cause significant refraction of the sound waves. The most remarkable feature is without doubt the SOFAR (SOund Fixing And Ranging) channel, a waveguide in which low-frequency waves transmitted and trapped around a minimum velocity depth can propagate over thousands of kilometres.

Refraction effects can be studied using geometrical acoustic theory. The acoustic energy is followed along its various propagation paths, accounting for refractions of the wave direction with the sound velocity gradient. Practically, this method is based on the very intuitive representation of ray tracing, and is the most commonly used, especially at high frequencies (as it very easily gives propagation times and angles). At low frequencies, one may prefer using wave methods for solving the propagation equation (normal modes, fast numerical transforms, parabolic equation). These methods are less intuitive than ray tracing and their physical meaning is less clear (except for normal modes), but they give more accurate results when computing detailed field amplitudes. The two types of approach (geometrical and wave) are used to process, in most cases of practical interest, the refraction effects commonly encountered and related to variations of velocity with depth, and to interpret the very special effects of the field structure caused by guided propagation.

Understanding the large-scale structure of the sound field implies modelling the structure of the multiple paths generated by refraction effects and by their confinement between surface and seabed. The need for this understanding is justified by many important practical applications; for example, in the military search for stealth zones, in bathymetric surveying to reconstruct very accurately the paths of the echoes from the seabed, and in acoustical oceanography to deduce hydrological characteristics from acoustic measurements.

9.1.2   Reflection and backscattering

The reflection of acoustic waves from the seabed or sea surface, or from obstacles in the water column, is a key aspect of many underwater acoustics applications. Two different physical modes must be considered: specular reflection from a regular surface, and scattering from an irregular surface. In the former, the incident wave is reflected symmetrically to its direction of arrival, with a reflection coefficient depending on the impedance contrast and the angle of incidence at the interface. This mode affects the transmission between distinct transmitters and receivers, and constrains the characteristics of multiple paths. An irregular surface will reflect incident energy in all angular directions: this is called scattering, and its useful component is backscattering towards the sonar. The backscatter level depends on the ratio between the signal wavelength and the characteristic dimensions of the target; a target that is small relative to the wavelength will not scatter very much energy (Rayleigh regime). At high frequencies, backscattering may be interpreted geometrically as a series of local reflections on elementary surfaces.

The study of reflection processes is, of course, essential to master the characteristics of propagation with multiple paths. But the reflection *per se* is usually not used directly as a measurement tool (because of the constraints it imposes, mostly bistatic configurations). However, backscattering is the physical value measured by the majority of sonar systems; its understanding and its modelling are therefore much used in the design of sonar systems (e.g., sounders, sidescan sonars, Doppler systems). They are primarily used to retrieve the target's characteristics from the backscattered energy. For example, the integrated measurement of backscattered intensity is used in fishery sonars to estimate the biomass in a given area quantitatively. The backscattering strength measured by mapping sonars is also a good indicator of the nature of the seabed.

9.1.3 Noise and fluctuations

The influence of the marine environment on signals transmitted or echoes back-scattered includes the many random perturbations that degrade the reception of these signals, resulting in noise being added to the useful signal. This noise can be of acoustic origin (present in the propagation medium and perceived by the receiver at the same time as the signal, more or less masking the signal), or of electrical origin (then affecting the receiver itself). There are many types of additive noise: the most common are caused by agitation of the sea surface and the intrinsic noise of the support platform. But other noise sources can be important too (e.g., shipping, rain, living organisms, thermal noise). The structure of ambient noise (i.e., its spatial and temporal structure) is quite complex, and follows the same propagation laws as the signals. Understanding noise and its generation is essential to the design of processes to remove as much noise as possible.

An important part of the noise affecting the receiver of an active sonar is due to reverberation, a generic term grouping all contributions from echoes that are different from the one desired. This phenomenon is very difficult to eliminate as it is generated with exactly the same characteristics as the useful signal: frequency filtering is therefore poorly suited to filter it out.

Finally, the signals themselves are perturbed by transmission (generation of multiple paths, because of refraction and reflections at the interfaces, or "micro-paths" associated with scattering), and by physical variations of the propagation medium and the targets. The signals received thus fluctuate, both in space and in time.

9.1.4 Transducers and array processing

To transmit or receive acoustic energy, sonar system engineering requires the design and use of electro-acoustic transducers. Sufficiently high levels at transmission are obtained using resonant mechanical systems, usually based on a piezoelectric engine controlled by the electrical signal, which reproduces the shape of the acoustic signal desired. The technologies used at high frequencies (Tonpilz) and at very high frequencies (ceramic blocks) are very general. At low frequencies (below 1 kHz), many variants are available, depending on the particular requirements of the type of application desired.

At very high transmission levels, particular phenomena appear. Cavitation (i.e., vaporisation of dissolved gas) is a limiting factor for acoustic projectors. Non-linearities in propagation can be exploited for certain types of transmission (e.g., parametric arrays).

An essential characteristic of antennas is their directivity pattern. Selection of a narrow beam in a favoured direction allows increasing the level transmitted, and improving the signal-to-noise ratio (SNR) on reception, as well as the spatial resolution of the measurements, whose angular selectivity varies with the ratio of wavelength to antenna dimension. The best performance of a sonar system is mostly attributable to the transducers used and the type of antenna processing. Highly directive antennas can measure the angular positions of targets[1] very accurately, or image a particular scene at high resolution.

Mechanically steered antennas suitable for use underwater are very difficult to produce, so sonar systems extensively use electronic steering. The physical array, made of many elementary transducers, remains fixed and so elementary transducer signals are phase-shifted or delayed to steer the directivity pattern along the direction desired. Beamforming is the processing technique most commonly used. It is based on the assumption that sound waves are plane at reception; it is used to steer the beams along directions that are highly inclined relative to the antenna's axis. Secondary lobes can be decreased by weighting to lessen the influence of the transducers at the array's ends. In the near field, the antenna must be focused to account for the curvature of the wave front. The Synthetic Aperture Sonar (SAS) technique replaces a long array with a physically short antenna, whose positions with time are measured very accurately. Despite some physical limitations due to underwater acoustic propagation, this technique offers some interesting perspectives to high-resolution imagery.

9.1.5   Signal processing

The acoustic performance of an underwater acoustic system (for detection, measurement, communication or imagery) mostly depends on the SNR it works at. Traditional theories of detection and estimation forecast that the probabilities of detection or false alarm for detection sonars, data errors for a transmission system, or variance of physical measurements (e.g., angle or time), will decrease rapidly in accordance with the SNR. The first function of a sonar receiver is therefore to improve this highly important parameter.

The fundamental processing operations used in underwater acoustics to extract the expected signal from the noise consist in:

- filtering in the frequency domain, to select the bandwidth of the signal expected;
- detection of the signal envelope, or quadrature to get the energy;
- summation of several signal realisations to improve the SNR.

Such a processing chain is intuitive for the detection of signals that are unknown and stable a priori (e.g., ship-radiated noise in passive sonar). For these, narrow or large bandpass filtering of the acoustic energy received (i.e., after quadrature) is completed by time integration, whose aim is the improvement of the SNR.

These receivers are also predominantly used in active sonar, when the signals transmitted are CW (Continuous Wave) pulses. These signals and their processing

---

[1] A variant on angular measurement is interferometry, using the phase difference between two transducers to measure very accurately the direction of arrival of a signal, without needing a highly directive antenna.

are remarkably simple. They are adequate at measuring times and angles (which is often the final goal of sonar systems), when the input SNR is appropriate.

Modulated signals (most often frequency-modulated, but sometimes phase-modulated too) are less frequently used. For detection and measurement applications, they imply the use of correlators at reception, to filter the signal received using parameters that perfectly match (including in phase) the time signal expected. The use of such signals and processing is particularly justified when the SNR is poor, since cross-correlation improves the processing gain substantially. The energy transmitted, sometimes for a relatively long time, is found in its entirety in a short output signal (pulse-compression technique). Communication systems, on the other hand, aim at reconstructing the information content of the modulation.

In practice, the operations of a sonar receiver use classic functions of demodulation, digitisation and filtering. An original processing is the Time-Varying Gain (TVG), whose aim is to keep the reception level relatively stable by compensating for the physical decrease of the echoes with distance (and therefore the time of propagation).

Unlike these elementary reception operations, found in most systems, post-processing modules vary widely with the systems considered. They use tools that are oriented toward the particular functions desired: measurement of immediately available physical parameters of the target echo (time, angle, phase, frequency), formatting and display of these values, most often graphically (and, for this purpose, using other types of data; e.g., from navigation, motion or environment sensors), and storing the values. Finally, specialised modules perform finer analyses of the signals received (time and/or frequency analyses) to characterise the target.

9.2   CURRENT TRENDS AND PERSPECTIVES

9.2.1   A major tool for monitoring and exploitation of the Earth

We shall not repeat here the detailed description of the different types of underwater acoustic system and application already presented at various levels in Chapters 1, 7 and 8. Let us instead simply recall that acoustics is used in navigation and safety instruments used on ships and in the offshore industry (sounding, marking, underwater positioning, speed measurement, obstacle detection, communication, remote control), in military operations (detection, tracking and characterisation of submarines and mines), in monitoring and exploitation of the biomass (detection, location, visualisation and measurement of fish shoals), in seabed mapping (bathymetry and imagery, for hydrographic, scientific or industrial aims), in geophysics and hydrocarbon exploration (subsurface sounding), in oceanography (hydrology variations, currents, . . .), etc. This list is long and imposingly varied.

Two main application fields, closely related, appear unavoidable and paramount at the beginning of the third millennium. The first is the control of environmental quality and the management of living natural resources. The second is the exploration and exploitation of energy resources. Both items will be increasingly important in the years to come, especially as far as the oceans are concerned. In this context, the role of underwater acoustics will be inevitably made stronger.

High-resolution mapping and monitoring of the seabed is an essential component of these realms, similar to space monitoring of the landmasses of the Earth. In this respect, the capabilities of specialised sonars are improving constantly. Although the ranges of different types of system, fixed by the physical characteristics of the underwater environment, cannot improve much further, resolution and accuracy of these systems are spectacularly improving. Bathymetric measurements can now reach relative accuracies of the order of 1/1,000, angular resolution can reach a fraction of a degree, and measurement point densities increase consequently. So does, correlatively, the quality of high-resolution acoustic imagery. This progress is supplemented by increasing coverage and data redundancy of mapping systems, and the emergence of new concepts using multiple beams not only across-track but also along-track or looking forward. The concept of SAS, although less necessary than in radar remote sensing, is justified by the need for very high-resolution imagery. But, in mapping, as well as in many other domains of application of underwater acoustics, the most important progress (and the less certain, as related to the physical conditions of the measurements) is expected in the absolute measurement of backscattered energy levels. There are very few sonar systems providing correctly calibrated backscattering strengths that are usable for target characterisation, whereas this capability has been available for years on airborne or spaceborne radars.

Biomass monitoring, management and exploitation are other domains where the demand for measurement or control systems is strong. Single-beam fishery echo sounders will need to evolve into instruments that give richer spatial information: multibeam sounders that explore instantaneously not just one but two[2] or three dimensions, enabling visualisation and quantitative assessment of fish shoals with more accuracy than is available today. The progress expected in the domain of qualitative species recognition, for which there is a strong demand, is, however, less easy to define, because of the complexity of the problem of identification of individual target strengths. The solution may lie in new tools for the visualisation of fish shoals. In a neighbouring domain, the development of aquaculture and fish farming will lead to the design of monitoring and counting systems that may be based on acoustics.

Monitoring the coastal environment and sediment fluxes implies the use of measurement systems to follow the evolution of surface deposits with better resolution than currently available with bathymetric systems. This will also require assessment of the structure and characteristics of the upper sedimentary layers and quantitative estimation of suspended particulate matter in the water column. Here again, acoustic instruments derived from modern sounders or ADCPs (Acoustic Doppler Current Profiles) will fulfil this demand.

---

[2] With a technology similar to that of mapping sonars.

9.2.2   Underwater acoustics in perspective

The progress of underwater acoustics has long been dependent on military research, which was the beneficiary of the largest funding (in this domain as in many others). Civilian developments have been therefore, technologically speaking, qualitatively lower than military ones. One explanation for this was the cost of developing innovative technologies and sophisticated processing systems, which traditional civilian users (hydrographers, fishermen and oceanographers) were not able to support. Still, today, economic support for the military applications of underwater acoustics is much larger than for civilian applications.

However, this situation began to change in the last two decades of the 20th century with the autonomous development of civilian underwater acoustics at an excellent technological level. This shift can be explained by:

- Changes in the needs of oceanography, implying an increasingly larger exploration of the oceans, generating new needs in tools for mapping, measuring, positioning and communicating. These in turn generated a strong R&D activity in this domain.
- The increasing need, for offshore industries (hydrocarbon exploration or exploitation, laying of pipelines or telecommunication cables), of survey and underwater intervention tools – with levels of funding in relation to the economical weight of this sector.
- Technological progress in relevant marine instrumentation (e.g., satellite positioning, motion sensors, communications, towed, tethered or autonomous underwater vehicles), making possible and facilitating the emergence of new acoustic transducer concepts.
- The spread of digital signal processing techniques, resulting in cheap and efficient tools that are more easily accessible to small teams. This led to the development of sonar systems that are independent of the expensive, specialised processors developed at high cost by the military industry. This favoured the appearance of many independent manufacturers and the remarkable progress of system performance in terms of processing capabilities, as well as lower costs.

In contrast, underwater acoustic transducer technologies have only indirectly benefited from the progress of other domains, and have evolved more slowly than electronics, or signal data processing. Electro-acoustic designs remain unavoidable in all acoustic systems, and very often determine the ultimate performance.[3] This area therefore remains very specific, and requires highly specialised and experienced personnel.

It is no exaggeration to say that underwater acoustics today performs most of the operations done by electromagnetic systems in the atmosphere and in space. The same basic functions can be achieved: detection, tracking, measurement and

---

[3] This fundamental principle is very well known in aerial acoustics: a genuine hi-fi buff will spend most of his budget on loudspeakers, whose impact largely exceeds the electronic elements for reading and amplifying.

characterisation of military targets, either from their echoes or from their intrinsic emissions; topography and reflectivity mapping with sidescan systems; transmission of digital data, etc. Considering mapping as an example, the orders of magnitude of bathymetric performance (accuracy or resolution) of multibeam sounders are, finally, comparable with those of airborne or spaceborne mapping radars. In terms of imagery, modern, high-resolution sidescan sonars provide images of quasi-photographic quality, with a lesser resolution but a much larger range. However, some characteristics of acoustic systems remain (and will remain) much behind those of electromagnetic systems, because of unavoidable constraints. For example, in mapping systems, sound velocity limits the pulse repetition frequency, hence the speed achievable by the support platform (itself limited by hydrodynamic constraints): coverage speed cannot be compared with that of a synthetic aperture radar installed on a satellite. In the domain of data transmission, the compulsory use of relatively low-frequency acoustic carrier frequencies limits information rates to values much smaller than what can be achieved with Hertzian transmission.

   Contrary to what can be observed in aerial acoustics, the techniques of underwater acoustics remain almost exclusively in the realm of marine professionals, and their commercial diffusion to a larger public remains very small. At another level, the impact of underwater acoustics in collective applications directly affecting the daily life of people is negligible compared with electromagnetic waves. Despite these fundamental limitations, our discipline has now entered a cycle of progress in which the widening of the market generates the economic potential for developing new systems that not only perform better but also cost less, these improvements generating a wider use. Sustained by user demand of increasing quantity and exigency, this process is accelerated even more by the rapid and sustained progress of associated technologies, especially in the fields of electronics and data processing. In such a context, the future of underwater acoustics will without doubt be as fascinating and exciting as its first century has been.

Appendices

A.1.1 THE UNITS OF UNDERWATER ACOUSTICS

The official SI or SI-derived units[1] are listed here for the main physical quantities encountered in underwater acoustics. They are followed by other common units, with the appropriate conversion factors (note the basic units are not defined here).

Length, surface and volume

SI unit

- metre (m);
- current multiples and submultiples: kilometre (1 km = 1,000 m), centimetre (1 cm = 0.01 m), millimetre (1 mm = 0.001 m), micrometer (1 μm = 10^{-6} m).

SI-derived

- square metre (m^2);
- cubic metre (m^3).

Others

- 1 foot ≈ 0.3048 m;
- 1 inch ≈ 0.0254 m;
- 1 yard ≈ 0.9144 m;
- 1 nautical mile (international) = 1,852 m;
- 1 fathom ≈ 1.8288 m (depth).

---

[1] SI: *Système International*. See Reports of the *Bureau International des Poids et Mesures*, France.

Time

SI unit

- second (s).

Speed

SI-derived unit

- metre/second (m/s).

Others

- 1 knot = 1 nautical mile/hour = 1,852 m/h ≈ 0.5144 m/s.

Mass

SI unit

- kilogram (1 kg = 1,000 g).

Others

- 1 pound ≈ 0.453 592 kg.

Density

SI unit

- kilogram/cubic metre (kg/m^3).

Force

SI-derived unit

- 1 Newton (N) = 1 m kg/s^2.

Energy

SI-derived unit

- 1 Joule (J) = 1 N m.

Power

SI-derived unit

- 1 watt (W) = 1 J/s.

Pressure

SI-derived unit

- 1 Pascal (Pa) $= 1\,\text{N/m}^2$.

Others

- 1 bar $= 10^5\,\text{Pa}$;
- 1 dyne/cm^2 = 1 μbar = 0.1 Pa;
- 1 atmosphere $\approx 1.013\,25 \times 10^5\,\text{Pa}$;
- 1 pound per square inch (p.s.i.) $\approx 6.8948 \times 10^3\,\text{Pa}$.

Acoustic impedence

Si-derived unit

- 1 Rayleigh (rayl) $= 1\,\text{kg/m}^2/\text{s}$

Temperature

SI unit

- degree Kelvin ($^\circ$K);
- degree Celsius ($^\circ$C): $t(^\circ\text{C}) = t(^\circ\text{K}) - 273.15$.

Others

- degree Fahrenheit ($^\circ$F): $t(^\circ\text{F}) = 1.8t(^\circ\text{C}) + 32$.

Electrical intensity

SI unit

- ampere (A).

Electrical voltage

SI-derived unit

- volt (V): $1\,\text{V} = 1\,\text{W/A}$.

Plane and solid angle

SI-coherent unit

- radian (rad);
- steradian (sr).

Logarithmic scales

SI-consistent units

- Neper (Np): unit of logarithmic decrement;
- bel (B): decimal logarithm of a power ratio;
- current submultiple: decibel (dB).

A.1.2   THE CARDINAL SINE FUNCTION

The sinc function (cardinal sine) is very extensively used in underwater acoustics modelling, where it is found to be, exactly or approximately, a convenient model for many current problems (directivity function of a linear array, frequency spectrum of a square envelope signal, autocorrelation function of a linear frequency modulation, ...). It is defined as:

$$\text{sinc}(x) = \frac{\sin(\pi x)}{\pi x} = \frac{\sin u}{u} \tag{A.1.1}$$

with $u = \pi x$. It is presented in Figure A.1.1 as a function of u, in natural values and in decibels $10\log(\sin u/u)^2$.

The central lobe width is given by $2u$ with $u \approx 1.0004$ at $-1.5\,$dB; $u \approx 1.3915$ at $-3\,$dB; $u \approx 1.8955$ at $-6\,$dB; $u \approx 2.3186$ at $-10\,$dB; $u = \pi$ at $-\infty\,$dB (first null).

The first sidelobe maximum is for $u \approx 4.4934$, with amplitude 0.2172 or $-13.26\,$dB. Some useful formulas are given below:

$$\int_{-\infty}^{\infty} \frac{\sin u}{u}\,du = \pi \tag{A.1.2}$$

$$\int_{-\infty}^{\infty} \left(\frac{\sin u}{u}\right)^2 du = \pi \tag{A.1.3}$$

$$\int_{-\infty}^{\infty} \left(\frac{\sin u}{u}\right)^3 du = \frac{3\pi}{4} \tag{A.1.4}$$

$$\int_{-\infty}^{\infty} \left(\frac{\sin u}{u}\right)^4 du = \frac{2\pi}{3} \tag{A.1.5}$$

$$\int_{-1.3915}^{1.3915} \left(\frac{\sin u}{u}\right)^2 du \Big/ \int_{-\infty}^{\infty} \left(\frac{\sin u}{u}\right)^2 du \approx 0.721 \tag{A.1.6}$$

$$\int_{-\pi}^{\pi} \left(\frac{\sin u}{u}\right)^2 du \Big/ \int_{-\infty}^{\infty} \left(\frac{\sin u}{u}\right)^2 du \approx 0.906 \tag{A.1.7}$$

$$\frac{\sin u}{u} \approx 1 - \frac{u^2}{6} + \frac{u^4}{120} + \varepsilon(u^6) \tag{A.1.8}$$

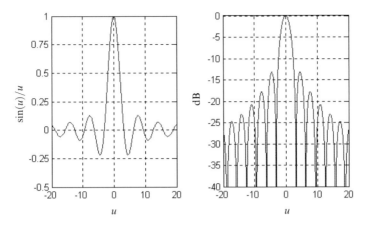

Figure A.1.1. The sinc function, in natural values (*left*) and in dB (*right*).

A.2.3 AVERAGE INTENSITY MODEL

When there are enough rays joining the source and the receiver, the multipath structure may be described as a continuum of angular beams, carrying on average the same intensity as the discrete summation of the rays (Weston, 1971, 1980a, b; Brekhovskikh and Lysanov, 1992). This approach is particularly suited to strongly guided propagation (e.g., in shallow water, surface or SOFAR [SOund Fixing And Ranging] channels: Lurton, 1992). An elementary beam [β; $\beta + \delta\beta$] propagating in a stratified medium (see Figure A.2.1) will be characterised by:

- Its cycle distance:

$$D_\beta = 2 \int_{z_{\min}}^{z_{\max}} \frac{1}{\tan \beta(z)} \, dz \qquad (A.2.6)$$

  where $z_{\min}$ and $z_{\max}$ are the depths of the turning points in the trajectories (Figure A.2.1), and $\beta(z)$ is the local angle.
- Its probability of crossing the depth z (i.e., the ratio between its local divergence and the cycle distance):

$$M_\beta(z) \, d\beta = \frac{4}{D_\beta} \frac{\partial r}{\partial \beta} \, d\beta \qquad (A.2.7)$$

  The factor 4 is explained by the fact that every beam β is transmitted up and down, and crosses the depth z twice in a cycle. $M_\beta(z) = 0$ if the beam cannot reach the depth z because of refraction.
- The beam-focusing factor (see Section 2.7.2 for more details):

$$F_\beta(z) = \cos \beta \left[\sin \beta(z) \frac{\partial r}{\partial \beta} \right]^{-1} \qquad (A.2.8)$$

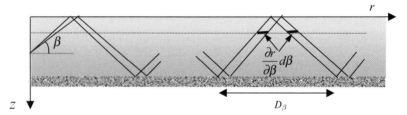

Figure A.2.1. Geometry and notation for the average intensity model.

The geometrical intensity loss for the beam is then:

$$P_\beta(r, z) = \cos\beta \left[r \sin\beta(z) \frac{\partial r}{\partial \beta} \right]^{-1} \tag{A.2.9}$$

Other effects should also be taken into account, such as water absorption and reflection losses. These are expressed as a loss term per cycle, converted into a continuous decrease with range. The total attenuation term is then:

$$A_\beta(r) = \exp(-\gamma_\beta r) \tag{A.2.10}$$

with:

$$\gamma_\beta = 2\alpha - \frac{\ln R_{F\beta}}{D_\beta} \tag{A.2.11}$$

where $R_{F\beta}$ is the intensity reflection coefficient on the seafloor for beam β. The loss $A_\beta(r)$ increases by an amount $R_{F\beta}$ every time r increases by D_β.

The average intensity at $(r; z)$ is finally obtained by integrating over the entire transmitted angular sector:

$$\langle p^2(r, z) \rangle = \frac{1}{r} \int_{\beta_{\min}}^{\beta_{\max}} M_\beta(z) F_\beta(z) A_\beta(r) \, d\beta$$

$$= \frac{4}{r} \int_{\beta_{\min}}^{\beta_{\max}} \frac{\cos\beta}{\sin\beta(z) D_\beta} \exp(-\alpha_\beta r) \, d\beta \tag{A.2.12}$$

This method is in fact relatively simple in principle. It gives the intensity field averaged over the entire waveguide considered, with the same description used for rays (and the same limitations in frequency range and caustic zones). Numerically, this technique gets free from the need of an accurate search for eigenrays. However, it only describes the *averaged field with range*; but this often proves to be sufficient for evaluating the transmission losses in strongly guided propagation configurations.

The averaged intensity model is also convenient for evaluation of the time and angle characteristics of the (averaged) acoustic field, and for modelling of ambient noise (in a way evidently similar to the noise model presented in Section 4.3.2). Finally, the averaged intensity model can be adapted to range-dependent environments.

A.2.4 NORMAL MODES IN ISOVELOCITY CONFIGURATIONS

A.2.4.1 Ideal waveguide

Let us consider a water layer with a constant sound velocity, and a perfectly rigid seafloor. The system of equations to solve for eigenmodes (of order $n = 1, 2, \ldots, N$) becomes:

$$\begin{cases} \dfrac{d^2\Phi_n}{dz^2} + (k^2 - k_n^2)\Phi_n = 0 \\[2mm] \Phi_n(0) = 0 \\[2mm] \dfrac{d\Phi_n}{dz}(H) = 0 \\[2mm] \displaystyle\int_0^\infty \Phi_n(z)\Phi_m(z)\,dz = \delta_{mn} \end{cases} \qquad (A.2.13)$$

Its general solution is under the form:

$$\Phi_{n(z)} = A_n \sin(k_{zn}z) + B_n \cos(k_{zn}z) \qquad (A.2.14)$$

with $k_{zn}^2 = k^2 - k_n^2$. The boundary conditions yield:

$$\begin{cases} B_n = 0 \\[2mm] A_n \cos(k_{zn}H) = 0 \quad \Rightarrow \quad k_{zn} = \left(\dfrac{n}{2} - 1\right)\dfrac{\pi}{H} \end{cases} \qquad (A.2.15)$$

The orthonormal condition yields:

$$A_n^2 \int_0^H \sin^2(k_{zn}z)\,dz = 1 \quad \Rightarrow \quad A_n = \sqrt{\dfrac{2}{H}} \qquad (A.2.16)$$

And the general solution is then:

$$p(r,z) = \dfrac{2}{H} \sum_n \sin(k_{zn}z)\sin(k_{zn}z_e)H_0^{(1)}(k_n r) \qquad (A.2.17)$$

Propagating modes

These modes correspond to a solution propagating along r:

$$k_n^2 = k^2 - k_{zn}^2 > 0 \qquad (A.2.18)$$

Therefore:

$$k^2 > \left(n - \dfrac{1}{2}\right)^2\left(\dfrac{\pi}{H}\right)^2 \quad \Rightarrow \quad n < \dfrac{kH}{\pi} + \dfrac{1}{2} \qquad (A.2.19)$$

Evanescent modes

These modes correspond to a solution that is evanescent along r. They verify:

$$k_n^2 < 0 \quad \Rightarrow \quad k_n = j\sqrt{k^2 - k_{zn}^2} \tag{A.2.20}$$

Their number is infinite. Their contribution is limited to small distances, because the Hankel function then degenerates into a negative exponential.

A.2.4.2   Isovelocity fluid model

This model is more realistic than the ideal waveguide model. It forms the basis of the development of modal theory in underwater acoustics (Pekeris, 1948). Let us consider a water layer with a constant sound velocity, above a bottom with higher velocity. Let us represent $\phi_1(z)$ and $\phi_2(z)$ as the wave functions of these modes, respectively, in water and at the bottom. The solution to $\phi_n(z)$ is of the form:

$$\phi_n(z) = A_n \sin(k_{zn}z) + B_n \cos(k_{zn}z) \tag{A.2.21}$$

And the boundary conditions write:

$$\begin{cases} \phi_{1n}(0) = 0 \quad \Rightarrow \quad B_n = 0 \\[2mm] \rho_1 \phi_{1n}(H) = \rho_2 \phi_{2n}(H) \\[2mm] \dfrac{d\phi_{1n}}{dz}(H) = \dfrac{d\phi_{2n}}{dz}(H) \\[2mm] \rho_1 \displaystyle\int_0^H \phi_{1n}^2(z)\, dz + \rho_2 \displaystyle\int_H^\infty \phi_{2n}^2(z)\, dz = 1 \end{cases} \tag{A.2.22}$$

By looking for a bottom-evanescent form for $\phi_{2n}(z)$ (for propagating modes), one can combine the preceding equations to get the transcendental equation:

$$\tan(k_{zn}H) = -j \frac{\rho_2}{\rho_1} \frac{k_{zn}}{\sqrt{k_2^2 - k_n^2}} \tag{A.2.23}$$

This relation is the *characteristic equation of eigenmodes*. It has real solutions in k_n (propagating modes) only for $k_2 \leq k_n \leq k_1$. The propagating modes are limited in number and fewer than for the corresponding ideal waveguide.

The eigenmodes can be interpreted physically as plane waves of wave vector (k_n, k_{2n}); that is, of angles β_n verifying:

$$\sin \beta_n = \frac{k_{zn}}{k_1}; \qquad \cos \beta_n = \frac{k_n}{k_1} \tag{A.2.24}$$

All incidence angles β do not, of course, correspond to eigenmodes, only those for which the wave is in phase again with itself after a cycle do so. The phase variation over a cycle writes:

$$2k_{zn}H + \pi + \chi = (n - \tfrac{1}{2})\pi \qquad n = 1, 2, 3, \ldots \tag{A.2.25}$$

where π is the phase shift at the reflection on the surface, and χ is the phase shift at the reflection on the bottom. It can be easily shown that this equation is equivalent to the characteristic equation given in Equation (A.2.23).

The condition $k_2 \leq k_n$ implies therefore that $\cos \beta_n \geq c_1/c_2$. The propagating modes thus correspond to grazing angles lower than the critical angle (acting as a cut-off angle for propagation in the waveguide). The portion of the angular spectrum where $\cos \beta_n < c_1/c_2$ corresponds to a loss of energy on the bottom. Rigorously speaking, one cannot then consider eigenmodes. But it is possible to accurately model the existence of *attenuated modes* corresponding to a propagation with energy loss. An imaginary part δ_n is then added to the wave number k_n. A good approximation consists in writing that the damping of a mode over a cycle (of horizontal length D_n) equals the reflection coefficient of the plane wave at the angle considered $V(\beta_n)$. The imaginary part δ_n of k_n is then given by:

$$\exp(-\delta_n D_n) = V(\beta_n) \tag{A.2.26}$$

Hence:

$$\delta_n = \frac{-1}{D_n} \ln V(\beta_n)$$

$$= -\frac{\tan \beta_n}{2H} \ln V(\beta_n)$$

$$= \frac{-1}{2H} \frac{k_{zn}}{k_n} \ln V(\beta_n) \tag{A.2.27}$$

The real part of the wave number for these attenuated modes is determined with a sufficient accuracy by:

$$\tan(k_{zn}H) = 0, \quad \text{with } k_{zn} > \sqrt{k_1^2 - k_2^2} \tag{A.2.28}$$

The normalisation term A_n of the wave functions in water $(\phi_n(z) = A_n \sin(k_{zn}z))$ writes:

$$A_n = \left[\rho_1 \left(\frac{H}{2} + \frac{\sin 2k_{zn}H}{4k_{zn}} \frac{k_1^2 - k_2^2}{k_2^2 - k_n^2} \right) \right]^{-1/2} \tag{A.2.29}$$

A.3.1   FLUID INTERFACE REFLECTION COEFFICIENTS – ADDITIONS

Reflection and transmission processes are conveniently described using plane wave expressions. Let (k_{ix}, k_{iz}), (k_{rx}, k_{rz}) and (k_{tx}, k_{tz}) be the wave vector coordinates, in the (x, z) plane, for the incident wave, the reflected wave and the transmitted wave, respectively. The reflection and transmission coefficients are V and W, respectively. The continuity conditions at the interface, for pressure and normal velocity, are:

$$\begin{cases} p_i + p_r = p_t \\ \dfrac{1}{\rho_1} \dfrac{\partial p_i}{\partial z} + \dfrac{1}{\rho_1} \dfrac{\partial p_r}{\partial z} = \dfrac{1}{\rho_2} \dfrac{\partial p_t}{\partial z} \end{cases} \tag{A.3.1}$$

From this, we can deduce:

$$
\begin{cases}
\exp j(k_{ix}x + k_{iz}z) + V \exp j(k_{rx}x + k_{rz}z) = W \exp j(k_{tx}x + k_{tz}z) \\
\dfrac{k_{iz}}{\rho_1} \exp j(k_{ix}x + k_{iz}z) + \dfrac{k_{rz}}{\rho_1} V \exp j(k_{rx}x + k_{rz}z) = \dfrac{k_{tz}}{\rho_2} W \exp j(k_{tx}x + k_{tz}z)
\end{cases}
\tag{A.3.2}
$$

Detailing the above for $z = 0$ and any x leads to $k_{ix} = k_{rx} = k_{tx} = k_x$. Furthermore:

$$
\begin{cases}
k_{ix} = \dfrac{2\pi f}{c_1} \sin \theta_1 \\[2mm]
k_{tx} = \dfrac{2\pi f}{c_2} \sin \theta_2
\end{cases}
\tag{A.3.3}
$$

The equality of the x-components is equivalent to the Snell–Descartes law:

$$
\frac{\sin \theta_1}{c_1} = \frac{\sin \theta_2}{c_2}
\tag{A.3.4}
$$

The other wave vector components are:

$$
\begin{cases}
k_{iz} = \sqrt{k_1^2 - k_x^2} = k_{1z} \\[2mm]
k_{rz} = -\sqrt{k_1^2 - k_x^2} = -k_{1z} \\[2mm]
k_{tz} = \sqrt{k_2^2 - k_x^2} = k_{2z}
\end{cases}
\tag{A.3.5}
$$

The reflection and transmission coefficients are finally:

$$
\begin{cases}
V = \dfrac{\rho_2 k_{1z} - \rho_1 k_{2z}}{\rho_2 k_{1z} + \rho_1 k_{2z}} \\[3mm]
W = \dfrac{2\rho_2 k_{1z}}{\rho_2 k_{1z} + \rho_1 k_{2z}}
\end{cases}
\tag{A.3.6}
$$

This type of notation is more convenient than using propagation vector angles, since it allows, with no change in the formulae, to deal with total refraction ($k_x > k_2$, and k_{2z} is then imaginary), or with absorption phenomena (then k_2 is complex). It should then be preferred in theoretical develoments and numerical programming.

A.3.1.1   Absorbing reflector

For an absorbing reflector, the reflection coefficient at grazing incidence may be obtained by accounting for an imaginary part in the wave number k_2. Finally, the result may be expressed approximately as:

$$
\begin{cases}
V \approx 1 - 2q + 2q^2 \\[3mm]
q = \dfrac{\rho_2}{\rho_1} \dfrac{c_1^2 c_2^2 \delta_2 \beta_1}{[c_2^2(1 - \beta_1^2) - c_1^2]^{1/2} \left[c_2^2 \left(1 + \beta_1^2 \left(\dfrac{\rho_2^2}{\rho_1^2} - 1\right)\right) - c_1^2 \right]}
\end{cases}
\tag{A.3.7}
$$

where β_1 is the grazing angle of the incident wave, and δ_2 the imaginary part of the wave number inside the reflecting medium.

For a rough first-order approximation, the above may be simplified into:

$$V \approx 1 - 2\frac{\rho_2}{\rho_1}\frac{c_1^2 c_2^2 \delta_2 \beta_1}{[c_2^2 - c_1^2]^{1/2}} \tag{A.3.8}$$

This expression is actually valid only for very small values of β_1.

A.3.1.2 "Slow" reflector

For a reflector with a sound velocity smaller than in the incident medium ($c_2 < c_1$), some particular features should be noted:

- the reflection coefficient is then strictly smaller than unity whatever the angle; no total reflection can be observed (except exactly at zero grazing incidence) since the wave in the reflecting medium can always be generated;
- a minimum value may be observed at a given angle; if $\rho_2 > \rho_1$, a null of the reflection coefficient may be observed if:

$$\rho_2 c_2 \cos\theta_1 = \rho_1 c_1 \sqrt{1 - \left(\frac{c_2}{c_1}\sin\theta_1\right)^2} \quad \text{or} \quad \sin\theta_1 = \frac{1}{c_2}\sqrt{\frac{\rho_2^2 c_2^2 - \rho_1^2 c_1^2}{\rho_2^2 - \rho_1^2}} \tag{A.3.9}$$

hence no acoustic energy is reflected by the second medium for this particular incidence.

Excepting the particular case of the water–air interface, this type of boundary condition is not commonly met in underwater acoustics. However, it corresponds to certain very soft sedimentary seafloors, mainly met in deep waters; note that the values of sediment velocity are not likely to be below 0.98 times the water velocity, while their density is around 1.2 to 1.3.

A.3.2 SIMPLIFIED MODEL OF FISH TARGET STRENGTH

The semi-heuristic model proposed here is a synthesis using two different approaches:

- the target strength of the swimbladder, considered as a gas bubble with its resonant characteristics, Rayleigh scattering at low frequencies and the geometric limit beyond resonance frequency;
- the average geometric scattering by the fish body at wavelengths shorter than the fish length.

The backscattering cross-section can be decomposed as a sum of two terms, corresponding to the two frequency ranges:

$$\sigma_{\text{fish}} = \sigma_{LF} + \sigma_{HF} \tag{A.3.10}$$

In low frequency:

$$\sigma_{LF} = \frac{a^2}{[(f_a/f)^2 - 1]^2 + \delta^2} \tag{A.3.11}$$

where a is the radius of the swimbladder, f_a is its resonance frequency and δ is the corresponding damping term. The resonance frequency is given by:

$$f_a = \frac{1}{2\pi a}\sqrt{\frac{3\gamma P_w}{\rho_w}} \approx \frac{3.25}{a}\sqrt{1 + 0.1z} \tag{A.3.12}$$

where ρ_w is the water density ($\approx 1{,}030\,\text{kg/m}^3$), P_w is the hydrostatic pressure ($\approx 10^5(1 + z/10)$ in Pa, z being the depth in metres) and γ is the gas adiabatic constant ($\gamma \approx 1.4$). Damping is due to the combined effects of radiation, shear viscosity and thermal conductivity. It may be approximated as:

$$\delta = \frac{2\pi f}{c_w}\frac{\rho_w}{\rho_f} + \frac{\varepsilon}{\pi f a^2 \rho_f} \tag{A.3.13}$$

ρ_f and ε are the fish flesh density and viscosity. Practically, ρ_f may be taken equal to ρ_w, and $\varepsilon \approx 50$. The swimbladder radius a may be related to the fish length L by $a \approx 0.04L$.

At the high-frequency limit, the backscattering cross-section of the fish body can be written for $f > f_1 \approx c/L$:

$$\sigma_{HF} = 0.0032L^2 \tag{A.3.14}$$

This expression is derived and simplified from the classical formula of Love (1978). In practice, to avoid a gap in the complete frequency response, one may use a smoothing factor describing a Rayleigh-like fall-off for this contribution below f_1:

$$\sigma_{HF} = 0.0032L^2 \frac{f^4}{f^4 + f_1^4} \tag{A.3.15}$$

A.3.3   COHERENT REFLECTION ON A ROUGH INTERFACE

Let us now consider a rough surface, whose height above the average plane surface is quantified by a random function $\xi(\vec{r})$ of mean 0 and variance h (Figure A.3.1). The phase perturbation of an acoustic wave reflecting on Point A, with height ξ relative to the ideal average plane surface, corresponds to the geometric distance ($\xi \cos\theta$) between A and the average surface:

$$\Delta\varphi = 2k\xi\cos\theta \tag{A.3.16}$$

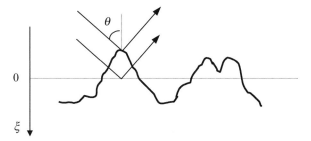

Figure A.3.1. Geometry of the rough surface and the incident wave.

The average value of the reflected field is obtained by integrating all reflections from the points ξ weighted by their probability $q(\xi)$, and using the pressure reflection coefficient V that would exist in the absence of the relief:

$$\langle p \rangle = V \int_{-\infty}^{\infty} \exp(-j\Delta\varphi)q(\xi)\,d\xi$$

$$= V \int_{-\infty}^{\infty} \exp(-2jk\xi\cos\theta)q(\xi)\,d\xi \qquad (A.3.17)$$

The probability $q(\xi)$ is assumed Gaussian:

$$q(\xi) = \frac{1}{h\sqrt{2\pi}}\exp\left(-\frac{\xi^2}{2h^2}\right) \qquad (A.3.18)$$

The integral in Equation (A.17) then has an exact solution:

$$\langle p \rangle = V\exp(-2k^2h^2\cos^2\theta) = V\exp\left(-\frac{P^2}{2}\right) \qquad (A.3.19)$$

This expression shows the appearance of the Rayleigh coefficient $P = 2kh\cos\theta$. Here, it is clear that P is the standard deviation of the phase fluctuation caused by the relief. The value of the reflected field thus obtained is the "coherent part" of the field reflected in the specular direction. It is valid when the standard deviation of the phase fluctuation remains smaller than $\pi/2$, which corresponds to $h = \lambda/(8\cos\theta)$. The loss associated with this extreme value is $\exp(-\pi^2/8) \approx 0.3$ (i.e., $-10.7\,\text{dB}$).

A.5.1 BASICS OF PIEZOELECTRICITY

A.5.1.1 Fundamental relations

The fundamental relations of piezoelectricity relate the mechanical parameters (*deformation* S or *stress* T) and the electrical parameters (*field* E or *induction* D). The direct piezoelectric effect writes:

$$\mathbf{D} = \mathbf{d}\cdot\mathbf{T} + \boldsymbol{\varepsilon}^{\mathbf{T}}\cdot\mathbf{E} \qquad (A.5.1)$$

And the inverse piezoelectric effect writes:

$$S = \mathbf{s}^E \cdot \mathbf{T} + \mathbf{d}^t \cdot \mathbf{E} \tag{A.5.2}$$

These equations use the following notation:

$\mathbf{D}$ = 3-dimensional *electric induction vector*, in C/m^2

$\mathbf{E}$ = 3-dimensional *electric field vector*, in V/m

$\mathbf{S}$ = 6-dimensional *relative deformation vector* (dimensionless)

$\mathbf{T}$ = 6-dimensional *mechanical stress vector*, in N/m^2

$\mathbf{d}$ = 3×6 *piezoelectricity matrix*, in C/N or in m/V ($\mathbf{d}^t$ is the transposition of $\mathbf{d}$)

ε^T = 3×3 electrical *permittivity matrix* (for $\mathbf{T}$ constant), in F/m

$\mathbf{s}^E$ = 6×6 *compliance matrix* (for $\mathbf{E}$ constant), in m^2/N

By convention, in the notation of vector and matrix indices, the index 3 corresponds to the axis giving the direction of the polarisation field. T_1, T_2, T_3, S_1, S_2, S_3 are the stresses and relative deformations of tension along the three axes 1, 2 and 3; T_4, T_5, T_6, S_4, S_5, S_6 are the stresses and relative deformations of shear along the three axes 1, 2 and 3.

A.5.1.2   Mechanical characteristics of piezoelectric ceramics

The *compliance matrix* (also called elasticity matrix) $\mathbf{s}^E$ relates the deformation $\mathbf{S}$ and the stress $\mathbf{T}$ through the following relation:

$$s_{ij}^E = \frac{S_i}{T_j} \quad \text{(in } m^2/N) \tag{A.5.3}$$

The *stiffness matrix* $\mathbf{c}$ is made of the inverse terms, and its coefficients are:

$$c_{ij} = \frac{T_i}{S_j} \quad \text{(in } N/m^2) \tag{A.5.4}$$

The *Young elasticity modulus* (in N/m^2) along the different directions is given by:

$$Y_{ii}^E = \frac{1}{s_{ii}^E} = c_{ii} \tag{A.5.5}$$

The (dimensionless) *Poisson coefficient* is given by:

$$\sigma^E = -\frac{s_{12}^E}{s_{11}^E} \tag{A.5.6}$$

The *frequency constants* N (in kHz mm) express, for the various vibration modes, the product of the resonance frequency and the characteristic dimension a of the ceramics for the mode considered:

$$N = f_0 a \tag{A.5.7}$$

These parameters are essential for dimensioning the ceramics of a transducer.

A.5.1.3 Electrical characteristics of piezoelectric ceramics

Dielectric permittivity relates induction to the electric field. If the dielectric environment is isotropic, the permittivity is scalar, and:

$$\mathbf{D} = \varepsilon \cdot \mathbf{E} \tag{A.5.8}$$

If the environment is anisotropic, the permittivity is a matrix ε and relation (A.5.8) becomes:

$$\mathbf{D} = \varepsilon \cdot \mathbf{E} \quad \text{with } \varepsilon^T = \begin{bmatrix} \varepsilon_{11}^T & 0 & 0 \\ 0 & \varepsilon_{22}^T & 0 \\ 0 & 0 & \varepsilon_{33}^T \end{bmatrix} \tag{A.5.9}$$

Dielectric stiffness expresses the breaking voltage of ceramics, limiting the electric field that can be applied (typically 3 to 4 kV/cm). The coercive field is the electric field that annihilates the polarisation of the ceramics. Finally, the dielectric losses can be represented by an equivalent resistance R_0 in parallel with the capacitance C_0 of the ceramics: the dielectric loss factor δ is defined as:

$$\tan \delta = \frac{1}{R_0 C_0 \omega} \tag{A.5.10}$$

A.5.1.4 Piezoelectric characteristics

The *charge piezoelectric coefficients* (in C/N, grouped in the matrix $\mathbf{d}$) express the relations between the charge densities appearing on the faces of the ceramics and the stresses applied (direct piezoelectric effect). Conversely (in m/V), it expresses the relations between the relative deformations obtained and the electric field applied (inverse effect). The matrix writes:

$$\mathbf{d} = \begin{bmatrix} 0 & 0 & 0 & 0 & d_{15} & 0 \\ 0 & 0 & 0 & d_{15} & 0 & 0 \\ d_{13} & d_{13} & d_{33} & 0 & 0 & 0 \end{bmatrix} \tag{A.5.11}$$

The *piezoelectric tension coefficients* (in V m/N, grouped in matrix $\mathbf{g}$, see Relation A.5.1.2) express the relations between the electric field appearing on the faces of the ceramics and the stresses applied (direct piezoelectric effect). Conversely (in m²/C), it expresses the relations between the relative deformations obtained and the charge densities (inverse effect).

The charge and tension piezoelectric constants are related by:

$$\mathbf{d} = \varepsilon^{\mathbf{T}} \cdot \mathbf{g} \tag{A.5.12}$$

For a submerged ceramic with dimensions small relative to the wavelength, the hydrostatic pressure generates an isotropic permanent stress on the entire surface:

$$d_h = d_{33} + 2d_{31} \tag{A.5.13}$$

The *Curie temperature* is the threshold above which the piezoelectric properties of the ceramic disappear: it is typically around 300°C.

A.5.1.5   Coupling coefficient

The *coupling coefficient* k_{ij} expresses the quality of conversion of electrical energy into mechanical energy, and vice versa. It depends on the shape of the sample, and the mode of vibration respective to the direction of polarisation. For thick samples vibrating along the direction of polarisation (direction "33"), it is written as:

$$k_{33} = \frac{d_{33}}{\sqrt{\varepsilon_{33}^T s_{33}^E}} \tag{A.5.14}$$

It corresponds to an electromechanical efficiency:

$$k_{33}^2 = \frac{Y d_{33}^2}{\varepsilon_{33}^T} \tag{A.5.15}$$

The coefficient k_{33} is of around 70% for PZT ceramics (10% for quartz).

A.5.2   PARAMETRIC ARRAYS

The radiation of a parametric array can be simply described with the model of Westervelt (1963). A line of virtual sources radiates at the frequency corresponding to the difference of the two primary frequencies, along the axis of the directivity lobe of the primary lobes. The attenuation of the primary waves (quantified by the decrement δ_m) limits the effective interaction length (conventionally $1/\delta_m$).

It can be shown (e.g., Medwin and Clay, 1998) that the pressure generated at the difference frequency, is expressed at distance R by:

$$p_d(R) = \frac{\pi \beta f_d^2 p_1 p_2 S_A \exp(-\alpha_d R)}{2 R \alpha_m \rho_0 c_0^4} \left[1 + \frac{k_d^2}{\delta_m^2} \sin^4 \left(\frac{\theta}{2} \right) \right]^{-1/2} \tag{A.5.16}$$

where:

$$\beta = \text{ the non-linearity coefficient } (\beta = 3.4 \text{ in water})$$

ρ_0 and $c_0 =$ the characteristics of water (respectively, its volume density and its celerity)

p_1 and $p_2 =$ the pressure amplitudes transmitted in the two primary beams

$S_A =$ the efficient section of the intersection of the two primary beams, and can be approximated by $\pi (2\theta_m/2)^2$ where $2\theta_m$ is the mean width of the beams

$\delta_m = \dfrac{\delta_1 + \delta_2}{2} =$ the average absorption coefficients of the primaries (the δ terms are amplitude decrements, in Np/m)

$$k_d = \frac{2\pi f_d}{c} = \frac{2\pi}{c}|f_2 - f_1| =$$ the wave number associated with the difference frequency (the primary frequencies being f_1 and f_2)

$\alpha_d =$ the absorption coefficient (in Np/m) of the difference frequency.

By expressing this pressure along the axis, and at a distance of 1 m from the transducer, one gets the equivalent level of the low-frequency secondary source:

$$p_d(R) = \frac{\pi \beta f_d^2 p_1 p_2 S_A}{2\alpha_m \rho_0 c_0^4} \tag{A.5.17}$$

We see that the secondary pressure increases with the square of the primary pressures, and the decrease by absorption is that of the secondary frequency.

The directivity function (in intensity) is then written:

$$D(\theta) = \left[1 + \frac{k_d^2}{\delta_m^2} \sin^4\left(\frac{\theta}{2}\right)\right]^{-1} \tag{A.5.18}$$

This function shows a narrow directivity lobe around $\theta = 0$, and a complete absence of side lobes. This is a remarkable property of parametric arrays.

The -3-dB aperture is given by:

$$2\theta_d = 4\sqrt{\frac{\delta_m}{k_d}} \tag{A.5.19}$$

Very simply, the Westervelt model gives results of very good quality, sufficient to assess directivity characteristics rapidly. A condition of applicability is to check a posteriori that the width of the primary beams is less than the width of the secondary beam, so that the model of a line of sources is meaningful.

The efficiency of a parametric array is defined as the ratio between the powers radiated in the secondary beam and the primary beams. It is approximated by:

$$\eta \approx \frac{\pi \beta^2 f_d^2 P_0}{2\rho_0 c_0^5 \theta_d^2} \approx \frac{\pi^2 \beta^2 f_d^3 P_0}{4\rho_0 c_0^6 \delta_m} \tag{A.5.20}$$

where P_0 is the total power of the primary waves ($P_0 = P_1 + P_2$). The efficiency of non-linear generation is very small, as is clearly seen in Figure A.5.1. The efficiency is better as the input power is higher, the secondary frequency higher and the secondary beam narrower.

A.5.3 ELECTRO-ACOUSTIC ANALOGIES

Electromechanical or electro-acoustic analogies enable the modelling of rather complex systems and processes with simple electric circuits. This approach is used to study the influences of different parameters in the working of a transducer, and the matching of a transducer and its associated electronics.

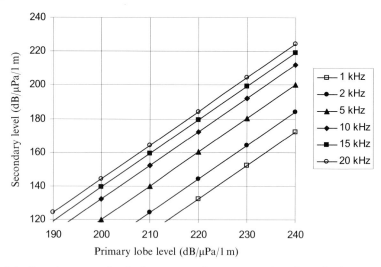

Figure A.5.1. Sound pressure level of a parametric array as a function of primary wave level and difference frequency. The primary frequencies are centred around 100 kHz, and the array diameter is 0.3 m.

The fundamental principle is the analogy between the electrical impedance (ratio between voltage and current) and the mechanical impedance (ratio between force and speed). A resonant mechanical system of the type mass + spring follows the movement equation:

$$F \sin \omega t = m\frac{dv}{dt} + R_m v + k\int v\,dt \qquad (A.5.21)$$

where $F \sin \omega t$ is the external force applied, m is the mass moving with speed v, R_m is the friction resistance and k is the elastic constant of the spring. This basic equation is similar to the equation of a resonant electrical circuit:

$$U \sin \omega t = L\frac{di}{dt} + Ri + \frac{1}{C}\int i\,dt \qquad (A.5.22)$$

Hence the correspondences between parameters are expressed in Table A.5.1.

With these correspondences, the electric circuit equivalent to a transducer can be represented as in Figure A.5.2. The leftmost part of the circuit is the blocked circuit, or the electric static circuit. R_0 is the dielectric loss resistance, C_0 is the capacitance of the dielectric plates. The transformer represents the electromechanical transduction of ratio ϕ. Therefore:

$$\begin{cases} F = \phi U \\[2mm] v = \dfrac{i}{\phi} \\[2mm] Z_m = \phi^2 Z_e \end{cases} \qquad (A.5.23)$$

Table A.5.1. Equivalence between mechanical values (*left column*) and electrical values (*right column*).

Force F	Voltage U
Speed v	Current i
Mass m	Inductance L
Friction R_m	Resistance R
Elasticity $1/k$	Capacitance C
Stiffness k	Inverse capacitance $1/C$
Impedance $Z_m = F/v$	Impedance $Z_e = U/i$

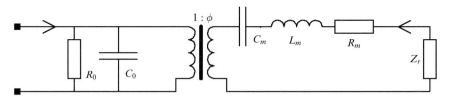

Figure A.5.2. Electrical equivalent circuit of an acoustic transducer.

Beyond the transformer, one finds the mechanical part of the transducer: R_m expresses the mechanical loss by damping in the material (part of the mechanical energy is dissipated through heat, like electric energy and the Joule effect); L_m expresses the dynamic mass of the transducer and C_m its elasticity.

At the end of the circuit, Z_r is the radiating impedance, the purely acoustic part of the system. Its expression depends on the geometry of the radiating part of the transducer. For a circular piston radiating into an infinite baffle, one can write:

$$Z_r = \pi a^2 \rho_0 c_0 \left[1 - \frac{J_1(2ka)}{ka} \right] + j \frac{\pi \rho_0 c_0}{2k^2} K_1(2ka) \qquad (A.5.24)$$

where ρ_0 and c_0 are, respectively, the density and the sound velocity of water; and J_1 and K_1 are the Bessel functions of order 1, respectively, of the first and second types. If the dimensions of the transmitting surface are large enough compared with the wavelength, the radiating impedance tends towards its limit value:

$$Z_r \rightarrow \pi a^2 \rho_0 c_0 \qquad (A.5.25)$$

or, more generally, $Z_r \rightarrow S\rho_0 c_0$ for a transducer of transmitting surface S.

The electric elements above the transformer form the *blocked impedance* of the transducer. Mechanical and acoustic terms are grouped into the *motional impedance*.

A.5.4   DIRECTIVITY PATTERN OF A LINEAR ARRAY

A.5.4.1   Directivity pattern

We want to calculate the field radiated, at a point A_0 in space, by a linear array of length L. The expression of this field as a function of the angle at a large distance, normalised relative to its maximum, forms the directivity pattern. It is obtained by integrating the contributions of all the points along the array:

$$D(\theta) = \left| \int_{-L/2}^{+L/2} \frac{\exp(-jkR(M))}{R(M)} dy \right|^2 \tag{A.5.26}$$

This is a problem with cylindrical symmetry (Figure A.5.3). The observation point A_0 can be located in space with its Cartesian coordinates (x_0, y_0) or with its cylindrical coordinates (R_0, θ_0).

In the following, we shall use a continuous transmitting array. The amplitude of the signal transmitted in a point is $1/L$, to get a level normalised to 1 after integration over the entire length L. These developments are also valid for a receiving array.

The acoustic pressure received at the observation point A_0 is the integral of the contributions from the spherical waves received at all the points of the antenna:

$$p_0 = \frac{1}{L} \int_{-L/2}^{+L/2} \frac{\exp(-jkR(y))}{R(y)} dy \tag{A.5.27}$$

where $R(y)$ is the oblique distance from the point M of abscissa y to A_0:

$$R(y) = \sqrt{x_0^2 + (y_0 - y)^2} = \sqrt{x_0^2 + y_0^2 - 2yy_0 + y^2} \tag{A.5.28}$$

A_0 is assumed to be far from the array, and therefore x_0, y_0 and R_0 are large compared with y and L. Therefore:

$$R(y) \approx \sqrt{x_0^2 + y_0^2 - 2yy_0} = R_0 \sqrt{1 - 2\frac{yy_0}{R_0^2}}$$

$$\approx R_0 \left(1 - \frac{yy_0}{R_0^2}\right) = R_0 - \frac{yy_0}{R_0} = R_0 - y \sin\theta_0 \tag{A.5.29}$$

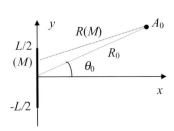

Figure A.5.3. Geometry and notation for computing the directivity of a linear array.

Substituting these developments into the expression of the pressure at A_0, we get:

$$p_0 = \frac{1}{L}\int_{-L/2}^{+L/2} \frac{\exp(-jkR(y))\,dy}{R(y)} \approx \frac{\exp(-jkR_0)}{LR_0}\int_{-L/2}^{+L/2} \exp(jky\sin\theta_0)\,dy$$

$$= \frac{\exp(-jkR_0)}{LR_0}\int_{-L/2}^{+L/2} [\cos(ky\sin\theta_0) - j\sin(ky\sin\theta_0]\,dy$$

$$= \frac{\exp(-jkR_0)}{LR_0}\frac{\sin\left(k\frac{L}{2}\sin\theta_0\right)}{k\frac{L}{2}\sin\theta_0} \tag{A.5.30}$$

Hence the directivity pattern:

$$D(\theta_0) = \left(\frac{\sin\left(k\frac{L}{2}\sin\theta_0\right)}{k\frac{L}{2}\sin\theta_0}\right)^2 \tag{A.5.31}$$

as $|\exp(-jkR_0)|^2 = 1$. Furthermore, the term $1/R_0$, considered to be a multiplicative constant, was omitted to normalise the function $D(\theta_0)$ to 1 at its maximum.

A.5.4.2   Directivity index

By definition, the directivity index (in natural units) is the inverse ratio of the directivity pattern integrated over the entire space and the solid angle of 4π of an array without directivity:

$$I_d = \frac{4\pi}{\int_{-\pi}^{+\pi} d\varphi \int_{-\pi/2}^{+\pi/2} D(\theta,\varphi)\cos\theta\,d\theta} \tag{A.5.32}$$

The denominator equals:

$$\int_{-\pi}^{+\pi} d\varphi \int_{-\pi/2}^{+\pi/2} D(\theta,\varphi)\cos\theta\,d\theta = \int_{-\pi}^{+\pi} d\varphi \int_{-\pi/2}^{+\pi/2} \left(\frac{\sin\left(k\frac{L}{2}\sin\theta\right)}{k\frac{L}{2}\sin\theta}\right)^2 \cos\theta\,d\theta$$

$$= 2\pi \int_{-1}^{+1} \left(\frac{\sin\left(k\frac{L}{2}u\right)}{k\frac{L}{2}u}\right)^2 du$$

$$= 2\pi\frac{2}{kL}\underbrace{\int_{-\infty}^{+\infty} \left(\frac{\sin v}{v}\right)dv}_{\pi} \approx \frac{4\pi^2}{kL} \tag{A.5.33}$$

This expression was obtained assuming that $kL/2$ is large ($L \gg \lambda$) and with the substitutions of variables $\begin{cases} u = \sin\theta \\ v = \dfrac{kL}{2}u \end{cases}$. The directivity index can finally be written as:

$$I_d \approx \frac{4\pi}{4\pi^2/kL} = \frac{kL}{\pi} \approx \frac{2L}{\lambda} \tag{A.5.34}$$

And in dB, this becomes:

$$ID = 10\log\left(\frac{2L}{\lambda}\right) \tag{A.5.35}$$

A.5.5 DIRECTIVITY PATTERN OF A CURVED ARRAY

Radiation from a cylindrical array is formally slightly more complex than for a linear array. The cylindrical array is characterised by its radius of curvature ρ_c, its curvilinear length L_c and its length h_c (Figure A.5.4). It can also be characterised by its angle of aperture γ_c, with $L_c = \rho_c \gamma_c$. Along the direction h_c, radiation is that of a linear array of length h_c and is not particularly interesting.

Along the transverse plane, radiation can be obtained by integrating the contributions from point sources distributed along the curvilinear abscissa. However, unlike the linear array, and even at large distances, the acoustic path differences between the contributions from these points will not converge toward a null value, but toward a constant imposed by the curvature of the array. Along the radiation axis, for a point of observation very far along the Ox axis (Figure A.5.4), the acoustic path difference between the central point O of the array and a point of the array at angle γ is given by:

$$\delta x(\gamma) = \rho_c(1 - \cos\gamma) = \rho_c \frac{\gamma^2}{2} \tag{A.5.36}$$

The field radiated by the entire curvilinear length is in fact an integral over the angle γ:

$$p(R) = \frac{1}{L_c}\int_{-L_c/2}^{+L_c/2}\frac{\exp(-jkR(\gamma))}{R(\gamma)}\,ds = \frac{\rho_c}{L_c}\int_{-\gamma_c/2}^{+\gamma_c/2}\frac{\exp(-jkR(\gamma))}{R(\gamma)}\,d\gamma$$

$$\approx \frac{\rho_c}{L_c}\frac{\exp(-jkR_0)}{R_0}\int_{-\gamma_c/2}^{+\gamma_c/2}\exp\left(jk\rho_c\frac{\gamma^2}{2}\right)d\gamma$$

$$\approx \frac{\rho_c}{L_c}\frac{\exp(-jkR_0)}{R_0}\int_{-\gamma_c/2}^{+\gamma_c/2}\cos\left(k\rho_c\frac{\gamma^2}{2}\right)d\gamma \tag{A.5.37}$$

The integral is akin to the integral of the Fresnel cosine $C(x) = \sqrt{2/\pi}\int_0^x \cos u^2\,du$.

In the case where the array curvature is large enough and the wavelength small enough that the phase of contributions from points along the array changes many

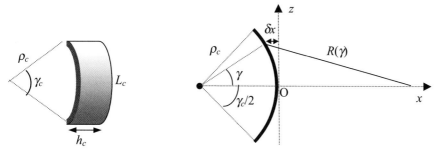

Figure A.5.4. Geometry and notation for radiation of a circular array in the Ox direction.

times, the Fresnel cosine integral tends toward the limit:

$$\int_0^{+\infty} \cos(au^2)\, du = \frac{1}{2}\sqrt{\frac{\pi}{2a}} \qquad (A.5.38)$$

The modulus of the radiated pressure is therefore:

$$|p(R)| \approx \frac{\rho_c}{L_c R_0}\sqrt{\frac{\pi}{k\rho_c}} = \frac{\rho_c}{L_c R_0}\sqrt{\frac{\lambda}{2\rho_c}} = \frac{1}{\gamma_c R_0}\sqrt{\frac{\lambda}{2\rho_c}} \qquad (A.5.39)$$

This order of magnitude is sufficient even if the size of the array does not fully justify the above approximation of an integration over $[0; +\infty]$. In fact, most of the energy contribution is radiated by the array portion bounding the first Fresnel zone (defined by $\delta x_F = \lambda/2$). The angle γ_F intercepted by this Fresnel zone equals:

$$\gamma_F = 2\sqrt{\frac{\lambda}{\rho_c}} \qquad (A.5.40)$$

The array points outside this sector only contribute little, making the resulting pressure fluctuate around the average value of Equation (A.5.39).

Consequently, if the Fresnel aperture γ_F is negligible compared with the physical aperture γ_c of the physical array, the width of the directivity diagram of a curvilinear array is close to γ_c. Because of the behaviour of the integral of the Fresnel cosine, oscillating around its asymptote, the directivity pattern is slightly fluctuating.

A.5.6   ARRAY-SHADING PERFORMANCE

We shall present here a simplified picture of the characteristics associated with some array weighting laws (Harris, 1978). The reference performances are those of the unweighted array (rectangular law). The parameters retained are:

- the maximal level of the side lobes A_{maxsec} (in dB);
- the decreasing slope of the side lobes ΔA_{sec} (in dB/octave);

- the widening of the main lobe (multiplicative coefficient relative to the unweighted width);
- the loss in directivity gain compared with the unweighted array ΔG_{dir} (in dB);
- the coherent summation gain of the array (summation of the receiver gains normalised by their number) G_{coh} (in dB).

They are presented in Table A.5.2, and their effects on the directivity pattern are graphically displayed in Figure A.5.5, for different kinds of weighting law.

The different shading laws used here are described below for $x \in [-L/2; +L/2]$:

- $\cos(x)$: $p(x) = \cos\left(\dfrac{\pi}{L}x\right)$;

- $\cos^2(x)$ or "raised cosine" or Hanning: $p(x) = \cos^2\left(\dfrac{\pi}{L}x\right) = 0.5 + \cos\left(\dfrac{2\pi}{L}x\right)$;

- Hamming: $p(x) = 0.54 + 0.46\cos\left(\dfrac{2\pi}{L}x\right)$;

- Gauss (with parameter α): $p(x) = \exp\left[-\dfrac{1}{2}\left(\alpha\dfrac{2}{L}x\right)^2\right]$;

- Kaiser–Bessel (with parameter α): $p(x) = \dfrac{I_0\left(\pi\alpha\sqrt{1-(2x/L)^2}\right)}{I_0(\pi\alpha)}$, where I_0 is the modified Bessel function of the first type and order 0: $I_0(x) = \sum\limits_{k=0}^{+\infty}\left[\dfrac{(x/2)^k}{k!}\right]^2$;

Table A.5.2. Parameters associated with the different weighting laws.

	A_{maxsec} (dB)	ΔA_{sec} (dB/octave)	$\Delta\theta_{lobe}$	ΔG_{dir} (dB)	G_{coh} (dB)
Rectangular (no weighting)	−13.5	−6	1	0	0
Triangle	−27	−12	1.44	−1.3	−6.0
$\cos(x)$	−23	−12	1.35	−0.9	−3.9
$\cos^2(x)$	−32	−18	1.62	−1.8	−6.0
Hamming	−43	−6	1.46	−1.3	−5.4
Gauss, $\alpha = 2.5$	−42	−6	1.49	−1.7	−5.8
Gauss, $\alpha = 3.0$	−55	−6	1.74	−2.1	−7.3
Kaiser–Bessel, $\alpha = 2$	−46	−6	1.61	−1.8	−6.2
Kaiser–Bessel, $\alpha = 2.5$	−57	−6	1.76	−2.2	−7.1
Dolph–Chebychev, $\alpha = 1.5$	−30	0	1.14	−0.5	−3.2
Dolph–Chebychev, $\alpha = 2.0$	−40	0	1.31	−0.9	−4.5
Dolph–Chebychev, $\alpha = 2.5$	−50	0	1.49	−1.4	−5.5

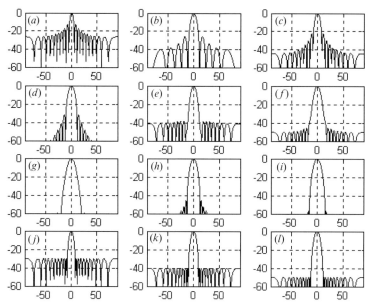

Figure A.5.5. Array-shading effect on directivity patterns, for different weighting laws: (*a*) no shading; (*b*) triangle; (*c*) cos *x*; (*d*) Hanning or cos² *x*; (*e*) Hamming; (*f*) Gauss, $\alpha = 2.5$; (*g*) Gauss, $\alpha = 3.0$; (*h*) Kaiser–Bessel, $\alpha = 2$; (*i*) Kaiser–Bessel, $\alpha = 2.5$; (*j*) Dolph–Chebychev, $\alpha = 1.5$; (*k*) Dolph–Chebychev, $\alpha = 2.0$; (*l*) Dolph–Chebychev, $\alpha = 2.5$.

• Dolph–Chebychev (parameterised at the level of the side lobes): the weighting coefficients are determined from the Chebychev polynomials of order $m = (N - 1)/2$, where N is the number of receivers. The details of these formulas will not be repeated here; the interested reader is instead directed to the classical references on the subject (Harris, 1978; Steinberg, 1975).

A.6.1   FOURIER TRANSFORM: MAIN PROPERTIES

Time signal	Frequency spectrum
$s(t)$	$S(f)$
$s_1(t) + s_2(t)$	$S_1(f) + S_2(f)$
$s(-t)$	$S(-f)$
$s^*(t)$ (complex conjugation)	$S^*(-f)$
$s(at)$	$\dfrac{1}{a} S\left(\dfrac{f}{a}\right)$

(cont.)

A.6.1   FOURIER TRANSFORM: MAIN PROPERTIES (*cont.*)

Time signal	Frequency spectrum
$s(t - t_0)$	$S(f)\exp(-j2\pi t_0 f)$
$s(t)\exp(j2\pi f_0 t)$	$S(f - f_0)$
$s_1(t)s_2(t)$	$S_1(f) \otimes S_2(f)$ (convolution product)
$s_1(t) \otimes s_2(t)$	$S_1(f)S_2(f)$
$\delta(t)$ (time domain Dirac)	1 (constant spectrum)
$\exp(j2\pi f_0 t)$	$\delta(f - f_0)$
$\cos(2\pi f_0 t)$	$\frac{1}{2}[\delta(f - f_0) + \delta(f + f_0)]$
$\sin(2\pi f_0 t)$	$\frac{1}{2j}[\delta(f - f_0) - \delta(f + f_0)]$
$\mathrm{rect}(t/T)$ (rectangle function upon $[-T/2, T/2]$)	$T\,\mathrm{sinc}(Tf)$ (cardinal sine)
$B\,\mathrm{sinc}(Bt)$	$\mathrm{rect}(f/B)$ (rectangle function upon $[-B/2, B/2]$)
$\exp\left(-\dfrac{t^2}{2\sigma^2}\right)$	$\sigma\sqrt{2\pi}\exp(-2\pi^2\sigma^2 f^2)$

A.6.2   ANALOGUE RECEIVERS FOR NARROWBAND PULSES AND PROCESSING GAIN

Analogue receivers used for narrowband pulses consist in:

- a bandpass filter as close as possible to the spectrum of the signal transmitted, possibly corrected for the expected Doppler shift;
- a square law envelope detector of the filtered signal;
- an integration of duration τ adapted to the duration of the signal length.

The main idea is, therefore, detection of the energy present in the bandwidth and the duration of the signal. This is called non-coherent processing (Figure A.6.1).

Letting E represent the energy of the signal received, of duration T, and $n/2$ the power spectral density of the noise, the input power SNR (Signal-to-Noise Ratio) r_{0P}

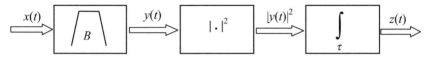

Figure A.6.1. Non-coherent processing of a narrowband pulse. From *left* to *right*: bandpass filtering, quadrature, integration.

is, by definition for a frequency band B including the signal spectrum:

$$r_{0P} = \frac{E/T}{nB} = \frac{E}{TnB} \tag{A.6.1}$$

If B is the width of the input bandpass filter, after integration over the time $\tau = T$, the output with noise only is a random variable of mean $\langle z_B(t) \rangle = nB$, and of variance $\sigma_n^2 = n^2 B / T$. The mean power of the output with signal + noise is the sum of the powers of signal and noise $\langle z_{s+n}(t) \rangle = E/T + nB$. And the output SNR r (ratio of the maximum output energy and the variance of noise only, after processing) is given, according to Definition (6.26) for a quadratic receiver, by:

$$r = \frac{\langle z_{s+n} \rangle - \langle z_n \rangle}{\sqrt{\sigma_n^2}} = \frac{E}{n\sqrt{BT}} = \sqrt{BT} r_{0P} \tag{A.6.2}$$

The processing gain PG can then be expressed in dB:

$$PG = 5 \log BT \tag{A.6.3}$$

PG is close to $0\,dB$, as the bandwidth B of the filter matches the duration T of the signal, and therefore is close to $1/T$. In fact, there is no real processing gain, since the energy of the signal is simply received inside the suitable frequency band.

A.8.1 PHASE-DIFFERENCE MEASUREMENT ERROR DUE TO NOISE

When measuring the phase difference $\Delta\phi$ between the two receivers of an interferometer, on a stable narrowband signal mixed with noise, the instantaneous standard deviation of $\Delta\phi$ is related to the SNR r by the following classical relation:

$$\delta\Delta\phi = \sqrt{\frac{2}{r}} \tag{A.8.1}$$

where $\delta\Delta\phi$ is in radians. This may be extended to low SNR values in (Quazi, 1981):

$$\delta\Delta\phi = \sqrt{\frac{2}{r} + \frac{4}{r^2}} \tag{A.8.2}$$

Practically, this error decreases when the phase difference is averaged over number N of samples, inversely to $\sqrt{N}$.

In the more realistic case where the signal amplitude fluctuates according to a Rayleigh law, it may be shown (Lurton, 2001) that the phase difference error is given, for $N = 1$, by

$$\delta\Delta\phi \approx \left[\frac{2.571 + \ln(r)}{r}\right]^{1/2} \tag{A.8.3}$$

and for $N > 2$, by:

$$\delta\Delta\phi \approx \left[\frac{1}{(N-1)r} + \frac{N}{2(N-1)(N-2)r^2}\right]^{1/2} \tag{A.8.4}$$

This expression is valid for values of $\delta\Delta\phi$ below ca. $30°$. It may be simplified into $\delta\Delta\phi \approx [1/Nr]^{1/2}$, for large values of both N and r.

Bibliography

Akal, T. and Berkson, J. M. *Ocean Seismo-Acoustics* (NATO Conference Series No. 16). New York: Plenum Press, 1986.

Allard, J. F. *Propagation of Sound in Porous Media*. London: Elsevier Science, 1993.

APL, *APL-UW High-Frequency Ocean Environmental Acoustic Models Handbook* (APL-UW TR 9407). Seattle, WA: Applied Physics Laboratory, University of Washington, 1994.

Babb, R. J. 'Feasibility of interferometric swath bathymetry using GLORIA, a long-range sidescan', *IEEE Journal of Oceanic Engineering*, **14**, 299–305, 1989.

Ben Menahem, A. and Singh, S. J. *Seismic Waves and Sources*. Berlin: Springer-Verlag, 1981.

Bendat, J. S. and Piersol, A. G. *Random Data: Analysis and Measurement Procedures*. New York: Wiley-Interscience, 1971.

Beranek, L. *Acoustics* (published for the Acoustical Society of America). New York: American Institute of Physics, 1954.

Blackington, J. G. 'Bathymetric resolution, precision and accuracy considerations for swath bathymetry mapping sonar systems', paper presented at *Oceans '91, Honolulu, HI*. Piscataway, NJ: IEEE, 1991.

Blondel, P. 'Segmentation of the Mid-Atlantic Ridge south of the Azores, based on acoustic classification of TOBI data', in C. J. MacLeod, P. Tyler and C. L. Walker (eds) *Tectonic, Magmatic and Biological Segmentation of Mid-Ocean Ridges* (Geological Society Special Publication No. 118). London: Geological Society, 1996, pp. 17–28.

Blondel, P. and Murton, B. J. *Handbook of Seafloor Sonar Imagery*. Chichester, UK: John Wiley & Sons/Praxis, 1997.

Boehme, H. and Chotiros, N. P. 'Acoustic backscattering at low grazing angles from the ocean bottom', *Journal of the Acoustical Society of America*, **84**, 1018–1029, 1988.

Bouvet, M. *Traitement des Signaux pour les Systèmes Sonar*. Paris: Masson, 1991.

Bouvet, M. and Bienvenu, G. *High Resolution Methods in Underwater Acoustics* (Lecture Notes in Control and Information Sciences Vol 155). Berlin: Springer-Verlag, 1991.

Boyle, F. A. and Chotiros, N. P. 'A model for acoustic backscatter from muddy sediments', *Journal of the Acoustical Society of America*, **98**, 525–530, 1995a.

Boyle, F. A. and Chotiros, N. P. 'A model for high-frequency acoustic backscatter from gas bubbles in sandy sediments at shallow grazing angles', *Journal of the Acoustical Society of America*, **98**, 531–541, 1995b.

Brekhovskikh, L. M. and Godin, O. A. *Acoustics of Layered Media I*. Berlin: Springer-Verlag, 1990.

Brekhovskikh, L. M. and Godin, O. A. *Acoustics of Layered Media II*. Berlin: Springer-Verlag, 1992.

Brekhovskikh, L. M. and Lysanov, Yu. P., *Fundamentals of Ocean Acoustics* (2nd edn). Berlin: Springer-Verlag, 1992.

Bruce, M. P. 'A processing requirement and resolution capability comparison of side-scan and synthetic aperture sonars', *IEEE Journal of Oceanic Engineering*, **17**(1), 1992.

Bruneau, M. *Manuel d'Acoustique Fondamentale*. Paris: Hermès, 1998.

Bucker, H. P. 'Sound propagation in a channel with lossy boundaries', *Journal of the Acoustical Society of America*, **48**, 1187–1194, 1970.

Bunchuk, A. V. and Zhitkovskii, Y. Y. 'Sound scattering by the ocean bottom in shallow-water regions (review)', *Soviet Physics Acoustics*, **26**, 363–370, 1980.

Burdic, W. S. *Underwater Acoustic System Analysis* (1st edn). Englewood Cliffs, NJ: Prentice Hall, 1984a.

Burdic, W. S. *Underwater Acoustic System Analysis* (2nd edn). Englewood Cliffs, NJ: Prentice Hall, 1984b.

Burns, D. R., Queen, C. B., Sisk, H., Mullarkey, W. and Chivers, R. C. 'Rapid and convenient sea-bed discrimination for fishery applications', *Proc. IOA.*, **11**, 169–178, 1989.

Cervenka, P. and de Moustier, C. 'Sidescan sonar image processing techniques', *IEEE Journal of Oceanic Engineering*, **18**, 108–122, 1993.

Chan, Y. T. *Underwater Acoustic Data Processing* (NATO ASI Series). Dordrecht, The Netherlands: Kluwer Academic, 1988.

Chapman, D. M. F. and Ellis, D. D. 'The group velocity of normal modes', *Journal of the Acoustical Society of America*, **74**, 973–979, 1983.

Chapman, R. P. and Harris, J. H. 'Surface backscattering strengths measured with explosive sound sources', *Journal of the Acoustical Society of America*, **34**, 1592–1597, 1962.

Chen, C. T. and Millero, F. J. 'Speed of sound in seawater at high pressures', *Journal of the Acoustical Society of America*, **62**, 1129–1135, 1977.

Cherkis, N. Z. 'Worldwide seafloor swath-mapping systems', online at http://www.iho.shom.fr/publicat/free/files/swathcur.pdf, 2001.

Chernov, L. A. *Wave Propagation in a Random Medium*. New York: McGraw-Hill, 1960.

Chotiros, N. P. 'High frequency acoustic bottom backscatter mechanisms at shallow grazing angles', in D. D. Ellis (ed.) *Ocean Reverberation*, Dordrecht, The Netherlands: Kluwer Academic Publishers, 1993.

Chotiros, N. P. 'Reflection and reverberation in normal incidence echosounding', *Journal of the Acoustical Society of America*, **96**, 2921–2928, 1994.

Chotiros, N. P. 'Acoustic penetration of a silty sand sediment in the 1–10 kHz band', *IEEE Journal of Oceanic Engineering*, **22**, 604–615, 1997.

Claerbout, J. *Fundamentals of Geophysical Data Processing*. New York: McGraw-Hill, 1976.

Clay, C. S. and Medwin, H. *Acoustical Oceanography: Principles and Applications*. New York: Wiley-Interscience, 1977.

Cohen, L. *Time-Frequency Analysis: Theory and Application*. Englewood Cliffs, NJ: Prentice Hall, 1995.

Collins, M. D. 'A split-step Padé solution for the parabolic equation method', *Journal of the Acoustical Society of America*, **93**, 1736–1742, 1993.

Coulon, F. *Théorie et traitement des signaux*. Paris: Dunod, 1984.

Cox, C. S. and Munk, W. H. 'Statistics of the sea surface derived from sun glitter', *Journal of Marine Research*, **13**, 198–227, 1954.

Cron, B. F. and Shaffer, L. 'Array gain for the case of directional noise', *Journal of the Acoustical Society of America*, **41**, 864, 1967.

Cron, B. F. and Sherman, C. H. 'Spatial correlation functions for various noise models', *Journal of the Acoustical Society of America*, **34**, 1732, 1962.

Cutrona, L. J. 'Comparison of sonar system performance achievable using synthetic aperture techniques with the performance achievable with conventional means', *Journal of the Acoustical Society of America*, **58**, 336–348, 1975.

Cutrona, L. J. 'Additional characteristics of synthetic-aperture sonar systems and a further comparison with nonsynthetic-aperture sonar systems', *Journal of the Acoustical Society of America*, **61**, 1213–1217., 1977.

Dashen, R. and Munk, W. 'Three models of global ocean noise', *Journal of the Acoustical Society of America*, **76**, 540, 1984.

de Moustier, C. 'Beyond bathymetry: Mapping acoustic backscattering from the deep seafloor with Sea Beam', *Journal of the Acoustical Society of America*, **79**, 316–331, 1986.

de Moustier, C. 'State of the art in swath bathymetry survey systems', *Int. Hydr. Rev.*, **65**, 25–54, 1988.

de Moustier, C. 'Signal processing for swath bathymetry and concurrent seafloor acoustic imaging', in J. M. F. Moura and I. M. G. Lourtie (eds) *Acoustic Signal Processing for Ocean Exploration*. Brussels: NATO Advance Study Institute, 1993, pp. 329–354.

Decarpigny, J. N., Hamonic, B. and Wilson, O. B. 'The design of low frequency underwater acoustic projectors: present status and future trends', *IEEE Journal of Oceanic Engineering*, **16**, 1991.

Del Grosso, V. A. 'New equation for the speed of sound in natural waters (with comparison to other equations)', *Journal of the Acoustical Society of America*, **56**, 1084–1091, 1974.

Denbigh, P. N. 'Swath bathymetry: Principles of operations and an analysis of errors', *IEEE Journal of Oceanic Engineering*, **14**, 289–298, 1989.

Desaubies, Y. and Dysthe, K. 'Normal-mode propagation in slowly varying ocean waveguides', *Journal of the Acoustical Society of America*, **97**, 933–946, 1995.

Desaubies, Y., Tarantola, A. and Zinn-Justin, J. *Oceanographic and Geophysical Tomography*. Amsterdam: Elsevier, 1990.

Di Napoli, F. R. and Deavenport, R. L. 'Theoretical and numerical Green's function field solution in a plane multilayered system', *Journal of the Acoustical Society of America*, **67**, 92–105, 1980.

Diner, N. and Marchand, Ph. *Acoustique et pêche maritime*. Brest, France: Editions IFREMER, 1995.

Dobrin, M. B. *Introduction to Geophysical Prospecting*. New York: McGraw-Hill, 1976.

Eckart, C. 'The scattering of sound from the sea surface', *Journal of the Acoustical Society of America*, **25**, 566–570, 1953.

Ehrenberg, J. E. 'A review of target strength estimation techniques', in Y. T. Chan (ed.) *Underwater Acoustics Data Processing*. Dordrecht, The Netherlands: Kluwer Academic, 1981, 161–176.

Elachi, C. *Introduction to the Physics and Techniques of Remote Sensing*. New York: Wiley-Interscience, 1987.

Etter, P. C. *Underwater Acoustic Modelling*. London: Elsevier Science, 1991.

Evans, R. B. 'A coupled mode solution for acoustic propagation in a waveguide with stepwise depth variations of a penetrable bottom', *Journal of the Acoustical Society of America*, **74**, 188–195, 1983.

Farmer, D. M. and Lemon, D. D. 'The influence of bubbles on ambient noise at high wind speeds', *J. Phys. Oceanogr.*, **14**, 1762–1778, 1984.

Flatté, S. *Sound in a Fluctuating Ocean*. Cambridge: Cambridge University Press, 1979.

Foote, K. G. 'Importance of the swimbladder in acoustic scattering by a fish: A comparison of gadoid and mackerel target strengths', *Journal of the Acoustical Society of America*, **67**, 2084–2089, 1980.

Foote, K. G. 'Optimizing copper spheres for precision calibration of hydroacoustic equipment', *Journal of the Acoustical Society of America*, **71**, 742–747, 1982.

Foote, K. G. 'Fish target strength for use in echo integrator surveys', *Journal of the Acoustical Society of America*, **82**, 981–987, 1987.

Francois, R. E. and Garrison, G. R. 'Sound absorption based on ocean measurements. Part I: Pure water and magnesium sulfate contributions', *Journal of the Acoustical Society of America*, **72**, 896–907, 1982a.

Francois, R. E. and Garrison, G. R. 'Sound absorption based on ocean measurements. Part II: Boric acid contribution and equation for total absorption', *Journal of the Acoustical Society of America*, **72**, 1879–1890, 1982b.

Frisk, G. V., *Ocean and Seabed Acoustics*. Englewood Cliffs, NJ: Prentice Hall, 1994.

Gardner, J. V., Field, M. E., Lee, H., Edwards, B. E., Masson, D. G., Kenyon, N. H. and Kidd, R. B. 'Ground-truthing 6.5-kHz sidescan sonographs: What are we really imaging?', *Journal of Geophysical Research*, **96**, 5955–5974, 1991.

Gensane, M. 'A statistical study of acoustic signals backscattered from the sea bottom', *IEEE Journal of Oceanic Engineering*, **14**, 84–93, 1989.

Gordon, G. D. 'The emergence of low-frequency active acoustics as a critical antisubmarine warfare technology', *Johns Hopkins APL Technical Digest*, **13**, 145–159, 1992.

Guillon, L. and Lurton, X. 'Backscattering from buried sediment layers: The equivalent input backscattering strength model', *Journal of the Acoustical Society of America*, **109**, 122–132, 2000.

Hall, M. V. 'A comprehensive model of wind-generated bubbles in the ocean and predictions of the effects on sound propagation at frequencies up to 40 kHz', *Journal of the Acoustical Society of America*, **86**, 1103–1117, 1989.

Hamilton, E. L. 'Compressional waves in marine sediments', *Geophysics*, **37**, 620–646, 1972.

Hamilton, E. L. 'Geoacoustic modelling of the sea floor', *Journal of the Acoustical Society of America*, **68**, 1313–1340, 1976.

Hamilton, E. L. and Bachman, R. T. 'Sound velocity and related properties of marine sediments', *Journal of the Acoustical Society of America*, **72**, 1891–1904, 1982.

Hampton, L. *Physics of Sound in Marine Sediments*. New York: Plenum Press, 1974.

Hare, R., Godin, A. and Mayer, L. *Accuracy Estimation of Canadian Swath (Multibeam) and Sweep (Multitransducer) Sounding Systems*. Canadian Hydrographic Service Internal Report, 1995.

Harris, F. J. 'On the use of windows for harmonic analysis with the Discrete Fourier Transform', *Proc. IEEE*, **66**, 51–83, 1978.

Haykin, S. *Advances in Spectrum Analysis and Array Processing*. Englewood Cliffs, NJ: Prentice Hall, 1994.

Helstrom, C. W. *Statistical Theory of Signal Detection*. New York: Pergamon Press, 1968.

Horton, Sr., W. 'A review of reverberation, scattering and echo structure', *Journal of the Acoustical Society of America*, **51**, 1049–1061, 1972.

Hughes-Clarke, J. 'Toward remote seafloor classification using the angular response of acoustic backscattering: A case study from multiple overlapping GLORIA data', *IEEE Journal of Oceanic Engineering*, **19**, 112–127, 1994.

IOA (Institute of Acoustics, UK), *Proceedings of the Institute of Acoustics: Sonar Transducers for the Nineties, University of Birmingham, 1990*.

IOA. *Proceedings of the Institute of Acoustics: Sonar transducers '95, University of Birmingham, 1995.*

IOA. *Proceedings of the Institute of Acoustics: Sonar transducers '99. Vol. 21, Pt 1, University of Birmingham, 1999.*

Isakovitch, M. A. and Kuryanov, B. F. 'Theory of low-frequency noise in the ocean', *Soviet Physics Acoustics*, **16**, 49, 1970.

ISEN, *Proceedings of the International Workshop on Power Sonic and Ultrasonic Transducers Design, ISEN, Lille.* Berlin: Springer-Verlag, 1988.

Ishimaru, A. *Wave Propagation and Scattering in Random Media*, Volume 1. *Single Scattering and Transport Theory.* San Diego, CA: Academic Press, 1978a.

Ishimaru, A. *Wave Propagation and Scattering in Random Media*, Volume 2. *Multiple Scattering, Turbulence, Rough Surfaces, and Remote Sensing.* San Diego, CA: Academic Press, 1978b.

Ivakin, A. N. 'Sound scattering by random inhomogeneities of stratified ocean sediments', *Soviet Physics Acoustics*, **32**, 492–496, 1987.

Ivakin, A. N. 'A unified approach to volume and roughness scattering', *Journal of the Acoustical Society of America*, **103**, 827–837, 1998.

Ivakin, A. N. and Lysanov, Y. P. 'Underwater sound scattering by volume inhomogeneities of a medium bounded by a rough surface', *Soviet Physics Acoustics*, **27**, 212–215, 1981.

Jackson, D. R. and Briggs, K. B. 'High-frequency bottom backscattering: Roughness versus sediment volume scattering', *Journal of the Acoustical Society of America*, **92**, 962–977, 1992.

Jackson, D. R., Winnebrenner, D. P. and Ishimaru, A. 'Application of the composite roughness model to high-frequency bottom backscattering', *Journal of the Acoustical Society of America*, **79**, 1410–1422, 1986.

Jackson, D. R., Briggs, K. B., Williams, K. L. and Richardson, M. D. 'Tests of models for high-frequency seafloor backscatter', *IEEE Journal of Oceanic Engineering*, **21**, 458–470, 1996.

Jensen, F. B., Kuperman, W. A., Porter, M. B. and Schmidt, H., *Computational Ocean Acoustics.* New York: AIP Press, 1994.

Keller, J. B. and Papadakis, J. S., *Wave Propagation and Underwater Acoustics.* Berlin: Springer-Verlag, 1977.

Kenneth, B. L. N. *Seismic Wave Propagation in Stratified Media.* Cambridge: Cambridge University Press, 1983.

Kerman, B. R. *Sea Surface Sound: Natural Mechanisms of Surface Generated Noise in the Ocean.* Boston: Kluwer Academic, 1988.

Kerman, B. R. *Natural Physical Sources of Underwater Sound.* Boston: Kluwer Academic, 1993.

Knudsen, V. O., Alford, R. S. and Emling, J. W. 'Underwater ambient noise', *Journal of Marine Research*, **7**, 410, 1948.

Knight, W. C., Pridham, R. G. and Kay, S. M. 'Digital signal processing for sonar', *Proc. IEEE*, **69**, 1451–1501, 1981.

Kuo, E. Y. T. 'Wave scattering and transmission at irregular surfaces', *Journal of the Acoustical Society of America*, **33**, 2135–2142, 1964.

Kuo, E. Y. T. 'Sea surface scattering and propagation loss: Review, update and new predictions', *IEEE Journal of Oceanic Engineering*, **13**, 1988.

Kuperman, W. A. and Jensen, F. B. *Bottom-interacting Acoustics.* New York: Plenum Press, 1980.

Lasky, M. 'Review of undersea acoustics to 1950', *Journal of the Acoustical Society of America*, **61**, 283–297, 1977.

Lavergne, M. *Méthodes sismiques*. Rueil-Malmaison, France: Institut Français du Pétrole, 1986.

LeBlanc, L. R., Mayer, L., Rufino, M., Schock, S. G. and King, J. 'Marine sediment classification using the chirp sonar', *Journal of the Acoustical Society of America*, **91**, 107–115, 1992a.

LeBlanc, L. R., Panda, S. and Schock, S. G. 'Sonar attenuation modelling for classification of marine sediments', *Journal of the Acoustical Society of America*, **91**, 116–126, 1992b.

LeChevalier, F. *Principes de Traitement des Signaux Radar et Sonar*. Paris: Masson, 1989.

Leduc, B. and Ayela, G. 'TIVA, a self-contained image/data transmission system for underwater applications', First French Congress of Acoustics, Lyon, France, *Journal de Physique*, **2**(Suppl.), 655–658, 1990.

Lee, D. and McDaniel, S. T., *Ocean Acoustic Propagation by Finite Difference Methods*. New York: Pergamon, 1988.

Leighton, T. G. *The Acoustic Bubble*. London: Academic Press, 1996.

Leroy, C. 'Formulas for the calculation of underwater pressure in acoustics', *Journal of the Acoustical Society of America*, **40**(2), 651–653, 1968.

Leroy, C. C. 'Development of simple equations for accurate and realistic calculations of the speed of sound in seawater', *Journal of the Acoustical Society of America*, **46**, 216–226, 1969.

Levitus, S. *Climatological Atlas of the World Ocean*. Washington, DC: NOAA Professional Paper 13, 1982.

Love, R. H. 'Resonant acoustic scattering by swimbladder-bearing fish', *Journal of the Acoustical Society of America*, **64**, 571–580, 1978.

Love, R. H. 'A model for estimating distributions of fish school target strengths', *Deep Sea Research*, **28A**, 705–725, 1981.

Lurton, X. 'The range-averaged intensity model: A tool for underwater acoustic field analysis', *IEEE Journal of Oceanic Engineering*, **17**, 138–149, 1992.

Lurton, X. and Pouliquen, E. 'Identification de la nature des fonds marins à l'aide de signaux d'échos-sondeurs: II. Méthode d'identification et résultats expérimentaux', *Acta Acustica*, **2**, 187–194, 1994.

Lurton, X. 'Swath bathymetry using phase difference: Theoretical analysis of acoustical measurement precision', *IEEE Journal of Oceanic Engineering*, **25**, 351–363, 2000.

Lurton, X. 'Précision de mesure des sonars bathymétriques en fonction du rapport signal à bruit', *Traitement du Signal*, **18**, 179–194, 2001.

Lyons, A. P., Anderson, A. L. and Dwan, F. S. 'Acoustic scattering from the seafloor: Modeling and data comparison', *Journal of the Acoustical Society of America*, **95**, 2441–2451, 1994.

McDaniel, S. T. 'Sea surface reverberation: a review', *Journal of the Acoustical Society of America*, **94**, 1905–1922, 1993.

McKinney, C. M. and Anderson, C. D. 'Measurements of backscattering of sound from the ocean bottom', *Journal of the Acoustical Society of America*, **36**, 158–163, 1964.

MacLennan, D. N. and Simmonds, E. J. *Fisheries Acoustics*. London: Chapman & Hall, 1992.

Marcos, S. *Les méthodes à haute-résolution – traitement d'antenne et analyse spectrale*. Paris: Ed. Hermès, 1997.

Mascle, J. and Huguen, C. 'Images may show start of European-African Plate Collision', *Transactions of American Geophysical Union*, **80**, 421–428, 1999.

Mascle, J. *et al.* 'Marine geologic evidence for a Levantine-Sinai plate, a new piece of the Mediterranean puzzle', *Geology*, **28**, 779–782, 2000.

Medwin, H. 'Speed of sound in water: A simple equation for realistic parameters', *Journal of the Acoustical Society of America*, **58**, 1318–1319, 1975.

Medwin, H. and Clay, C. S. *Fundamentals of Acoustical Oceanography*. Boston: Academic Press, 1998.

Miasnikov, E. V. 'What is known about the character of noise created by submarines?', *The Future of Russia's Strategic Nuclear Forces – Discussions and Arguments*. Moscow: Center for Arms Control, Energy and Environmental Studies, Moscow Institute of Physics and Technology, Appendix 1, 1998.

Mitson, R. B. *Fisheries Sonar*. Farnham, UK: Fishing New Books, 1983.

Mitson, R. B. *Underwater Noise of Research Vessels*, Vol. 1. Copenhagen: International Council for the Exploration of the Sea, 1995, p. 61.

Moura, J. M. F. and Lourtie, I. M. G. *Acoustic Signal Processing for Ocean Exploration* (NATO ASI Series). Dordrecht, The Netherlands: Kluwer Academic, 1993.

Munk, W. H. and Forbes, A. M. G. 'Global ocean warming: An acoustic measure?', *Journal of Physical Oceanography*, **19**, 1765–1777, 1989.

Munk, W. H. and Wunsh, C. 'Ocean acoustic tomography: A scheme for large scale monitoring', *Deep Sea Research*, **26A**, 123–160, 1979.

Munk, W. H. and Wunsh, C. 'Ocean acoustic tomography: Rays and modes', *Review of Geophysics and Space Physics*, **21**, 777–793, 1983.

Munk, W., O'Reilly, W. and Reid, J. 'Australia-Bermuda sound transmission experiment (1960) revisited', *J. Phys. Oceanogr.*, **18**, 1876–1898, 1988.

Munk, W. H., Spindel, R. C., Baggeroer, A. B. and Birdsall, T. G. 'An overview of the Heard Island feasibility experiment', *Journal of the Acoustical Society of America*, **96**, 2330–2332, 1994.

Munk, W. H., Worcester, P. F. and Wunsch, C. *Ocean Acoustic Tomography*. Cambridge: Cambridge University Press, 1995.

Nagl, A., Uberall, H., Haug, A. J. and Zarur, G. L. 'Adiabatic mode theory of underwater sound propagation in range-dependent environment', *Journal of the Acoustical Society of America*, **63**, 739–749, 1978.

NDRC (National Defense Research Committee), *Physics of Sound in the Sea*. Los Altos, CA: Peninsula Publishing, 1945.

Neumann, G. and Pierson, Jr, W. J. *Principles of Physical Oceanography*. Englewood Cliffs, NJ: Prentice Hall, 1966.

Nielsen, R. O. *Sonar Signal Processing*. Boston: Artech House, 1991.

Officer, C. B. *Introduction to the Theory of Sound Transmission*. New York: McGraw-Hill, 1958.

Ol'shevskii, V. V. *Characteristics of Sea Reverberation*. New York: Consultants Bureau, 1967.

Oliver, C. and Quegan, S. *Understanding Synthetic Aperture Radar Images*. Boston: Artech House, 1998.

Pace, N. G. *Acoustics and the Sea Bed*. Bath, UK: Bath University Press, 1983.

Pace, N. G. and Gao, H. 'Swathe seabed classification', *IEEE Journal of Oceanic Engineering*, **13**, 83–90, 1988.

Pace, N. G., Pouliquen, E., Bergem, O. and Lyons, A. P. 'High-frequency acoustics in shallow water', *SACLANTCEN Conference Proceedings CP-45*. La Spezia, Italy: NATO SACLANT Undersea Research Centre, 1997.

Papoulis, A. *Signal Analysis*. New York: McGraw-Hill, 1977.

Papoulis, A. *Probability, Random Variables and Stochastic Processes*. New York: McGraw-Hill, 1991.

Pekeris, C. L. 'Theory of propagation of explosive sound in shallow water', *Geological Society of America Memoirs*, **27**, 1948.

Peterson, W. W., Birdsall, T. G. and Fox, W. C. 'The theory of signal detectability', *IEEE Trans. Information Theory*, **4**, 171–212, 1954.

Pierce, A. D. 'Extension of the method of normal modes to sound propagation in an almost stratified medium', *Journal of the Acoustical Society of America*, **37**, 19–27, 1965.

Pouliquen, E. and Lurton, X. 'Identification de la nature des fonds marins à l'aide de signaux d'échos-sondeurs: I. Modélisation d'échos réverbérés par le fond', *Acta Acustica*, **2**, 113–126, 1994.

Proakis, J. G. *Digital Communications* (2nd edn). New York: McGraw-Hill, 1989.

Proakis, J. G. and Manolakis, D. G. *Digital Signal Processing Principles, Algorithms, and Applications*. Macmillan, 1988.

Quazi, A. H. 'An overview of the time delay estimate in active and passive systems for target localization', *IEEE Transactions on Acoustics, Speech and Signal Processing*, **29**, 527–533, 1981.

Rayleigh, Lord [John William Strutt]. *The Theory of Sound*. New York: Dover Publications, 1945.

Renard, V. and Allenou, J. P. 'Sea Beam multibeam echo sounding on Jean Charcot: Description, evaluation and first results', *Int. Hydr. Rev.*, **56**, 36–57, 1979.

Richardson, M. D. and Davis, A. M. 'Modelling gassy sediment structure and behavior', special issue of *Continental Shelf Research*, **18**, Nos 14–15, 1998.

Rihaczek, A. W. *Principles of High-resolution Radar*. New York: McGraw-Hill, 1969.

Robber, R. J. *Underwater Electroacoustics Measurements*. Los Altos, CA: Peninsula Publishing, 1988.

Ross, D. *Mechanics of Underwater Noise*. New York: Pergamon Press, 1976.

Rossi, M. *Acoustics and Electroacoustics*. Lausanne, Switzerland: Presses Polytechniques Romandes, 1986.

Shepard, F. P. 'Nomenclature based on sand-silt-clay ratios', *J. Sediment Petrol.*, **24**, 151–158, 1954.

Sheriff, R. E. and Geldart, L. P. *Exploration Seismology*. Cambridge: Cambridge University Press, 1982.

Somers, M. L. 'Sonar imaging of the seabed: techniques, performances, applications', in J. M. F. Moura and I. M. G. Lourtie (eds) *Acoustic Signal Processing for Ocean Exploration*. Dordrecht, The Netherlands: Kluwer Academic, 1993, pp. 355–369.

Stanton, T. K. 'Volume scattering: Echo peak PDF', *Journal of the Acoustical Society of America*, **77**, No. 4, 1358–1366, 1985.

Steinberg, B. D. *Principles of Aperture and Array System Design*. New York: John Wiley & Sons, 1975.

Stoica, P. and Moses, R. *Introduction to Spectral Analysis*. Upper Saddle River, NJ: Prentice Hall, 1997.

Stoll, R. D. 'Marine sediment acoustics', *Journal of the Acoustical Society of America*, **77**, 1789–1799, 1985.

Tappert, F. D., 'The parabolic approximation method', in J. B. Keller and J. S. Papadakis (eds) *Wave Propagation in Underwater Acoustics*. New York: Springer-Verlag, 1977, pp. 224–287.

Tavolga, W. H. *Marine Bioacoustics*. New York: Pergamon Press, 1964.

Thomas, H. 'New advanced underwater navigation techniques based on surface relay buoys', *Oceans '94, Brest, France*, 1994.

Thomson, D. J. and Chapman, N. R. 'A wide-angle split-step algorithm for the parabolic equation', *Journal of the Acoustical Society of America*, **74**, 1848–1854, 1983.

Tindle, C. T. 'The equivalence of bottom loss and mode attenuation per cycle in underwater acoustics', *Journal of the Acoustical Society of America*, **66**, 250–255, 1979.

Tolstoy, I. and Clay, C. S. *Ocean Acoustics*. New York: American Institute of Physics, 1987.

Tyce, R. C. 'Deep seafloor mapping systems – A review', *MTS Journal*, **20**, 4–16, 1988.

Tyler, G. D. 'The emergence of low-frequency active acoustics as a critical antisubmarine warfare technology', *Johns Hopkins APL Technical Digest*, **13**, 145–159, 1992.

Urick, R. J. 'The backscattering of sound from a harbor bottom', *Journal of the Acoustical Society of America*, **26**, 231–235, 1954.

Urick, R. J. *Sound Propagation in the Sea*. Los Altos, CA: Peninsula Publishing, 1982.

Urick, R. J. *Principles of Underwater Sound* (3rd edn). New York: McGraw-Hill, 1983.

Urick, R. J. *Ambient Noise in the Sea*. Los Altos, CA: Peninsula Publishing, 1986.

Van Trees, H. L. *Detection, Estimation and Modulation Theory*. New York: John Wiley & Sons, 1968.

Vogt, P., Jung, W. Y. and Nagel, D. 'GOMAP: A matchless resolution to start the new millennium', *EOS Trans. AGU*, **81**, 254–258, 2000.

Voronovitch, G. *Wave Scattering from Rough Surfaces*. Berlin: Springer-Verlag, 1994.

Wenz, G. M. 'Acoustic ambient noise in the sea: spectra and sources', *Journal of the Acoustical Society of America*, **34**, 1936, 1962.

Wenz, G. M. 'Review of underwater acoustics research: noise', *Journal of the Acoustical Society of America*, **51**, 1010, 1972.

Westervelt, P. J. 'Parametric acoustic arrays', *Journal of the Acoustical Society of America*, **35**, 535-537, 1963.

Weston, D. E. 'Intensity-range relations in oceanographic acoustics', *Journal of Sound and Vibration*, **18**, 271–287, 1971.

Weston, D. E. 'Acoustic flux methods for oceanic guided waves', *Journal of the Acoustical Society of America*, **68**, 287–296, 1980a.

Weston, D. E. 'Ambient noise depth-dependence models and their relation to low-frequency attenuation', *Journal of the Acoustical Society of America*, **67**, 530, 1980b.

Wilson, O. B. 'An introduction to the theory and design of sonar transducers' (Thesis). Washington, DC: Naval Post-Graduate School. US Government Printing Office, 1985.

Wilson, W. D. 'Equation for the speed of sound in seawater', *Journal of the Acoustical Society of America*, **32**, 1357, 1960.

Winder, A. A. 'Sonar system technology', *IEEE Transactions on Sonics and Ultrasonics*, **SU-22**, 291–332, 1975.

Wong, H. K. and Chesterman, W. D. 'Bottom backscattering near grazing incidence in shallow water', *Journal of the Acoustical Society of America*, **44**, 1713–1718, 1968.

Woodward, P. M. *Probability and Information Theory with Application to Radar*. New York: Pergamon Press, 1953.

Worcester, P. F., Cornuelle, B. and Spindel, R. C. 'A review of oceanic acoustic tomography: 1987–1990', *Review of Geophysics*, 557–570, 1991.

Zhitkovskii, Y. Y. and Lysanov, Y. P. 'Reflection and scattering of sound from the ocean bottom (Review)', *Soviet Physics Acoustics*, **30**, 1–13, 1965.

Index

Printing: Mercedes-Druck, Berlin
Binding: Stein+Lehmann, Berlin